VOLUME ONE HUNDRED AND FIFTY EIGHT

PROGRESS IN MOLECULAR BIOLOGY AND TRANSLATIONAL SCIENCE

Neuroepigenetics and Mental Illness

CONTENTS

VOLUME ONE HUNDRED AND FIFTY EIGHT

PROGRESS IN MOLECULAR BIOLOGY AND TRANSLATIONAL SCIENCE

Neuroepigenetics and Mental Illness

Edited by

BART P.F. RUTTEN

Department of Psychiatry and Neuropsychology,
School for Mental Health and Neuroscience (MHeNs),
European Graduate School of Neuroscience (EURON),
Faculty of Health, Medicine and Life Sciences,
Maastricht University Medical Centre +, Maastricht,
The Netherlands

ELSEVIER

ACADEMIC PRESS
An imprint of Elsevier

Academic Press is an imprint of Elsevier
50 Hampshire Street, 5th Floor, Cambridge, MA 02139, United States
525 B Street, Suite 1650, San Diego, CA 92101, United States
The Boulevard, Langford Lane, Kidlington, Oxford OX5 1GB, United Kingdom
125 London Wall, London EC2Y 5AS, United Kingdom

First edition 2018

ISBN: 978-0-12-812592-2
ISSN: 1877-1173

Publisher: Zoe Kruze
Acquisition Editor: Ashlie M. Jackman
Editorial Project Manager: Joanna Collett
Production Project Manager: James Selvam
Cover Designer: Victoria Pearson

Typeset by SPi Global, India

CONTRIBUTORS

Ning An
School for Mental Health and Neuroscience (MHeNS), Department of Psychiatry and Neuropsychology, Faculty of Health, Medicine and Life Sciences, Maastricht University, Maastricht, The Netherlands

Katherine C. Bassil
School for Mental Health and Neuroscience (MHeNS), Department of Psychiatry and Neuropsychology, Faculty of Health, Medicine and Life Sciences, Maastricht University, Maastricht, The Netherlands

Raoul Belzeaux
McGill Group for Suicide Studies, Douglas Mental Health University Institute, Department of Psychiatry, McGill University, Montreal, QC, Canada

Rahul Bharadwaj
Lieber Institute for Brain Development, Johns Hopkins Medical Campus, Baltimore, MD, United States

Stephan Claes
Department of Neuroscience, KU Leuven, Leuven, Belgium

Laurence de Nijs
School for Mental Health and Neuroscience (MHeNS), Department of Psychiatry and Neuropsychology, Faculty of Health, Medicine and Life Sciences, Maastricht University, Maastricht, The Netherlands

Alec Dick
Max Planck Institute of Psychiatry, Munich, Germany

Kathinka Evers
Centre for Research Ethics & Bioethics, Department of Public Health and Caring Sciences, Uppsala University, Uppsala, Sweden

Laura M. Fiori
McGill Group for Suicide Studies, Douglas Mental Health University Institute, Department of Psychiatry, McGill University, Montreal, QC, Canada

Mats G. Hansson
Centre for Research Ethics & Bioethics, Department of Public Health and Caring Sciences, Uppsala University, Uppsala, Sweden

Ali Jawaid
Laboratory of Neuroepigenetics, Medical Faculty of the University of Zurich and Department of Health Science and Technology of the Swiss Federal Institute of Technology, Neuroscience Center Zurich, Zurich, Switzerland

Chelsey Ju
McGill Group for Suicide Studies, Douglas Mental Health University Institute, Department of Psychiatry, McGill University, Montreal, QC, Canada

Ulrik Kihlbom
Centre for Research Ethics & Bioethics, Department of Public Health and Caring Sciences, Uppsala University, Uppsala, Sweden

Grace S. Kim
Neuroscience Program, University of Illinois at Urbana–Champaign; Medical Scholars Program, University of Illinois College of Medicine, Urbana, IL, United States

Ilona Koupil
Department of Public Health Sciences, Stockholm University; Department of Public Health Sciences, Karolinska Institutet, Stockholm, Sweden

Laurence Lanfumey
Inserm UMR S894, Centre de Psychiatrie et Neuroscience, Université Paris Descartes, Paris, France

Roy Lardenoije
Department of Psychiatry and Neuropsychology, Maastricht University Medical Centre, Maastricht, The Netherlands

Rixing Lin
McGill Group for Suicide Studies, Douglas Mental Health University Institute, Department of Psychiatry, McGill University, Montreal, QC, Canada

Katie Lunnon
University of Exeter Medical School, University of Exeter, Exeter, United Kingdom

Isabelle M. Mansuy
Laboratory of Neuroepigenetics, Medical Faculty of the University of Zurich and Department of Health Science and Technology of the Swiss Federal Institute of Technology, Neuroscience Center Zurich, Zurich, Switzerland

Raymond Mongeau
EA 4475, Pharmacologie de la Circulation Cérébrale, Faculté de Pharmacie, Université Paris Descartes, Paris, France

Bertram Mueller-Myhsok
Department of Translational Research in Psychiatry, Max Planck Institute of Psychiatry; Munich Cluster for Systems Neurology (SyNergy), Munich, Germany; University of Liverpool, Liverpool, United Kingdom

Caroline M. Nievergelt
Department of Psychiatry, University of California, San Diego, La Jolla, CA, United States

Ehsan Pishva
Department of Psychiatry and Neuropsychology, Maastricht University Medical Centre, Maastricht, The Netherlands; University of Exeter Medical School, University of Exeter, Exeter, United Kingdom

Nadine Provencal
Faculty of Health Sciences, Simon Fraser University, Burnaby; Healthy Starts Theme, BC Children's Hospital Research Institute, Vancouver, BC, Canada

Giovanna Punzi
Lieber Institute for Brain Development, Johns Hopkins Medical Campus, Baltimore, MD, United States

Elizabeth J. Radford
Department of Pediatrics, University of Cambridge, Cambridge, United Kingdom

Mathilde Règue-Guyon
Inserm UMR S894, Centre de Psychiatrie et Neuroscience, Université Paris Descartes, Paris, France

Chiara Renzi
Applied Research Division for Cognitive and Psychological Science, European Institute of Oncology, Milan, Italy

Martin Roszkowski
Laboratory of Neuroepigenetics, Medical Faculty of the University of Zurich and Department of Health Science and Technology of the Swiss Federal Institute of Technology, Neuroscience Center Zurich, Zurich, Switzerland

Bart P.F. Rutten
Department of Psychiatry and Neuropsychology, School for Mental Health and Neuroscience (MHeNs), European Graduate School of Neuroscience (EURON), Faculty of Health, Medicine and Life Sciences, Maastricht University Medical Centre +, Maastricht, The Netherlands

Alicia K. Smith
Department of Gynecology and Obstetrics; Department of Psychiatry and Behavioral Sciences, Emory University, Atlanta, GA, United States

Clara Snijders
School for Mental Health and Neuroscience (MHeNS), Department of Psychiatry and Neuropsychology, Faculty of Health, Medicine and Life Sciences, Maastricht University, Maastricht, The Netherlands

Alix Thomson
School for Mental Health and Neuroscience (MHeNS), Department of Psychiatry and Neuropsychology, Faculty of Health, Medicine and Life Sciences, Maastricht University, Maastricht, The Netherlands

Gustavo Turecki
McGill Group for Suicide Studies, Douglas Mental Health University Institute, Department of Psychiatry, McGill University, Montreal, QC, Canada

Monica Uddin
Neuroscience Program; Department of Psychology; Carl R. Woese Institute for Genomic Biology, University of Illinois at Urbana–Champaign, Urbana, IL, United States

Gianluca Ursini
Lieber Institute for Brain Development, Johns Hopkins Medical Campus; Department of Psychiatry and Behavioral Sciences, Johns Hopkins University School of Medicine, Baltimore, MD, United States

Daniel L. van den Hove
Department of Psychiatry and Neuropsychology, Maastricht University Medical Centre, Maastricht, The Netherlands

Ramona A.J. Zwamborn
Department of Neurology and Neurosurgery, Brain Center Rudolf Magnus, University Medical Center Utrecht, Utrecht, The Netherlands

CHAPTER ONE

Neuroepigenetics of Mental Illness: The Inside Outs of the Outside Within

Bart P.F. Rutten[1]
Department of Psychiatry and Neuropsychology, School for Mental Health and Neuroscience (MHeNs), European Graduate School of Neuroscience (EURON), Faculty of Health, Medicine and Life Sciences, Maastricht University Medical Centre +, Maastricht, The Netherlands
[1]Corresponding author: e-mail address: b.rutten@maastrichtuniversity.nl

Contents

Abstract

Epigenetics refers to environmentally sensitive modifications to DNA and chromatin that regulate gene transcription without altering the genetic sequence itself. Because of the brain's central role in a person's adaptation to dynamic changes in the environment, the field of epigenetic research is particularly pertinent for the neurosciences and mental health and illness. "Neuroepigenetics" refers to the field of epigenetics, as applied to research of the nervous system and related functional abilities. The onset and course of mental disorders revolve around person–environment interactions, i.e., the interplay between environmental factors with people (and their brain) throughout life and on the background of an individual's inherited genomic make-up. With respect to the nervous system and its functional abilities, neuroepigenetic research aims to turn the inside outs by *explicating* the epigenetic profiles that regulate gene expression *within* cells while epigenetic profiles themselves may reflect the imprints of external factors (the outside of the organism). Neuroepigenetic research is of prime importance for elucidating the molecular underpinnings of gene–environment interplay in relation to functional abilities of the brain and mental health throughout life, and thus for identifying the mechanisms underlying onset and course of mental disorders. However, neuroepigenetic research is still in its infancy and many conceptual and methodological challenges are apparent. In addition, there are considerable complexities and major

Progress in Molecular Biology and Translational Science, Volume 158
ISSN 1877-1173
https://doi.org/10.1016/bs.pmbts.2018.04.003

challenges in conceptualizing the nature and diagnosis of mental disorders, and these need to be taken into account in studying neuroepigenetics of mental disorders. The current manuscript provides an overview and conceptual framework for neuroepigenetic research in mental disorders.

Epigenetics refers to environmentally sensitive modifications to DNA and chromatin that regulate gene transcription without altering the genetic sequence itself. The field of epigenetic research is particularly pertinent for the neurosciences and mental health and illness.

The onset and course of mental disorders revolve around person–environment interactions, i.e., the interplay between environmental factors with people (and their brain) throughout life and on the background of an individual's inherited genomic make-up. Thus, research on neuroepigenetics seems of prime importance for elucidating the molecular underpinnings of gene–environment interplay in relation to functional abilities of the brain and mental health throughout life as well as for elucidating the mechanisms underlying onset and course of mental disorders. As (i) neuroepigenetic research aims to *explicate* the epigenetic profiles that regulate gene expression *within* cells and (ii) these epigenetic profiles within cells are subject to external/environmental impact in relation to mental health and illness, this chapter is entitled *Neuroepigenetics of mental illness: The inside outs of the outside within.*

Through providing an overview and conceptual framework for neuroepigenetic research in mental disorders, the current chapter also serves as an introduction to the series of manuscripts comprising the book "Neuroepigenetics of Mental Illness."

1. THE EPIGENETIC MACHINERY

The epigenetic machinery covers multiple levels of regulation of gene expression and includes modification to DNA such as methylation and hydroxymethylation, remodeling of chromatin and non-coding RNA (ncRNA).[1] In contrast to genetic variations which usually reflect permanent characteristics of the DNA sequence, epigenetic changes are in principle reversible.[2] Further information on the various epigenetic modification and measurements of epigenetic modifications are provided in elsewhere (refer Snijders et al.[2a] and Radford[2b]). Although DNA methylation, chromatin remodeling, and ncRNAs represent the most well-known epigenetic mechanisms, recent work has also identified other epigenetic mechanisms such as RNA editing,

RNA methylation and mitochondrial epigenetics and knowledge in these areas is rapidly evolving. It should furthermore be noted that epigenetic mechanisms of DNA (de)methylation, chromatin remodeling, and ncRNAs do not act independently, but closely interact with each other to form a complex regulatory system that dynamically fine-tunes gene expression in response to local or external factors, see, e.g., Refs. 3,4.

While having the opportunity to make use of a constantly improving toolbox of techniques for measuring epigenetic modifications (refer Snijders et al.[2a]), the field of epigenetic epidemiology is challenged by several conceptual and methodological aspects.[5] These comprise (among others) the required sample sizes for detecting significant effects (particularly for genome-wide datasets requiring corrections for multiple comparisons), tissue and cell-type specificity of epigenetic profiles, the need for replication and validation approaches, the need for longitudinal analyses in order to develop our research efforts from association studies using cross-sectional designs to (statistical mediation) analyses in longitudinal genetically-sensitive designs, and the plethora of possible confounders (environmental and genetic). For example, genetic effects on inter-individual variation in DNA methylation, methylation quantitative trait loci, complicate the interpretation of data on epigenetic profiles, and availability of information on genetic variation of the same subjects is therefore of great value for appropriate interpretation of epigenetic data. The Genetics of DNA Methylation Consortium (GoDMC) (www.godmc.org.uk/) provides a framework for bringing together genome-wide datasets of DNA methylation and genetic variants and may serve as reference source and a basis for integrating genomic and epigenomic datasets. In addition, genome-wide analyses of epigenetic profiles also require innovative developments in generating a bioinformatic toolbox to analyze and interpret the generated datasets, such as described elsewhere.[6]

Besides this, the analysis and interpretation of epigenetic data is currently hampered by limited availability of reference datasets. However, recent developments in projects such as ENCODE2, the US National Institutes of Health Epigenomics Roadmap, the Accessible Resource for Epigenomic Studies and the International Human Epigenome Consortium are generating important reference data sets, also for developmental and aging trajectories in a range of tissue types, including brain.

Evidence accumulates that epigenetic mechanisms play crucial roles in regulating gene expression during and after exposures to environmental factors across the entire life span of mammals. The current evidence that epigenetic mechanisms may mediate the impact of environmental factors on the phenotype of interest mainly stems from experimental animal studies while a first wave of human epigenetic studies, see, e.g., Refs. 7–10, has provided further support for the concept of epigenetic mediation of environmental influences.

As epigenetic mechanisms mediate proliferation and tissue and cell-type specification and differentiation, it is logical that different cell and tissue types also harbor differences in epigenetic imprints. The central nervous system is the site of primary interest for mental disorders, the limited availability of brain tissue as well as the timing of collection of brain samples in humans challenge neuroepigenetic research considerably. High-standard brain biobanks, such as, for example, the Netherlands Brain Bank, the National Institute of Mental Health, the Stanley Foundation and the tissue bank of the Banner Sun Health Research Institute besides human brain data repositories such as the Allen Brain Atlas as well as Brain Cloud of the Lieber Institute for Brain Development are therefore of pivotal importance. Besides limited availability of well-characterized and high-quality brain samples, it should be noted that epigenetic read-outs from post-mortem analyses of brain samples represent an end stage of the often chronic course of mental ill-health in these individuals and that the read-outs are thus likely influenced by numerous possible co-variates and confounding variables including a range of factors which do not necessarily reflect the actual course of illness, e.g., social isolation, use of medication, drugs, alcohol and/or nicotine. Besides these, measured epigenetic variation can also arise from stochastic factors or be the consequence of methodological factors in using post-mortem samples such as those related to long post-mortem interval periods and long-term storage of material (e.g., repeated freezing and thawing).

In order to assess epigenetic modifications across life and in close temporal relation to changes in mental health phenotypes, it would be very attractive if epigenetic profiles in accessible tissue such as blood or buccal swab samples would reflect (at least part) of the epigenetic profiles of the same individual in individual in the brain or if they would correlate with the diagnosis or symptom severity. Some recent studies have suggested that epigenetic profiles in peripherally accessible cell types such as blood, buccal epithelial cells and saliva may indeed be worthwhile (see below). Non-coding RNAs such as microRNAs (miRNAs) are of particular importance in this respect. For example, these miRNAs can inhibit expression of several target genes, act as signaling molecules throughout the body and be secreted by neuronal and non-neuronal cells in response to an event thereby influencing gene expression in other cell types in the organism. miRNAs and other non-coding RNAs can, also be secreted by non-neuronal and neuronal cells in the brain, freely circulating in the cerebrospinal fluid or encapsulated within extracellular vesicles, and can cross the blood–brain barrier, circulate within the blood, influence cells at distance or being secreted into the urine.[11]

With respect to DNA methylation, studies have furthermore indicated that a proportion of the blood-based DNA methylation profiles may be linked with DNA methylation profiles in the brain. Although the general consensus is that different cell types (and tissue types) vary strongly in their DNA methylation profiles, a recent study reported that inter-individual differences in DNA methylation do show some correlations across brain and blood.[12] For the majority of DNA methylation sites in this study using Infinium 450K Bead Chip arrays, inter-individual variation in whole blood did not correlate strongly with inter-individual variation in the brain. However, variation at a subset of the 450K probes did correlate across tissues, even in instances when the actual level of DNA methylation was significantly different between them. While these latter correlations may co-depend on genetic variation, they also open the possibility that brain-localized epigenetic modifications (at least for some probes) may be reflected in more accessible samples such as blood, saliva and/or buccal epithelial cells.

Another line of argumentation for possible relevance of epigenetic profiling in peripheral tissues (including blood) for understanding mental disorders is that the biology underlying the pathophysiology of many mental and neurodegenerative disorders not only impacts or localizes within the brain, but that tissues and cell types outside of the brain are also affected and may even mediate some of the links between risk factors and mental phenotypes. For example, evidence is accumulating for a prime role of the immune system in a range of mental disorders and that changes of immune cell types can be traced in blood samples.

An interesting further suggestion on the possibility that severe stress impacts on a range of cell types throughout the body and that epigenetic imprint can be traced via peripheral samples, is found in a recent longitudinal epigenetic study on the impact of traumatic stress on mental health in relation to DNA methylation profiles measured in whole blood samples.[13] It should be noted studying whole blood samples does not allow for (blood) cell type specific investigations, which seem relevant given different functions of blood-based cell types. Nonetheless, this prospective methylome-wide study reported that longitudinal changes in the level of symptoms of post-traumatic stress disorder (PTSD) 1 month before and 6 months after a deployment period were significantly associated with longitudinal changes in DNA methylation profiles.[13] Thus, based on blood-based findings on longitudinal changes across a period of exposure to robust trauma, it is tempting to speculate that (1) these blood-based DNA methylation changes may reflect similar changes in DNA methylation in the brain, and/or (2) that perhaps DNA methylation profiles of peripheral stem cells such as hematopoietic stem cells are impacted by severe, long-term stress. Thus, while the findings require further replications, validations and extrapolation to other populations and stress exposures, they do provide a first basis to explore whether it may indeed be possible to tap into the mechanisms underlying mental ill-health via accessible biological samples in living humans. However, significant progress in establishing tissue, and cell-type specificity of epigenetic profiles across the life span, and in response to environmental exposures is needed in order to improve the interpretation of research findings.

Another window of opportunities and challenges in the field of epigenetic research lies in the revitalized concept of epigenetic inheritance. The classical concept that the molecular biology of transgenerational inheritance only involves genetic factors has been challenged by findings in animal studies that environmentally induced epigenetic profiles can also be transmitted to several generations of offspring after the original exposure. This field has re-opened the Lamarckian concept of heritability of acquired traits. Rodent studies have provided evidence that maternal separation and subsequent stressors during development impact on behavior and biology across multiple generations of offspring. Although there is some suggestive evidence for inheritance of the impact of environmental exposures across generations in humans, it needs to be established whether epigenetic inheritance is indeed relevant for humans. The current status of epigenetic inheritance of stress is reviewed elsewhere (refer Jawaid et al.[13a]).

2. THE NATURE OF MENTAL DISORDERS

The onset and course of mental disorders revolve around person–environment interactions, and neuroepigenetics thus is of prime importance for elucidating the molecular underpinnings of gene–environment interplay in relation to functional abilities of the brain and mental health throughout life as well as for elucidating the mechanisms underlying onset and course of mental disorders. However, considerable complexities and challenges have been identified in conceptualizing the nature and diagnosis of mental disorders. The complex and debatable nosology of mental disorders and related classification systems has a tremendous impact on scientific attempts in trying to identify links between (neuro)biology and expression of mental ill-health, and developments in this area off course need to be taken into account in research on neuroepigenetic underpinning of mental disorders.

2.1 Complexity of Developmental Programming and the Brain

It is widely accepted that development of an individual with a distinct genetic make-up occurs under the constant influence of experiences and exposures during life culminating into a person with distinct characteristics such as self-esteem, intellectual capacities, stability of regulation of emotion, thoughts and behavior and habitual styles for coping with significant challenges in life. One can thus assume that all these complex developmental trajectories underlying the huge range of functional abilities are mediated by the brain and are subject to regulation through epigenetic mechanisms. Nevertheless, evidence on epigenetic mechanisms across normal developmental trajectories of the mammalian brain and in response to developmental experiences is limited, and mainly stems from developmental disturbances in monogenic disorders with mutation in epigenetically relevant genes, see also elsewhere (refer Radford[2b]). In order to better understand links between neuroepigenetics and functional abilities of a person, it is thus of prime essence to deepen and broaden the knowledge on the theme of developmental programming and environmental epigenetics during development. In order to relate measurable epigenetic profiles with functional abilities of the brain during development and aging of living humans, the field is in need of adequately designed longitudinal studies preferably starting from (or before) birth and collecting accessible tissue samples in humans as well as measuring behavioral and experiential phenotypes and monitoring environmental exposures across life.

The majority of current epigenetic studies in relation to mental disorders in humans involve cross-sectional studies comparing tissue samples of people diagnosed with a certain mental disorder with samples from people without a diagnosis. Over the last decades diagnoses of mental disorders have been categorized in classification systems such as the Diagnostic Statistical Manual (DSM) and International Statistical Classification of Diseases and Related Health Problems (ICD), and these bimodal (i.e., diagnosis "yes" or "no") classification systems have been used also in epigenetic studies on mental disorders. Whereas such studies do provide a first step in attempts to identify neuroepigenetic mechanisms underlying onset and course of mental disorders, the interpretation and relevance of them does suffer from the same critical notions about these diagnostic systems as also apparent in the wider psychiatry community and in research on mental health and illness. It is clear that these bimodal categorical classification systems do not provide an adequate basis for describing the biopsychosocial complexities of the nature of mental disorders nor the dynamic changes in quantitative phenotypes (including biological) across life,[14] and alternative notions have been proposed. It has, for example, been acknowledged that the "comorbidities" of categorized classifications of mental disorders actually reflect variations in underlying, transdiagnostic dimensions which may be reflected variation in human behavior, genetic variation, and/or neurobiological function, see, e.g., for psychosis.[15,16] Thus, clinicians and researchers are increasingly focusing on dimensional approaches to explain variation in functional abilities of the brain and related psychopathology, which cut across classical (DSM- or ICD-based) diagnostic entities. This shift toward dimensional approaches intends to more adequately assess clinically and scientifically relevant phenomena and to increase attention to and understanding of the causes and consequences of such variation. Initiated by the National Institute of Mental Health in the United States, the Research Domain Criteria (RDoC) provides a further alternative attempt to innovate research in mental health and illness.[17] While DSM and ICD approaches conceptualize mental disorders as largely discrete entities that are characterized by distinctive signs, symptoms, and natural histories, the RDoC approach conceptualizes psychopathology as the product of dysfunctions in neural circuitries, which can be traced at molecular, cellular and behavioral levels.[18]

2.2 Complexity of Dynamic, Rather Than Static, Concepts of Mental Illness

Recent research efforts to predict which individuals will progress to full-threshold (i.e., DSM or ICD diagnosable) clinically expressed mental illness have focused on early diagnosis of subthreshold, prodromal stages.[19,20] In this respect, it should be noted that phenotypes such as affective state, cognitive abilities and behavior of a certain individual are known to change dynamically across different periods of life but also across time periods of weeks, days, hours or even minutes. It thus seems unlikely that cross-sectional sampling of biological, clinical or psychological data of principally dynamic features (e.g., a snapshot of an epigenetic profile) can thus fully predict future emergence of mental illness or progression to more advanced stages of a mental disorder. Thus, conceptual models and analytical techniques need to incorporate the dynamic nature of life, biology and mental (ill)health and it seems evident that neuroepigenetic studies on mental illness should incorporate dynamic network models on changes in variables (including epigenetic) and networks in relation to transdiagnostic, dimensional changes of experiences and symptoms.

Another noteworthy issue is that RDoC, DSM and ICD are all based on natural classification systems, which are built on the assumption of causality between variables. Alternative notions have suggested that mental disorders can possibly also be understood using latent variable analyses or network models.[21] For example, it has been proposed that a range of childhood behaviors could be characterized into broad dimensions such as introversion and extraversion and that "comorbidity" among common mood, anxiety, and substance use disorders in life could be better explained by the latent dimensional constructs rather than by traditional DSM-based diagnostic categories.[22]

In addition to this notion, it has been noted that dimensions of mental phenotypes and their underlying biological mechanisms may act as complex systems, and that each system may have a different type of constitutive architecture and different characteristics which may give rise to important differences, particularly in dynamics of transitional phases during onset and course of mental disorders.[23] While these concepts on system dynamics can be applied to longitudinal changes in symptoms and momentary experiences and behavior in the flow of daily life,[24] it would be very interesting (though challenging) to conduct corresponding dynamic system analyses also using longitudinally collected measures of biological read-out parameters, such as, e.g., epigenetic and gene expression profiles.

3. AN INTEGRATIVE AND DYNAMIC CONCEPTUAL MODEL

Accumulating evidence that gene-by-environment interactions mediate and moderate vulnerability to and dynamic changes in mental illness phenotypes has changed the conventional understanding of the etiology of mental disorders. Besides genetic variants that moderate the impact of environmental influences on gene expression and behavioral phenotypes, recent evidence has shown that epigenetic mechanisms may moderate as well as

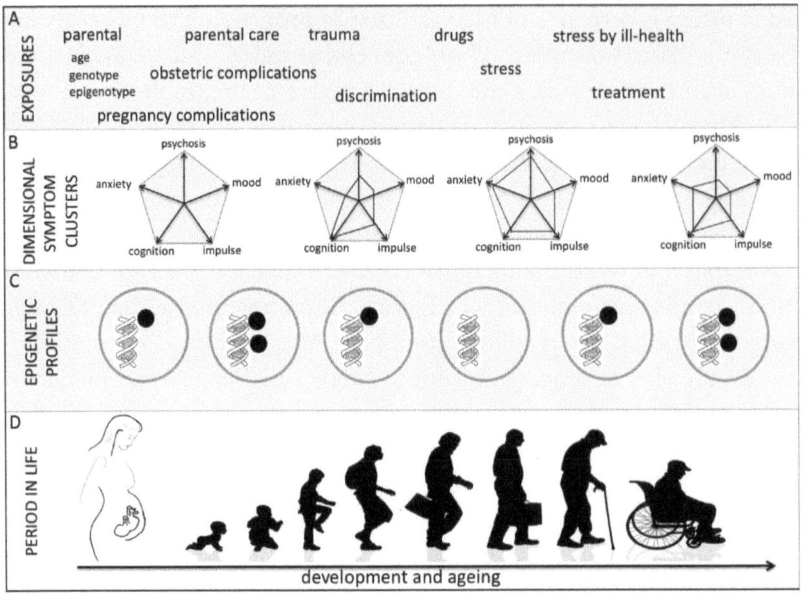

Fig. 1 Conceptual model of the dynamics of gene–environment interplay and neuroepigenetics in relation to dimensional changes in symptoms of mental disorders.

mediate environmental influences across the lifespan. Fig. 1 illustrates a current model integrating genetic, environmental and epigenetic factors in the dynamic changes of symptom dimension involved in mental ill-health. Inherited genetic and epigenetic profiles constitute the molecular basis of an individual to develop over the life course, during which environmental exposures interact with genetic and epigenetic profiles and may furthermore lead to long-term alterations of the epigenome, altering expression of genes influencing behavioral and experiential dimensional phenotypes of individuals.

4. FUTURE PERSPECTIVES

While epigenetic mediation of environmental factors in the development of mental disorders has generated considerable recent interest, little evidence is currently available and, when available, epigenetic findings should be interpreted cautiously. In this respect, it is important to realize that (1) most of the findings mentioned above have not (yet) been replicated and were obtained with limited samples sizes, (2) molecular methodologies assessing epigenetic profiles are developing rapidly and the "normal" epigenome is still largely unknown, (3) most studies on environmentally induced epigenetic changes were performed on animals, and (4) the current wave of human studies has mainly been performed on patients with chronic forms of mental disorders.

While struggling with aligning etiological and clinical nosological challenges and being cautious in interpreting the current findings, the current state of the literature has allowed researchers to summarize and review the published findings in comprehensive manners, e.g., connected to distinct neurobiological processes such on the neurotrophin signaling (refer Règue-Guyon et al.[24a]), the hypothalamus–pituitary–adrenal axis (refer Dick and Provencal[24b]) and the WNT signaling pathways (refer Zwamborn et al.[24c]), timing of environmental exposures during life (refer Claes[24d]), to aging (refer Lardenoije et al.[24e]), to current classifications of diagnostic categories of mental disorders such as psychotic disorders (refer Punzi et al.[24f]) and PTSD (refer Kim et al.[24g]), to treatment response (refer Fiori et al.[24h]) as well as on transgenerational epigenetic inheritance (refer Jawaid et al.[13a]).

The field is currently in great need of prospective studies on longitudinal changes of transdiagnostic (latent) liability phenotypes combined with assessments of multiple relevant (adverse and protective/beneficial) environmental exposures as well as multi-level biological profiling (including genetic and epigenetic) across life. Given the confounding effects of several disorder-related exposures such as medication and social disadvantage, it will be crucial to conduct such studies by capturing (and following) the relevant dynamic and network-based factors involved (clinical, social, psychological and biological) in a prospective manner in the general population and populations at high-risk for robust exposures. In addition to this, it should be overlooked that much progress is also needed in the more fundamental neurosciences. By conducting experimental studies; i.e., after the phase of discovery (and replication), candidate epigenetic marks need to be experimentally manipulated in appropriate cell or animal model systems in order to start testing for causality.

Taken together, the current state of the literature points out that genes and environment do not act independently and that the nature–nurture dichotomy is much less rigorous as previously thought. Given the possibility of great social and ethical implications, current research efforts should incorporate social, ethical and philosophical concepts (see Renzi et al.[24i]). Such integration of what has often been seen as separate research approaches holds new promise for studies of neuroepigenetics of mental disorders.

REFERENCES

1. Adwan L, Zawia NH. Epigenetics: a novel therapeutic approach for the treatment of Alzheimer's disease. *Pharmacol Ther.* 2013;139(1):41–50. https://doi.org/10.1016/j.pharmthera.2013.03.010.
2. Henikoff S, Matzke MA, Mauro ED. Exploring and explaining epigenetic effects. *Trends Genet.* 1997;13(8):293–295. https://doi.org/10.1016/S0168-9525(97)01219-5. (a) Snijders C, Bassil KC, de Nijs L. Methodologies of neuroepigenetic research: background, challenges and future perspectives. *Prog Mol Biol Transl Sci.* 2018;158:15–27. (b) Radford EJ. An introduction to epigenetic mechanisms. *Prog Mol Biol Transl Sci.* 2018;158:29–48.
3. Sun G, Yu RT, Evans RM, Shi Y. Orphan nuclear receptor TLX recruits histone deacetylases to repress transcription and regulate neural stem cell proliferation. *Proc Natl Acad Sci USA.* 2007;104(39):15282–15287. https://doi.org/10.1073/pnas.0704089104.
4. Wu H, Coskun V, Tao J, et al. Dnmt3a-dependent nonpromoter DNA methylation facilitates transcription of neurogenic genes. *Science.* 2010;329(5990):444–448.
5. Mill J, Heijmans BT. From promises to practical strategies in epigenetic epidemiology. *Nat Rev Genet.* 2013;14(8):585–594. https://doi.org/10.1038/nrg3405.
6. Angarica VE, del Sol A. Bioinformatics tools for genome-wide epigenetic research. *Adv Exp Med Biol.* 2017;978:489–512. https://doi.org/10.1007/978-3-319-53889-1_25.
7. Heijmans BT, Mill J. Commentary: the seven plagues of epigenetic epidemiology. *Int J Epidemiol.* 2012;41(1):74–78. https://doi.org/10.1093/ije/dyr225.
8. Houtepen LC, Vinkers CH, Carrillo-Roa T, et al. Genome-wide DNA methylation levels and altered cortisol stress reactivity following childhood trauma in humans. *Nat Commun.* 2016;7:10967. https://doi.org/10.1038/ncomms10967.
9. Klengel T, Mehta D, Anacker C, et al. Allele-specific FKBP5 DNA demethylation mediates gene-childhood trauma interactions. *Nat Neurosci.* 2013;16(1):33–41. https://doi.org/10.1038/nn.3275.
10. Tobi EW, Lumey LH, Talens RP, et al. DNA methylation differences after exposure to prenatal famine are common and timing- and sex-specific. *Hum Mol Genet.* 2009;18(21):4046–4053. https://doi.org/10.1093/hmg/ddp353.
11. Pegtel DM, Peferoen L, Amor S. Extracellular vesicles as modulators of cell-to-cell communication in the healthy and diseased brain. *Philos Trans R Soc Lond B Biol Sci.* 2014;369(1652) pii: 20130516. https://doi.org/10.1098/rstb.2013.0516. PubMed PMID: 25135977; PubMed Central PMCID: PMC4142037.
12. Hannon E, Lunnon K, Schalkwyk L, Mill J. Interindividual methylomic variation across blood, cortex, and cerebellum: implications for epigenetic studies of neurological and neuropsychiatric phenotypes. *Epigenetics.* 2015;10(11):1024–1032. https://doi.org/10.1080/15592294.2015.1100786.
13. Rutten BPF, Vermetten E, Vinkers CH, et al. Longitudinal analyses of the DNA methylome in deployed military servicemen identify susceptibility loci for post-traumatic stress disorder. *Mol Psychiatry.* 2018;23(5):1145–1156. https://doi.org/10.1038/mp.2017.120. Epub 2017 Jun 20. PubMed PMID: 28630453. (a) Jawaid A, Roszkowski M, Mansuy IM. Transgenerational epigenetics of traumatic stress. *Prog Mol Biol Transl Sci.* 2018;158:273–298.
14. Kendler KS. The nature of psychiatric disorders. *World Psychiatry.* 2016;15(1):5–12. https://doi.org/10.1002/wps.20292. PubMed PMID: 26833596; PubMed Central PMCID: PMC4780286.
15. Barch DM, Bustillo J, Gaebel W, et al. Logic and justification for dimensional assessment of symptoms and related clinical phenomena in psychosis: relevance to DSM-5. *Schizophr Res.* 2013;150(1):15–20. https://doi.org/10.1016/j.schres.2013.04.027. Epub 2013 May 22. Review. PubMed PMID: 23706415.

16. van Os J, Kenis G, Rutten BP. The environment and schizophrenia. *Nature.* 2010;468(7321):203–212. https://doi.org/10.1038/nature09563. PubMed PMID: 21068828.
17. Insel T, Cuthbert B, Garvey M, et al. Research domain criteria (RDoC): toward a new classification framework for research on mental disorders. *Am J Psychiatry.* 2010;167(7):748–751. https://doi.org/10.1176/appi.ajp.2010.09091379. PubMed PMID: 20595427.
18. Lilienfeld SO, Treadway MT. Clashing diagnostic approaches: DSM-ICD versus RDoC. *Annu Rev Clin Psychol.* 2016;12:435–463. https://doi.org/10.1146/annurev-clinpsy-021815-093122. Epub 2016 Feb 3. Review. PubMed PMID: 26845519; PubMed Central PMCID: PMC5154554.
19. Yung AR, Nelson B. The ultra-high risk concept—a review. *Can J Psychiatry.* 2013;58(1):5–12. PubMed PMID: 23327750.
20. Yung AR, Fusar-Poli P, Nelson B. The ultra high risk approach to define psychosis risk. *Curr Pharm Des.* 2012;18(4):346–350. Review. PubMed PMID: 22239565.
21. Zachar P, Kendler KS. The philosophy of nosology. *Annu Rev Clin Psychol.* 2017;13:49–71. https://doi.org/10.1146/annurev-clinpsy-032816-045020. Review. PubMed PMID: 28482691.
22. Kim H, Eaton NR. The hierarchical structure of common mental disorders: connecting multiple levels of comorbidity, bifactor models, and predictive validity. *J Abnorm Psychol.* 2015;124(4):1064–1078. https://doi.org/10.1037/abn0000113. PubMed PMID: 26595482.
23. Nelson B, McGorry PD, Wichers M, Wigman JTW, Hartmann JA. Moving from static to dynamic models of the onset of mental disorder: a review. *JAMA Psychiatry.* 2017;74(5):528–534. https://doi.org/10.1001/jamapsychiatry.2017.0001. Review. PubMed PMID: 28355471.
24. Hasmi L, Drukker M, Guloksuz S, et al. Network approach to understanding emotion dynamics in relation to childhood trauma and genetic liability to psychopathology: replication of a prospective experience sampling analysis. *Front Psychol.* 2017;8:1908. https://doi.org/10.3389/fpsyg.2017.01908. eCollection 2017. PubMed PMID: 29163289; PubMed Central PMCID: PMC5673657. (a) Règue-Guyon M, Lanfumey L, Mongeau R. Neuroepigenetics of neurotrophin signaling: neurobiology of anxiety and affective disorders. *Prog Mol Biol Transl Sci.* 2018;158:159–193. (b) Dick A, Provencal N. Central neuroepigenetic regulation of the hypothalamic–pituitary–adrenal axis. *Prog Mol Biol Transl Sci.* 2018;158:105–127. (c) Zwamborn RAJ, Snijders C, An N, et al. Wnt signaling in the hippocampus in relation to neurogenesis, neuroplasticity, stress and epigenetics. *Prog Mol Biol Transl Sci.* 2018;158:129–157. (d) Claes S. Neuroepigenetics of prenatal psychological stress. *Prog Mol Biol Transl Sci.* 2018;158:83–104. (e) Lardenoije R, Pishva E, Lunnon K, et al. Neuroepigenetics of aging and age-related neurodegenerative disorders. *Prog Mol Biol Transl Sci.* 2018;158:49–82. (f) Punzi G, Bharadwaj R, Ursini G. Neuroepigenetics of schizophrenia. *Prog Mol Biol Transl Sci.* 2018;158:195–226. (g) Kim GS, Smith AK, Nievergelt CM, et al. Neuroepigenetics of post-traumatic stress disorder. *Prog Mol Biol Transl Sci.* 2018;158:227–253. (h) Fiori LM, Lin R, Ju C, et al. Using epigenetic tools to investigate antidepressant response. *Prog Mol Biol Transl Sci.* 2018;158:255–272. (i) Renzi C, Provencal N, Bassil KC, et al. From epigenetic associations to biological and psychosocial explanations in mental health. *Prog Mol Biol Transl Sci.* 2018;158:299–323.

CHAPTER TWO

Methodologies of Neuroepigenetic Research: Background, Challenges and Future Perspectives

Clara Snijders, Katherine C. Bassil, Laurence de Nijs[1]
School for Mental Health and Neuroscience (MHeNS), Department of Psychiatry and Neuropsychology, Faculty of Health, Medicine and Life Sciences, Maastricht University, Maastricht, The Netherlands
[1]Corresponding author: e-mail address: laurence.denijs@maastrichtuniversity.nl

Contents

Abstract

Over the last years, interest in epigenetic mechanisms has strongly increased in the field of neuroscience. Neuroepigenetics has intensely evolved and now refers to the assessment of a variety of epigenetic marks which can be found across several regions of the healthy or diseased brain. These marks include DNA (hydroxy)methylation, a large diversity of post-translational histone modifications and an increasing number of non-coding RNAs. The present chapter aims to concisely summarize the techniques used to study these mechanisms in the brain and provides an overview of their current challenges along with future perspectives that will allow the field to move forward.

Progress in Molecular Biology and Translational Science, Volume 158
ISSN 1877-1173
https://doi.org/10.1016/bs.pmbts.2018.04.009

1. INTRODUCTION

The field of epigenetics only recently emerged, challenging the discussion of *nature vs. nurture*. These two phenomena ceased to be regarded as dichotomous entities and epigenetics formed the joining entity allowing for an active interplay between the genome and the environment.[1] It is now clearly established that environmental experiences such as psychological or physical stress, education, and drug usage have a lasting influence on neuronal function through epigenetic alterations of gene expression.[1,2] Nowadays, the impact of life-events on the regulation of epigenetic mechanisms in the central nervous system (CNS) is extensively studied under the rapidly expanding field of neuroepigenetics.[1]

Neuroepigenetics, the name initially opted for by Jeremy Day,[3] focuses on epigenetic mechanisms that play a crucial role in both healthy and pathological brain functions, not only throughout an individual's lifespan, but also across generations[4] and evolution.[5] Some of the major mechanisms involved are covalent modifications of DNA, of which DNA methylation and hydroxymethylation are the most prominently studied, histone post-translational modifications (PTMs), and non-coding RNAs (ncRNAs). Recent discoveries have shed light on a range of epigenetic mechanisms and their key role in developmental and behavioral processes, synaptic plasticity, neurotoxicity, drug dependency and CNS disorders like Alzheimer's disease, psychiatric disorders or epilepsy.[1] These findings highlight the involvement of epigenetic mechanisms in the progression from a "healthy" to a "diseased" brain state, or *vice versa*. Over the past years, the methods used to study these mechanisms have strongly evolved. However, these tools should be viewed in light of their own strengths and limitations in order to optimize their utilization and promote their constant evaluation and revision. Therefore, this introductory chapter will discuss the methodologies used to study epigenetic mechanisms in the CNS by providing an overview of the most commonly used methods, the challenges they bring and suggestions on how these can be overcome.

2. METHODOLOGIES OF NEUROEPIGENETIC RESEARCH

2.1 DNA Methylation and Hydroxymethylation

Initial epigenetic studies have mainly examined alterations in DNA methylation which is believed to be the most stable epigenetic modification. DNA methylation is the process through which a methyl group is covalently added to the carbon-5 position of cytosine residues to form 5-methylcytosine (5-mC).[6] This process is catalyzed by DNA methyltransferases and predominantly occurs in CpG dinucleotides.[6] As a result, and depending on its exact location outside or within a gene, this chemical modification can influence gene expression by triggering transcriptional repression, activating gene expression or alternative splicing.[7–9] More recently, the 5-hydroxymethylcytosine (5-hmC) modification, the oxidized form of 5-mC, which is catalyzed by enzymes of the ten-eleven translocation (Tet) family, has been given increasing attention, especially because of its abundance in the brain.[10,11]

The analysis of the chemical modifications of DNA is well documented and can be categorized into approaches that detect global DNA methylation, those that yield information about the methylation status of specific CpG sites, and more advanced, high-coverage, epigenome-wide methods. The first studies that examined the chemical modifications of DNA in the brain belonged to the first category and mainly performed immunohistochemical (IHC) stainings in brain tissue by using antibodies raised against 5-mC and later 5-hmC. Other techniques that fall under this category include enzyme-linked immunosorbent assays (ELISAs), and various restriction enzyme-based methods that rely on polymerase chain reaction (PCR) amplification to differentiate between differently methylated sequences. While these types of studies can highlight the tissue and disease-state specific nature of epigenetic alterations and are among the most cost-efficient methods used to study DNA methylation, they lack single site resolution to identify precise and cell-type specific epigenetic alterations and only yield rough estimates of the global methylation status of the brain region being studied.

In order to identify exactly which genomic sites hold epigenetic alterations, more advanced techniques have been developed. Within these tools, candidate gene studies examine methylation profiles in a pre-defined set of genes and include techniques such as pyrosequencing, quantitative reverse transcription PCR (qRT-PCR), and methylation-specific PCR. Although specific, the hypothesis-driven nature of these methods does not allow one to identify new or genome-wide epigenetic variations.

Today's most commonly used methods for profiling genome-wide DNA methylation involve three steps; (1) the fragmentation of genomic DNA (e.g., by restriction enzymes or sonication) followed by (2) bisulfite conversion (combined or not with an oxidation step) or an affinity enrichment-based method, and (3) next generation sequencing (NGS) or microarray analysis.[12]

Bisulfite conversion refers to the treatment of DNA with sodium bisulfite which allows one to differentiate between methylated and unmethylated C residues, and between 5-mC and 5-hmC when an oxidation step is incorporated (e.g., in oxidative bisulfite sequencing and Tet-assisted bisulfite sequencing).[12,13] Alternatively, affinity enrichment-based methods enrich methylated DNA regions by using either methyl-CpG-binding domain (MBD) proteins that bind to methylated DNA, or 5-mC-specific antibodies (such as in methylated DNA immunoprecipitation; MeDIP). Regardless of the method being used, the obtained products can then be analyzed using a high-coverage approach such as NGS or an array-based method. Although bisulfite treatment of DNA combined with NGS (whole-genome bisulfite sequencing; WGBS) is considered to be the gold-standard technique used to provide genome-wide single-base resolution of methylated cytosines,[14–16] obtaining such data still requires a large investment of time, budget, and specific technical expertise. Alternatively, reduced-representation bisulfite sequencing (RRBS) combines the digestion of DNA by restriction enzymes with bisulfite conversion and NGS for the analysis of methylation profiles at a lower cost than WGBS. However, bisulfite treatment is more often combined with methylation arrays which are more user-friendly and thus, used more extensively. The Illumina Infinium Human Methylation 450K array (450K array) spans more than 450,000 methylation sites covering close to 99% of the CpG islands[17,18] while the newly developed Illumina Infinium Methylation EPIC BeadChip goes even further by covering more than 850,000 CpG methylation sites across the genome.[19] The main advantages and limitations of the above-mentioned techniques can be found in Table 1 while more detailed overviews of the different techniques available to analyze DNA methylation can be found elsewhere.[12,20]

2.2 Histone Modifications

Histone PTMs include a large variety of dynamic modifications such as methylation, phosphorylation, acetylation, ubiquitylation and sumoylation.[21] The modifications of the N-terminal tail residues of histone proteins have been most-studied and have been shown to, among others, alter chromatin structure and impact a variety of processes including transcriptional activation and DNA repair.[22]

Focusing on single histone modifications has led to the false believe that single histone marks could influence downstream processes on their own. However, it is now recognized that several histone PTMs mostly function

Table 1 Advantages and limitations of the most common experimental techniques used to profile DNA methylation.

	Advantages	**Limitations**
Global		
IHC, ELISA	Cost- and time-efficient	Provide rough estimates of the global methylation status, no single site resolution
Restriction enzyme-based methods + PCR	Relatively fast and cost-efficient	Limited to restriction enzymes digestion sites, moderate resolution, prone to PCR bias
Gene-specific		
Pyrosequencing	Single site resolution, quantitative, high throughput	Only short reads, design of assay can be difficult in CG dense areas, need of specific equipment
Methylation-specific PCR	Sensitive, fast, inexpensive and relatively easy to perform	Not quantitative, no singly site resolution, only one or two CpG sites at a time
qRT-PCR	Relatively easy to perform, equipment easily accessible	Low precision, no single site resolution
Epigenome-wide		
WGBS	Whole-genome mapping with single site resolution	DNA degradation following bisulfite treatment, high sequencing cost, data storage, time-consuming, requires high technical expertise
MBD-Seq	Relatively cost-effective, detects DMRs within CpG regions of high and low density, discriminates 5-mC from 5-hmC	Relatively low resolution, biased toward hypermethylated regions
MeDIP-Seq	Relatively cost-effective, specific to 5-mC/5-hmC depending on the specificity of the antibody	Limited by the quality/specificity of antibody, biased toward hypermethylated regions, no identification of single 5mC sites
RRBS	High sensitivity, relatively cost-effective	DNA degradation following bisulfite treatment, limited to regions base on enzymes' recognition sites

ELISA, enzyme-linked immunosorbent assay; *IHC*, immunohistochemistry; *MBD-Seq*, methyl-CpG-binding domain sequencing; *MeDIP-Seq*, methylated DNA immunoprecipitation sequencing; *PCR*, polymerase chain reaction; *qRT-PCR*, quantitative reverse transcription PCR; *RRBS*, reduced-representation bisulfite sequencing; *WGBS*, whole-genome bisulfite sequencing.

Table 2 Main advantages and limitations of antibody- and MS-based techniques used to profile histone modifications.

	Advantages	Limitations
Antibody-based Western blot, ELISA, ChIP	Relatively fast, no need for expensive equipment, highly sensitive for the detection of low-abundant proteins when specific antibodies are available	Cross-reactivity of most antibodies to highly similar PTMs, low-throughput; inability to assess several modifications at a time, requires *a priori* knowledge of the modification and its position
MS-based Bottom-up, top-down, middle-down	Can be used in high-throughput fashion; simultaneously monitor several PTMs, discovery of new modifications, detection of previously unknown modifications, investigation of combinatorial marks	Sample preparation and data analysis are more time-consuming, limited in high sensitivity as compared to the antibody-based techniques

ChIP, chromatin immunoprecipitation; *ELISA*, enzyme-linked immunosorbent assay; *MS*, mass spectrometry; *PTMs*, post-translational modifications.

together to bring about structural or functional change. Therefore, although well-suited for low-throughput studies, studying PTMs using antibody-based techniques such as western blotting, ELISA and chromatin immunoprecipitation (ChIP) followed by qRT-PCR (ChIP-qRT-PCR) or sequencing (ChIP-Seq) holds several limitations which can be found in Table 2. A growing interest in analyzing both combinatorial and new PTMs allowed for the development of techniques such as mass spectrometry (MS) of which the bottom-up approach is now one of the most well-characterized, suitable and sensitive tools used to study these marks (reviewed in Refs. 23 and 24 together with the top-down and middle-down approaches). Using this approach, a recently published paper proposes an entire workflow to prepare histones for the efficient analysis of PTMs in which the protein is enzymatically digested into shorter peptides which facilitates further separation by liquid chromatography and detection by MS.[25] In view of its own advantages and drawbacks, it is generally recommended to validate results obtained by MS with antibody-based techniques.[24] Alternatively, the combination of *in situ* hybridization (ISH) with proximity ligation assays (ISH-PLA) can assess histone modifications of specific cell types at specific gene loci.[26] Finally, gaining insights into human *in vivo* neuroepigenetics is becoming possible through the use of positron emission tomography imaging which yields unprecedented epigenetic information on the presence or absence of histone deacetylases, the well-studied class of enzymes that remove acetyl groups from histones.[27]

2.3 Non-Coding RNAs

After not being considered for a long time, ncRNAs such as short interfering RNAs, piwi-interacting RNAs and microRNAs (miRNAs) are now receiving increasing attention. These RNAs are not translated into proteins but instead are mainly involved in the transcriptional and post-transcriptional regulation of gene expression.[28]

ncRNAs can be found and studied in most biological samples including neuronal tissue and body fluids. In body fluids specifically, miRNAs are increasingly being recognized for their potential to serve as relatively stable biomarkers. Large-scale profiling of ncRNAs isolated from neural tissue samples is currently mostly done by using microarrays or NGS. Following this first discovery phase, the validation of the obtained findings is mainly done using approaches such as qRT-PCR and Northern blotting.[29] Since microarrays, NGS and qRT-PCR are among the most-used techniques to profile ncRNAs, their main advantages and limitations can be found in Table 3. In addition, fluorescence ISH can be used to specifically characterize the RNA's origin within tissue samples.[30]

Table 3 Advantages and limitations of the most common experimental techniques used to profile ncRNAs.

	Advantages	**Limitations**
Microarray	Relatively easier than NGS with less time-intensive sample preparation and data analysis, well-established bioinformatics	Potential for cross-hybridization, low dynamic range, relatively higher signal to noise ratio; limits the detection of rare/low-abundance RNAs
NGS	Ability to identify novel ncRNAs, multiplex capacity, increased specificity and sensitivity as compared to arrays, depth can be adapted to identify rare/low-abundance RNAs	Newer than arrays so relatively less well-established and more complex bioinformatics
qRT-PCR	Relatively easy to perform, equipment easily accessible, cheaper than microarrays or NGS	Low precision, no single site resolution

3. CHALLENGES AND FUTURE PERSPECTIVES

3.1 Technological Issues

Next to the technological challenges mentioned throughout the previous paragraphs, additional issues currently remain unsolved. First, regarding the study of DNA methylation, variants such as 5-formylcytosine (5-fC) and 5-carboxylcytosine (5-caC) currently remain understudied. Both modifications are thought to be the result of Tet enzymes sequentially oxidizing 5-mC to 5-hmC, 5-fC and 5-caC and are believed to be involved in the active DNA demethylation process.[31] However, given their low abundance, there is a strong need for the development of more sensitive and accurate methods to detect their presence.[32,33] Additionally, and as highlighted in Ref. 34, today's techniques used to study DNA methylation primarily assess cytosine methylation within CpG-rich promoter areas whereas recent studies have shed light on the importance of these modifications occurring outside of promoter regions and even outside of CpG islands.[35,36]

ncRNAs have not been studied as extensively as other epigenetic marks. The study of these small RNAs can be challenging which is partly due to their low abundance in the brain and body fluids. Moreover, it is not uncommon to detect inconsistencies in miRNA expression levels specifically when using different approaches to analyze the same samples.[37,38] One study in particular compared the ability of different platforms (microarray, NGS and qRT-PCR) from different vendors to detect miRNAs across several biological samples and provides an overview of practical criteria, reproducibility and reliability of these methods.[39] Finally, as a single miRNA is known to target hundreds of different messenger RNAs, unraveling and validating these complex interactions poses yet an additional challenge.

3.2 Tissue and Cell-Type Specificity

Several studies have now provided evidence for the tissue-specific nature of DNA methylation along with its region-specificity within the brain.[35,40] The observed methylation differences between brain regions could help explain the area-specific functionalities and should be taken into account when comparing epigenetic profiles between brain samples from different individuals. Another question that applies for all epigenetic studies aims at understanding how to deal with the cell-type specific nature of epigenetic modifications. One should be aware of the highly heterogeneous mix of cells present in the mammalian brain and the limitations of certain techniques which lack cellular resolution.[41] To tackle this issue, researchers have turned to single-cell analyses by using techniques such as fluorescence activated cell sorting or laser capture microdissection which make it possible to obtain homogenous cell populations or even single cells and go on with studying the epigenetic landscape of these specific cells.[41,42] It is worth noting that *in silico* approaches have been developed in order to distinguish neuronal from non-neuronal DNA methylation signals.[43,44] Such analytic tools could thus be applied to adjust for cell-type composition and (re-)analyze existing datasets that did not take cellular heterogeneity into account.

3.3 Causality

One of the main limitation in the field of (neuro)epigenetics is the issue of causality. Today's most widely used techniques prevent the understanding of whether the observed specific epigenetic alterations are causal to the disease process at play or rather the consequence of several pathological changes associated with the disease. As recently stated by Roubroeks et al., the use of novel and more targeted (epi)genome editing techniques such as the clustered regularly interspersed short palindromic repeats (CRISPR)-associated endonuclease Cas9 system will enable the experimental manipulation of DNA methylation and demethylation through adding or removing methyl groups to specific DNA loci.[45,46] Alternatively, approaches such as engineered zinc finger proteins and sequence-specific transcription activator-like effectors further hold the potential to overcome said limitation. Using such approaches will enable one to assess the functional consequences not only of specifically altered methylation states, but also of precise histone modifications and ncRNAs. Indeed, recent findings have shown that the CRISPR-Cas9 system can be useful in uncovering the regulation and function of clustered miRNAs[47] and knocking-down or knocking-out specific ncRNAs both *in vitro* and *in vivo*.[48,49] In the future, these approaches hold the potential to further reveal the involvement of a variety of epigenetic mechanisms in the human brain and its associated disease states. Alternatively, quasi-experimental designs such as (two-step) Mendelian randomization can further help in gaining understanding of epigenetic causal pathways to disease. While epigenetic applications are still relatively rare, its usage could further aid in the development of epigenetic-based interventions. More detailed insights on the applications and limitations of this approach can be found elsewhere.[50,51] Finally, future research will also need to aim at gaining understanding of the nature and complexity of numerous modifications working together to bring about functional or structural change.

3.4 Study Design

Several remarks can also be made regarding the requirements for study designs used in neuroepigenetics. First, one has to keep the evidently limited or even inexistent options to study *in vivo* neuroepigenetics in mind along with the limited availability of high quality human post-mortem brain tissue. Second, since factors such as the disease-stage and age of the subject can have a significant impact on the observed epigenetic profiles of certain areas in post-mortem brain material, care must be taken to properly balance these factors across study groups. Finally, since males are currently overrepresented in animal studies and human studies often lack power to compare epigenetic profiles between genders, there is a need for future research to focus on understanding epigenetic similarities as well as differences between males and females in order to allow for the generalization of obtained findings to both populations.

3.5 Analytical Techniques

As mentioned, many sources of variability can potentially affect the interpretation of epigenetic findings such as age, gender, medication use, smoking habits, and even the analytical tools and normalization methods used to process the generated data. Therefore, it is recommended to comprehensively gather and document as many potential sources of confounding or covariability as possible, in order to account for these influences during the data analysis stage. It is also worth noting that several databases (see, e.g., http://epigenie.com/epigenetic-tools-and-databases/ for both) provide information on the normal and pathological epigenome, allele-specific methylation and miRNA expression profiles.[52]

4. CONCLUSION

The present chapter provides a brief overview of some of the most-used techniques in the field of neuroepigenetics along with their main challenges and suggestions for improvement. Being aware of the methodologies used in this field will allow neuroscientists to acknowledge their limitations, understand their particular usage and hence help in improving the research outcomes necessary not only for a better understanding of brain disorders, but also for more efficient prevention, diagnosis and treatment.

REFERENCES

1. Sweatt JD. The emerging field of neuroepigenetics. *Neuron.* 2013;80(3):624–632.
2. Griffiths BB, Hunter RG. Neuroepigenetics of stress. *Neuroscience.* 2014;275:420–435.
3. Day JJ, Sweatt JD. DNA methylation and memory formation. *Nat Neurosci.* 2010;13(11):1319–1323.
4. Klengel T, Dias BG, Ressler KJ. Models of intergenerational and transgenerational transmission of risk for psychopathology in mice. *Neuropsychopharmacology.* 2016;41(1):219–231.
5. Schmidt B. Epigenomics: from chromatin biology to therapeutics. *Yale J Biol Med.* 2013;86(1):112–113.
6. Gopalakrishnan S, Van Emburgh BO, Robertson KD. DNA methylation in development and human disease. *Mutat Res.* 2008;647(1–2):30–38.
7. Ziller MJ, Gu H, Muller F, et al. Charting a dynamic DNA methylation landscape of the human genome. *Nature.* 2013;500(7463):477–481.
8. Maunakea AK, Chepelev I, Cui K, Zhao K. Intragenic DNA methylation modulates alternative splicing by recruiting MeCP2 to promote exon recognition. *Cell Res.* 2013;23(11):1256–1269.
9. Varley KE, Gertz J, Bowling KM, et al. Dynamic DNA methylation across diverse human cell lines and tissues. *Genome Res.* 2013;23(3):555–567.
10. Kriaucionis S, Heintz N. The nuclear DNA base 5-hydroxymethylcytosine is present in Purkinje neurons and the brain. *Science.* 2009;324(5929):929–930.
11. Tahiliani M, Koh KP, Shen Y, et al. Conversion of 5-methylcytosine to 5-hydroxymethylcytosine in mammalian DNA by MLL partner TET1. *Science.* 2009;324(5929):930–935.
12. Yong WS, Hsu FM, Chen PY. Profiling genome-wide DNA methylation. *Epigenetics Chromatin.* 2016;9:26.
13. Fukuzawa S, Takahashi S, Tachibana K, Tajima S, Suetake I. Simple and accurate single base resolution analysis of 5-hydroxymethylcytosine by catalytic oxidative bisulfite sequencing using micelle incarcerated oxidants. *Bioorg Med Chem.* 2016;24(18):4254–4262.
14. Berman BP, Weisenberger DJ, Aman JF, et al. Regions of focal DNA hypermethylation and long-range hypomethylation in colorectal cancer coincide with nuclear lamina-associated domains. *Nat Genet.* 2011;44(1):40–46.
15. Heyn H, Vidal E, Sayols S, et al. Whole-genome bisulfite DNA sequencing of a DNMT3B mutant patient. *Epigenetics.* 2012;7(6):542–550.
16. Li N, Ye M, Li Y, et al. Whole genome DNA methylation analysis based on high throughput sequencing technology. *Methods.* 2010;52(3):203–212.
17. Bibikova M, Barnes B, Tsan C, et al. High density DNA methylation array with single CpG site resolution. *Genomics.* 2011;98(4):288–295.
18. Bibikova M, Fan JB. Genome-wide DNA methylation profiling. *Wiley Interdiscip Rev Syst Biol Med.* 2010;2(2):210–223.
19. Pidsley R, Zotenko E, Peters TJ, et al. Critical evaluation of the Illumina MethylationEPIC BeadChip microarray for whole-genome DNA methylation profiling. *Genome Biol.* 2016;17(1):208.
20. Kurdyukov S, Bullock M. DNA methylation analysis: choosing the right method. *Biology.* 2016;5(1) https://doi.org/10.3390/biology5010003.
21. Kouzarides T. Chromatin modifications and their function. *Cell.* 2007;128(4):693–705.
22. Dong X, Weng Z. The correlation between histone modifications and gene expression. *Epigenomics.* 2013;5(2):113–116.
23. Sidoli S, Cheng L, Jensen ON. Proteomics in chromatin biology and epigenetics: elucidation of post-translational modifications of histone proteins by mass spectrometry. *J Proteomics.* 2012;75(12):3419–3433.

24. Onder O, Sidoli S, Carroll M, Garcia BA. Progress in epigenetic histone modification analysis by mass spectrometry for clinical investigations. *Expert Rev Proteomics*. 2015;12(5):499–517.
25. Sidoli S, Bhanu NV, Karch KR, Wang X, Garcia BA. Complete workflow for analysis of histone post-translational modifications using bottom-up mass spectrometry: from histone extraction to data analysis. *J Vis Exp*. 2016;(111):e54112.
26. Gomez D, Shankman LS, Nguyen AT, Owens GK. Detection of histone modifications at specific gene loci in single cells in histological sections. *Nat Methods*. 2013;10(2):171–177.
27. Wey HY, Gilbert TM, Zurcher NR, et al. Insights into neuroepigenetics through human histone deacetylase PET imaging. *Sci Transl Med*. 2016;8(351):351ra106.
28. Santosh B, Varshney A, Yadava PK. Non-coding RNAs: biological functions and applications. *Cell Biochem Funct*. 2015;33(1):14–22.
29. Huang Q, Mao Z, Li S, Hu J, Zhu Y. A non-radioactive method for small RNA detection by northern blotting. *Rice*. 2014;7:26.
30. Nielsen BS, Holmstrom K. Combined microRNA in situ hybridization and immunohistochemical detection of protein markers. *Methods Mol Biol*. 2013;986:353–365.
31. Klungland A, Robertson AB. Oxidized C5-methyl cytosine bases in DNA: 5-Hydroxymethylcytosine; 5-formylcytosine; and 5-carboxycytosine. *Free Radic Biol Med*. 2017;107:62–68.
32. Guo M, Li X, Zhang L, et al. Accurate quantification of 5-methylcytosine, 5-hydroxymethylcytosine, 5-formylcytosine, and 5-carboxylcytosine in genomic DNA from breast cancer by chemical derivatization coupled with ultra performance liquid chromatography-electrospray quadrupole time of flight mass spectrometry analysis. *Oncotarget*. 2017;8(53):91248–91257.
33. Song C-X, Szulwach KE, Dai Q, et al. Genome-wide profiling of 5-formylcytosine reveals its roles in epigenetic priming. *Cell*. 2013;153(3):678–691.
34. Lunnon K, Mill J. Epigenetic studies in Alzheimer's disease: current findings, caveats, and considerations for future studies. *Am J Med Genet B Neuropsychiatr Genet*. 2013;162B(8):789–799.
35. Davies MN, Volta M, Pidsley R, et al. Functional annotation of the human brain methylome identifies tissue-specific epigenetic variation across brain and blood. *Genome Biol*. 2012;13(6):R43.
36. Lister R, Pelizzola M, Dowen RH, et al. Human DNA methylomes at base resolution show widespread epigenomic differences. *Nature*. 2009;462(7271):315–322.
37. Saugstad JA. MicroRNAs as effectors of brain function. *Stroke*. 2013;44(6 0 1):S17–S19.
38. Chen Y, Gelfond JA, McManus LM, Shireman PK. Reproducibility of quantitative RT-PCR array in miRNA expression profiling and comparison with microarray analysis. *BMC Genomics*. 2009;10:407.
39. Git A, Dvinge H, Salmon-Divon M, et al. Systematic comparison of microarray profiling, real-time PCR, and next-generation sequencing technologies for measuring differential microRNA expression. *RNA*. 2010;16(5):991–1006.
40. Ladd-Acosta C, Pevsner J, Sabunciyan S, et al. DNA methylation signatures within the human brain. *Am J Hum Genet*. 2007;81(6):1304–1315.
41. Mitchell A, Roussos P, Peter C, Tsankova N, Akbarian S. The future of neuroepigenetics in the human brain. *Prog Mol Biol Transl Sci*. 2014;128:199–228.
42. Datta S, Malhotra L, Dickerson R, Chaffee S, Sen CK, Roy S. Laser capture microdissection: big data from small samples. *Histol Histopathol*. 2015;30(11):1255–1269.
43. Montano CM, Irizarry RA, Kaufmann WE, et al. Measuring cell-type specific differential methylation in human brain tissue. *Genome Biol*. 2013;14(8):R94.
44. Guintivano J, Aryee MJ, Kaminsky ZA. A cell epigenotype specific model for the correction of brain cellular heterogeneity bias and its application to age, brain region and major depression. *Epigenetics*. 2013;8(3):290–302.

45. Roubroeks JAY, Smith RG, van den Hove DLA, Lunnon K. Epigenetics and DNA methylomic profiling in Alzheimer's disease and other neurodegenerative diseases. *J Neurochem*. 2017;143(2):158–170.
46. Liu XS, Wu H, Ji X, et al. Editing DNA methylation in the mammalian genome. *Cell*. 2016;167(1):233–247: e217.
47. Lataniotis L, Albrecht A, Kok FO, et al. CRISPR/Cas9 editing reveals novel mechanisms of clustered microRNA regulation and function. *Sci Rep*. 2017;7(1):8585.
48. Chang H, Yi B, Ma R, Zhang X, Zhao H, Xi Y. CRISPR/cas9, a novel genomic tool to knock down microRNA in vitro and in vivo. *Sci Rep*. 2016;6:22312.
49. Ho TT, Zhou N, Huang J, et al. Targeting non-coding RNAs with the CRISPR/Cas9 system in human cell lines. *Nucleic Acids Res*. 2015;43(3):e17.
50. Relton CL, Davey Smith G. Two-step epigenetic Mendelian randomization: a strategy for establishing the causal role of epigenetic processes in pathways to disease. *Int J Epidemiol*. 2012;41(1):161–176.
51. Latvala A, Ollikainen M. Mendelian randomization in (epi)genetic epidemiology: an effective tool to be handled with care. *Genome Biol*. 2016;17(1):156.
52. Panwar B, Omenn GS, Guan Y. miRmine: a database of human miRNA expression profiles. *Bioinformatics*. 2017;33(10):1554–1560.

CHAPTER THREE

An Introduction to Epigenetic Mechanisms

Elizabeth J. Radford[1]
Department of Pediatrics, University of Cambridge, Cambridge, United Kingdom
[1]Corresponding author: e-mail address: ejr41@cam.ac.uk

Contents

Abstract

The astonishing array of cellular phenotypes required to make a complex organism such as a human is generated from an identical genetic sequence in the nucleus of each cell. The central nervous system is a highly ordered, complex system composed of multiple neuronal and glial cell types. Both neurons and glia are derived from neural stem/precursor cells (NPCs) in a temporally and spatially patterned process of cellular division and differentiation, migration, maturation and the establishment of neuronal connectivity. In part, this is testament to the power and flexibility of epigenetic mechanisms to direct transcription, producing a myriad of outcomes. The progressive restriction of developmental potential involves the acquisition of a hierarchy of epigenetic regulation which restricts and directs the developmental program of gene expression. Differentiation into each cell type is associated with the acquisition of a distinct epigenetic profile. The importance of epigenetic mechanisms for neurodevelopment is exemplified by the many neurodevelopmental disorders caused by the genetic disruption of epigenetic readers, writers or modifiers. In this chapter I will introduce the different classes of

Progress in Molecular Biology and Translational Science, Volume 158
ISSN 1877-1173
https://doi.org/10.1016/bs.pmbts.2018.04.002

epigenetic mechanisms, and I will briefly discuss some of the known neurodevelopmental disorders caused by genetic disruption of these pathways to illustrate the importance of epigenetic processes in brain development and function.

1. INTRODUCTION

The astonishing array of cellular phenotypes required to make a complex organism such as a human is generated from an identical genetic sequence in the nucleus of each cell. In part, this is testament to the power and flexibility of epigenetic mechanisms to direct transcription, producing a myriad of outcomes. Waddington first defined epigenetics as "*the branch of biology which studies the causal interactions between genes and their products, which bring the phenotype into being.*"[1] A more mechanistic definition is provided by Bird: "*the structural adaptation of chromosomal regions so as to register, signal or perpetuate altered activity states.*"[2] The progressive restriction of developmental potential involves the acquisition of a hierarchy of epigenetic regulation which restricts and directs the developmental program of gene expression. These epigenetic mechanisms are generally heritable upon mitosis, resulting in cellular phenotypic stability which is critical for tissue identity and integrity under normal conditions.[3] The reacquisition of developmental potential, as occurs naturally during the formation of the germ cells and following fertilization, or experimentally during cloning or derivation of induced pluripotent stem cells, involves genome-wide epigenetic reprogramming. The plastic development of an organism in response to environmental conditions during early life involves the modulation of the developmental program of gene expression, and is therefore hypothesized to involve epigenetic mechanisms. It is important to note that the term "epigenetic" has historically signified a variety of things, and continues to be used in different ways today.[4] For the purposes of this chapter, we consider epigenetics to mean "*the structural adaptation of chromosomal regions so as to register, signal or perpetuate altered activity states.*"[2]

1.1 Epigenetic Mechanisms and Neurodevelopment

The central nervous system is a highly ordered, complex system composed of multiple neuronal cell types in addition to the glial cell lineages: astrocytes and oligodendrocytes. Both neurons and glia are derived from neural stem/precursor cells (NPCs) in a temporally and spatially patterned process of cellular division and differentiation, migration, maturation and the establishment of neuronal connectivity.[5] During embryonic neurodevelopment, NPCs first produce neurons, with the different neuronal cell types being generated in a tightly ordered sequence. In the neocortex, the sequential generation and migration of these neuronal cell types produces the cortical layers. Subsequently differentiation of the progeny of NPCs switches to glia, and the ordered production of the different oligodendrocyte and astrocyte lineages occurs.[6,7] Differentiation into each cell type is associated with the acquisition of a distinct epigenetic profile. The epigenetic remodeling associated with neural development has been studied in detail using both in vitro and in vivo models of neural stem cell differentiation.[8–10] The importance of epigenetic mechanisms for neurodevelopment is exemplified by the many neurodevelopmental disorders caused by the genetic disruption of epigenetic readers, writers or modifiers. As most neurons, particularly in the human brain, are post-mitotic, there is also considerable interest in characterizing the role of epigenetic mechanisms in this context. Neuronal activity has been shown to result in dynamic changes in DNA methylation,[11] and there is evidence that DNA methylation is involved in processes such as long term potentiation, critical for learning and memory.[12] In this chapter I will introduce the different classes of epigenetic mechanisms, and I will briefly discuss some of the known neurodevelopmental disorders caused by genetic disruption of these pathways to illustrate the importance of epigenetic processes in brain development and function.

2. EPIGENETIC MECHANISMS OF REGULATION

Two general classes of epigenetic regulation can be defined: covalent modifications to DNA and covalent modifications to the histones around which the DNA is wound. In addition, non-coding RNAs and the three-dimensional structure and arrangement of chromatin within the nucleus are both regulated by and contribute to the establishment and maintenance of epigenetic states (reviewed in Bird[2]). These different classes of epigenetic modifications are intricately and intimately related, resulting in multiple layers of control. Histone modifications associated with repressive or active chromatin have been shown, respectively, to recruit and exclude DNA methylation (reviewed in Ref. 13) while some histone modifying enzymes contain DNA methyl-binding domains.[14,15] Non-coding RNAs have been shown to target both histone modifying enzymes and DNA methylation (for example: Refs. 16–18). Local changes in chromatin structure can have an impact on distant genes as they affect chromosomal looping and the accessibility of regulatory elements. Furthermore, there is increasing evidence that the organization of chromatin into distinct domains within the nucleus is important for transcriptional regulation. Topologically associated domains (TADs) are regions of the genome which preferentially physically associate with each other.[19] An elegant recent study has demonstrated the importance of TADs in mouse neurodevelopment.[10] The disruption of TAD boundaries has been shown to alter transcription of nearby genes, and has been observed in disease processes, including cancer and developmental disorders.[20] Euchromatic, actively transcribed regions are brought together in the nucleus, while repressive domains help to maintain silencing at heterochromatic regions.[19,21,22] However, much remains unknown regarding the processes involved in establishing and maintaining nuclear organization, and their impact on cellular phenotype. The 4D nucleome consortium is an international collaboration aiming to understand the principles behind the three-dimensional organization of the nucleus in space and time, the role nuclear organization plays in gene expression and cellular function, and how changes in the nuclear organization affect normal development and disease.[23] It is hoped and anticipated that this project will significantly advance our understanding of how the nuclear architecture is established and its importance in health and disease.

2.1 DNA Methylation

DNA methylation involves the addition of a methyl group to the fifth carbon in the cytosine pyrimidine ring. In vertebrate somatic tissues this largely occurs symmetrically on the two strands in CpG dinucleotides, although appreciable levels of non-CpG methylation have been identified in embryonic stem cells[24] and mouse oocytes.[25] All further discussion here will refer to CpG methylation. 70%–80% of mammalian CpGs are thought to be methylated[26] with the majority of unmethylated CpGs being found in high-density clusters, called CpG islands, which often have regulatory functions and are frequently found at gene promoters. Two mammalian de novo methyltransferases, DNMT3A and DNMT3B are responsible for the establishment of methylation patterns, while DNMT1 has a preference for hemi-methylated DNA and thus effectively maintains CpG methylation on DNA replication.[27,28] The targeting of de novo methylation is thought to be mediated by the interaction of the de novo methyltransferases DNMT3A and DNMT3B with factors such as DNMT3L (DNMT3Like, a non-catalytic paralog) and Np95 (reviewed in Refs. 29,30). DNMT3L has been shown to bind to unmethylated H3K4 tails in vitro, and thus it has been proposed that the absence of H3K4 methylation, a histone modification which is often associated with open chromatin and actively transcribed areas (see below), may result in the targeting of de novo methylation through DNMT3A to specific loci, including imprinted genes. Alternatively the presence of H3K4 methylation may protect from the action of DNA methyltransferases.[31,32] Indeed, multiple studies have found a genome-wide negative correlation between H3K4 methylation and DNA methylation (for example: Refs. 33–36). Consistent with this, a homozygous germline mutation in the H3K4 demethylase AOF1/KDM1B has been shown to perturb the acquisition of some maternal methylation imprints during oogenesis. This indicates a requirement for histone demethylation in order to establish germline CpG methylation at some imprinting control regions, ICRs.[37] Genomic imprinting and the epigenetic mechanisms which control it will be discussed further in Section 3.2.

2.1.1 DNA Methylation and Transcriptional Control

DNA methylation is enriched at heterochromatic regions and at repetitive elements, where it is thought to promote a closed chromatin structure and aid the prevention of expression and/or recombination (reviewed in Ref. 2). While DNA methylation undeniably has a role in the suppression of transcription and the maintenance of heterochromatin, in certain genomic contexts DNA methylation may have a more nuanced role.

In general, the methylation of promoter regions is associated with down-regulation of transcription. It is often assumed that gain of methylation results in transcriptional suppression, however, the acquisition of methylation may in fact be as a consequence of transcriptional suppression, although it is possible that promoter methylation may then contribute to the maintenance of a silent state.[38–40] The relationship between promoter methylation and transcription depends in part on the CpG density of the promoter. Promoters can be divided according to their CpG content into high, intermediate and low density classes, HCP, ICP and LCP.[35,36] The vast majority of HCPs are unmethylated, even if they are transcriptionally silent. Methylation of the CpG island core of HCPs tends to result in exclusion of trithorax and polycomb groups normally associated with the promoter, the formation of local heterochromatin and gene silencing. Such HCP silencing is generally only employed at imprinted or germline associated genes and during X-chromosome inactivation.[35] DNA methylation of such promoters is not thought to be the initial silencing signal but instead "locks in" and maintains the stably silent state. In contrast, the rarer CpGs in LCPs tend to be methylated, regardless of the activity state of the promoter.[36] This confirms that a low density of methylated CpGs is not incompatible with transcriptional activity, as has previously been observed.[41] At ICPs there is a negative correlation between degree of methylation and transcriptional activity. Methylation associated with the silencing of ICPs is much more commonly observed, implying that either these promoters are more susceptible to de novo methylation, and/or that CpG methylation has a mechanistic function in the control of transcription at these promoters. Methylation is thought to impair transcriptional activity largely through the recruitment of methyl-binding domain proteins, although CpG methylation can also directly inhibit transcription factor binding.[42]

Recent studies have revisited the role of DNA methylation of non-5′ CpG islands in the regulation of transcriptional control. In the human and mouse brain, intragenic CpG islands have been shown to have a much higher frequency of tissue-specific methylation than proximal promoter CpG islands (34% and 2%, respectively). It has been proposed that these intragenic CpG islands may be alternative promoters, and thus that intragenic CpG methylation may play a role in the regulation of alternative transcript usage between different cell types.[43] However, such a correlation does not imply causality, and the absence of transcriptional initiation from an intragenic CpG island may be permissive for the acquisition of DNA methylation.

DNA methylation over the gene body is known to correlate positively with the level of gene transcription in the mouse and human genomes.[33,44,45] Indeed, in female mammals some regions of the active X show much higher overall levels of DNA methylation than the inactive X. This methylation is concentrated in gene bodies.[46] X inactivation leads to a reduction in such gene-body methylation.[47] In *Arabidopsis thaliana*, such gene-body methylation has been proposed to inhibit spurious transcriptional initiation, but the function, if any, in mammalian genomes remains unknown.[42] Indeed, it is possible that the accumulation of methylation on the active X is simply symptomatic of the greater accessibility of DNA methyltransferases to this chromosome.

Wu et al. have suggested a role for DNMT3A and non-promoter methylation in the control of gene expression in postnatal neurogenesis. *DNMT3A*$^{-/-}$ mice demonstrate approximately an equal number of downregulated as upregulated transcripts in neural stem cells. This is inconsistent with a primary role for DNA methylation during neuronal differentiation as a mechanism of promoter silencing. Wu et al. suggest that DNA methylation outside of the proximal promoter prevents Polycomb mediated gene repression through interfering with binding of the PRC2 complex. Loss of DNMT3A results in the accumulation of H3K27 and gene silencing.[48] Consistent with an important role for DNMT3A in brain development, in humans the loss of function of one allele results in an autosomal dominant developmental disorder characterized by intellectual disability, tall stature and a distinctive facial appearance.[49] DNMT3A has recently also been shown to be important in motor neuron development.[9] Mice lacking CNS DNMT3A are healthy at birth, but have fewer motor neurons, movement defects, and die prematurely.

Another neurodevelopmental disorder which illustrates the importance of brain DNA methylation is Rett syndrome. Rett syndrome is an X-linked dominant autism spectrum disorder caused mutations in MECP2 which encodes a methyl-CpG binding domain protein.[50] Affected children usually have normal development until they are 6–12 months of age, followed by regression and the loss of acquired developmental skills, leading to severe mental retardation and motor impairment. Although MECP2 is expressed in many tissues, in mouse models the brain-specific loss of *MeCP2* recapitulates the conventional phenotype.[51] Remarkably, in the mouse phenotypic rescue can obtained by inducing normal MeCP2 expression in adult MeCP2 mutant animals. Importantly, such rescue is observed in animals in which the disease is already manifest. It was initially thought that MeCP2 binds to methylated regions and acts as a transcriptional repressor.[52,53] However, more recent data suggest that MeCP2 is expressed at very high levels in postnatal neurons, with a genome-wide profile of binding. Consequently it has been hypothesized that it may act more as a structural element of chromatin than a transcriptional regulator of specific genes.[54] A further re-examination of the role of MeCP2 has been precipitated by recent data which suggest that MeCP2 binds to hydroxymethylated cytosines (5hmC, see below), in neurons with similar affinity to 5mC, and that the Rett syndrome causing mutation R133C preferentially inhibits MeCP2 binding to 5hmC.[55,56]

2.2 DNA Hydroxymethylation

Cytosine hydroxymethylation (5hmC) is produced by the oxidation of 5-methyl-cytosine, a reaction catalyzed by the TET family of enzymes.[57,58] 5hmC is not recognized by DNMT1 nor most methyl-binding domain containing proteins, and is thought to functionally resemble unmodified cytosine to the cellular apparatus.[58] All three mammalian TET enzymes can convert 5mC to 5hmC, further TET activity can convert 5hmC to 5-formylcytosine (5fC) and 5-carboxylcytosine (5caC), which are subsequently recognized and excised by thymine DNA glycosylase (TDG) to be replaced with unmethylated cytosine.[59,60] This is thought to be the mechanism through which regions are actively demethylated in mammals. 5hmC is relatively abundant in mouse embryonic stem cells, ESCs, the early embryo and in adult brain.[57,61–63] Hydroxymethylation of pluripotency genes in ESCs is important in the maintenance of pluripotency.[64] However, the full role of hydroxymethylation in the brain, as well as other tissues, remains to be elucidated.

2.3 Histone Modifications

A histone octamer of two H2A–H2B dimers and an H3–H4 tetramer acts as a scaffold around which 147 bp of DNA is wrapped to form a nucleosome, producing the famous "beads on a string" structure (reviewed in Ref. 65). Covalent modifications to key residues in the histone tails, especially those of H3 and H4, alter the interaction between adjacent nucleosomes and/or between histones and the DNA, changing the three-dimensional chromatin structure on both a global and a local scale. The resulting higher-order chromatin structures can be divided into tightly packed heterochromatic, transcriptionally silent domains, and euchromatic domains where the DNA is more accessible to nuclear protein complexes and thus the potential for transcription is maintained.

The dynamic modulation of histone modifications plays an important role in transcriptional control. Efforts have been made to divide histone modifications into repressive and active marks according to how they correlate with levels of transcriptional activity. Acetylation neutralizes the negative charge of lysine (K), disrupting tight packaging of chromatin, and is consistently associated with transcriptional activity. In contrast, the effect of lysine methylation is dependent on the residue and histone modified and has been associated with both transcriptional activity: H3K4, H3K36, H3K79 and transcriptional repression: H3K9, H3K27, H4K20.[66] However, this is an oversimplification and increasing evidence suggests that the impact of a particular histone mark may be dependent on its genomic context.[67]

The close association between certain histone modifications and particular regulatory elements or sequence features has been identified by multiple studies. For example, high CpG density, unmethylated promoters are usually marked by H3K4me3, in association with other characteristics of open chromatin such as histone acetylation, DNAseI hypersensitivity and histone variant H3.3 incorporation; while enhancers are associated with H3K4me1 and histone 3 acetylation.[68–73] The Polycomb mediated modification H3K27me3 is generally found in euchromatic regions, and is associated with transcriptional repression (reviewed by Bernstein et al.). In ESCs H3K27 is found in conjunction with H3K4me3 at "bivalent" promoters.[69] H3K9 methylation, although found at inactive promoters and enhancers, is also associated with heterochromatic regions, and is enriched at repetitive elements.[68–72] The combinatorial nature of this "histone code" suggests the potential for great flexibility and multiple layers of regulation. Further evidence of this has been provided by large international collaborations aiming to map the epigenomes of multiple cell types in both mouse and human such as the Roadmap Epigenomics and the Encylopaedia of DNA elements (ENCODE) projects.[34,74–76]

The importance of chromatin in brain development and function is underlined by the large and expanding number of neurodevelopmental disorders caused by mutations in the genes encoding chromatin remodelers, histone modifiers or proteins which "read" such marks. These can be divided into "writers," "readers," "erasers," and remodelers.[77] For example, mutations in "writers" can cause Rubinstein–Taybi syndrome (the histone acetyltransferases CBP and p300); Kabuki syndrome type 1 (KMT2D), Wiedemann–Steiner syndrome (KMT2A); Sotos syndrome (NSD1); Weaver syndrome (EZH2) and Sotos-like syndrome (SET2D), among others. Examples of syndromes caused by mutations in "erasers" include Kabuki syndrome type 2 (KDM6A) and BDMR syndrome (HDAC4); while examples of mutations in remodelers include the ATRX (ATRX) and CHARGE syndromes (CHD7) reviewed in Ref. 77. Many of the recently identified developmental disorders are caused by the genetic disruption of chromatin structure or function, and as well-powered unbiased projects investigating the genetic etiology of developmental disorders such as the Deciphering Developmental Disorders project report back, the number of chromatin genes implicated in human neurodevelopment is likely to rise.

2.4 Non-coding RNA

Pioneering work in yeast identified RNA-mediated mechanisms crucial for the control of the targeting of epigenetic modifications (reviewed in Ref. 78) and the heritability of histone modifications during replication.[79]

Although these studies illustrate RNA-mediated processes occurring in fission yeast that lack DNA methylation, small RNAs have also been shown to direct DNA methylation and are required for the silencing of transposable elements in mammalian germ cell development.[16,80–82] Piwi-interacting RNAs, piRNAs are 25–30 nucleotide long small RNAs produced by the cleavage of transposon-derived transcripts in the male mammalian germline. piRNAs associate with the Argonaute proteins MIWI and MILI and guide the further cleavage of transposon-derived transcripts, resulting in a ping-pong cycle of piRNA amplification which is required for the de novo DNA methylation and silencing of these transposable elements.[81,83] DNMT3L is also required for the methylation of transposable elements in the germ line, and piRNAs are present in DNMT3L mutant germ cells, suggesting that piRNAs act upstream of and may target DNMT3L. However, preliminary studies suggest that MIWI and MILI do not interact directly with DNMT3L.[84] *Dnmt3c*, a recently discovered functional paralog of *DNMT3B* in the Muroidea families including mouse, rat and hamster, has been shown to be required for the silencing of evolutionarily recent transposable elements in the male germline of these animals.[85] As no known human paralog exists, it is currently thought that this role in humans is fulfilled redundantly by DNMT3A and DNMT3B.

Small non-coding RNAs have also been implicated in dosage control of unique sequences. MicroRNAs bind to targets in the 3′UTR of mRNAs, suppressing expression either by translational inhibition, or through triggering mRNA decay (reviewed in Ref. 86). At the imprinted Dlk1/Dio3 locus, maternally expressed microRNAs control the dosage of the paternally expressed Rtl1 gene post-transcriptionally.[87] However, a single microRNA can have multiple targets, which increases the complexity of dosage control mediated by small RNA populations.

Long non-coding RNAs, lncRNAs, sometimes called macro non-coding RNAs, also play a role in regional and *trans*-acting dosage modulation, and are involved with imprinting control at some loci. Imprinted expression of lncRNAs is required for long-distance bidirectional *cis*-acting control of transcription at several imprinted clusters, such as the *Igf2r* domain, the *Kcnq1* domain, the *Snrpn* locus and the *Gnas* cluster.[17,88–91] Targeted truncation or deletion of these large non-coding transcripts results in the de-repression in *cis* of protein-coding genes normally repressed on that allele. This is exemplified by the *Igf2r* locus, where the ncRNA *Airn* accumulates in *cis* at the promoter of the *Slc22a3* gene in the placenta, recruiting the histone H3K9 methyltransferase G9a, resulting in the silencing of the paternally inherited allele of *Slc22a3*.[17] Individual lncRNAs have also been shown to regulate the dosage of a variety of mRNA transcripts. Staufen-1 mediated decay acts on translationally active mRNAs. Gong and Maquat demonstrate that imperfect base pairing between Alu elements in lncRNAs and Alu elements in the 3′UTR of mRNAs enables the binding of Staufen-1, stimulating decay.[92] However, additional, currently unknown factors contribute to targeting specificity.

3. GENOMIC IMPRINTING

Genomic imprinting is of critical importance to brain development and normal behavior. Furthermore, the study of imprinted regions has been extremely informative regarding how epigenetic marks alter genome function and thus the phenotype of an organism. Diploid organisms inherit two copies of each gene from their parents, a maternal and a paternal allele. Most genes are expressed from both alleles, however, a small group of about 100 genes are subject to genomic imprinting. They are functionally monoallelic and are expressed from either the maternally or the paternally inherited chromosome. This is controlled by multiple layers of epigenetic regulation.[93] Imprinted genes are often found in clusters where the parental allele-specific pattern of gene expression is coordinately regulated by ICRs through long range *cis*-acting epigenetic mechanisms. All ICRs so far identified have regions of differential DNA methylation (differentially methylated regions—DMRs) between the two parental chromosomes.[93] ICRs can be divided into those which are methylated on the paternally inherited copy and those with maternally inherited methylation. The role of histone modifications in imprinting control is less clear, however DMRs are characterized by the asymmetrical accumulation of different histone modifications on the two parental chromosomes and a "*tri-mark*" comprising the trimethylation of H3K4 and H3K9 and the trimethylation of H4K20.[94] As mentioned above, many imprinted *loci* contain non-coding RNAs, some of which are of critical importance in the maintenance of genomic imprinting.[93] The role of hydroxymethylation at imprinted *loci* remains unclear, but often appears to be asymmetrically distributed at DMRs.

Most imprinted genes are expressed in the brain and several exhibit brain-specific imprinting. Recent data have demonstrated the importance of the imprinted gene *Dlk1* and *Igf2* as postnatal neurogenic niche factors in the mammalian brain. The dosage of these two genes is regulated by a remarkable temporal and spatially specific selective loss of imprinting.[95,96] The monoallelic expression of imprinted loci renders them vulnerable to genetic dysregulation, as mutation of the sole expressed allele can compromise expression and lead to severe developmental defects. The human congenital imprinting syndromes, including Angelman (AS) and Prader–Willi syndromes (PWS), are all characterized by neurological and behavioral impairments and learning difficulties,[97] indicating the importance of genomic imprinting in brain development and function. Rarely, human imprinting syndromes are caused by a failure of epigenetic reprogramming events (see below) resulting in an epigenotype switch that can phenocopy genetic deletion or duplication events.

4. REPROGRAMMING OF EPIGENETIC MARKS DURING DEVELOPMENT

There are two phases of global epigenetic reprogramming during the mammalian life cycle: prior to implantation in the early embryo and during the development of primordial germ cells (PGCs). These are required for viability and fertility, respectively, and involve the reprogramming of both DNA methylation and histone modifications.

4.1 Histone Modifications and Epigenetic Reprogramming

Our knowledge of chromatin dynamics during early embryonic and PGC reprogramming is limited due to the technical difficulty of conducting chromatin immunoprecipitation on the tiny amounts of cellular material that can be obtained at these stages. However, it is known that a dramatic reprogramming of histone modifications and exchange of histone variants takes place in PGCs immediately following, and perhaps initiated by, active DNA demethylation.[98] Some data suggest that this reprogramming of chromatin may in turn be necessary for the subsequent de novo DNA methylation of germ cells, as a homozygous germline mutation in the H3K4 demethylase AOF1/KDM1B has been shown to perturb the acquisition of some maternal methylation imprints during oogenesis.[37]

In the early embryo, little is known about histone modification dynamics during passive demethylation of the maternal pronucleus. However, upon fertilization, the protamines of the paternal pronucleus are replaced by hyper-acetylated histones, and immediately following histone incorporation, histone methylation marks such as H3K4me1, H3K9me1 and H3K27me1 are detectable. This rapid remodeling of paternal pronuclear chromatin occurs prior to the onset of DNA demethylation and it is not known if these processes are related, or whether particular sequence elements are associated with certain histone modifications (reviewed in Ref. 99). Indeed, the finding that modified histones are retained at developmentally important loci in human and, to a lesser extent, mouse sperm, has refocused attention on the role of histone modifications in the male germ cell and the zygote.[100,101] The fate of such modifications post-fertilization, and their impact on early embryonic development is entirely unknown.

4.2 Reprogramming of DNA Methylation

Erasure of DNA methylation in PGCs begins following specification at E7.5 and continues throughout the migration of the PGCs to the genital ridge. A second wave of demethylation occurs around E11.5 and includes the dramatic and rapid erasure of methylation at imprinted loci (reviewed by Ref. 102). Dogma dictates that this demethylation is complete, and that there is no epigenetic inheritance through meiosis, however, some elements, such as intra-cisternal A particles and rare unique regions, appear to partially escape this methylation reprogramming.[103–106] This has also been shown to be the case in human PGC development.[107–109] This will be discussed further in Section 3.4 in the context of transgenerational epigenetic inheritance.

DNA demethylation can be active or passive. Passive demethylation is replication dependent, and requires the absence or inactivity of the maintenance methyltransferase DNMT1 and the consequent failure to copy methylation onto the daughter strand at replication. Active mammalian DNA demethylation is observed both in the paternal zygotic pronucleus,[110,111] and in PGCs.[104] Direct removal of the methyl group involves breaking a highly stable carbon–carbon bond, therefore indirect pathways to DNA demethylation, which involve the DNA repair machinery, are thought to be more likely.[104] In plants active DNA demethylation is achieved through the combined action of DNA glycosylases and the base excision repair, BER, pathway (reviewed in Ref. 112). In mammals, the oxidation of 5mC to 5hmC catalyzed by the TET family of enzymes is thought to precede BER in active demethylation.[58,113,114] However, the observation of significant levels of hydroxymethylated cytosine in some cellular and genomic contexts suggests that 5hmC does not solely exist as an intermediate in active demethylation. In the early embryo DNA methylation is reprogrammed by a combination of active and passive mechanisms. Active TET mediated demethylation of the paternal pronucleus is observed, with global hydroxymethylation persisting into the cleavage stage embryo, indicating that there is no rapid removal of 5hmC by base excision repair or related pathways at this stage.[61,115]

The maternal genome also undergoes reprogramming of CpG methylation in the early embryo, but this is a passive, replication dependent process.[110,111,116] It is not fully understood how ooplasmic factors distinguish between the maternal and paternal pronuclei in the zygote. However, PGC7/Stella has been shown to contribute to the protection of the maternal genome from demethylation, as evident from the behavior of repeats and some methylation imprints in *Stella* mutants.[115,117] Reprogramming in the zygote is thought to be less extensive than in PGCs, as DNA methylation at centromeric and pericentromeric heterochromatin and ICRs (see below) appears largely unaffected.

5. EPIGENETIC INHERITANCE

There has been significant interest and speculation regarding the possibility of inheritance of environmentally induced epigenetic information between generations. It has been hypothesized that such transgenerational epigenetic inheritance may result in altered development of the offspring, dependant on the environmental factors their parents were exposed to. The extent to which intergenerational (between parents to offspring) and transgenerational (across multiple generations) epigenetic inheritance occurs will depend on the degree of epigenetic reprogramming occurring in the germline, and the extent to which these reprogramming mechanisms are vulnerable to environmental stress. There is clear evidence of epigenetic inheritance in plants and lower metazoans such as the worm *C. elegans*.[109] In mammals, although many examples of non-Mendelian inheritance have been observed,[118] the exact mechanisms responsible, and the extent to which they contribute to human health and disease remain to be fully determined. The inheritance of environmentally induced epigenetic information may be more advantageous in organisms with short lifespans and the limited capacity to change their environment, such as plants. It is theoretically possible that the evolutionary cost of epigenetic inheritance increases with an organism's lifespan and its ability to modulate its environment, as the parental experience of the environment may be less predictive of that of their offspring. Therefore it is possible that epigenetic inheritance may be significantly less prevalent in higher order mammals than in lower metazoans and plants.[109] However, further high quality, well powered, prospective studies in humans and animal models will be required to investigate this.

6. CONCLUDING REMARKS

A wealth of experimental data from patients and animal models demonstrates that epigenetic modifications in a mitotic context are crucial for normal brain development and adult neurogenesis. However there is increasing appreciation that epigenetic modifications in a post-mitotic context are also important for brain function, including learning and memory. As epigenetic marks are theoretically reversible they hold great therapeutic promise, particularly if treatments can be developed which modulate them in a site-specific manner. However, there remain significant technical challenges to be overcome in this endeavor, although progress has been made. The continued investigation of epigenetic phenomena in patients, cell, *in silico* and animal models, will contribute significantly to our understanding of brain development and function.

ACKNOWLEDGMENT

I apologise to the researchers whos work was not cited in this chapter due to space limitations.

REFERENCES

1. Waddington CH. The epigenotype. 1942. *Int J Epidemiol.* 2012;41:10–13.
2. Bird A. Perceptions of epigenetics. *Nature.* 2007;447:396–398.
3. Almouzni G, Cedar H. Maintenance of epigenetic information. *Cold Spring Harb Perspect Biol.* 2016;8.
4. Lappalainen T, Greally JM. Associating cellular epigenetic models with human phenotypes. *Nat Rev Genet.* 2017;18:441–451.
5. Hirabayashi Y, Gotoh Y. Epigenetic control of neural precursor cell fate during development. *Nat Rev Neurosci.* 2010;11:377–388.
6. Kriegstein A, Alvarez-Buylla A. The glial nature of embryonic and adult neural stem cells. *Annu Rev Neurosci.* 2009;32:149–184.
7. Bayraktar OA, Fuentealba LC, Alvarez-Buylla A, Rowitch DH. Astrocyte development and heterogeneity. *Cold Spring Harb Perspect Biol.* 2015;7.
8. Ziller MJ, et al. Dissecting neural differentiation regulatory networks through epigenetic footprinting. *Nature.* 2015;518:355–359.
9. Ziller MJ, et al. Dissecting the functional consequences of de novo DNA methylation dynamics in human motor neuron differentiation and physiology. *Cell Stem Cell.* 2018;22:559–574.e9. https://doi.org/10.1016/j.stem.2018.02.012.
10. Bonev B, et al. Multiscale 3D genome rewiring during mouse article multiscale 3D genome rewiring during mouse neural development. *Cell.* 2017;171557.e1–557.e24.
11. Guo JU, et al. Neuronal activity modifies the DNA methylation landscape in the adult brain. *Nat Neurosci.* 2011;14:1345–1351.

12. Day JJ, Sweatt JD. DNA methylation and memory formation. *Nat Neurosci.* 2010;13:1319–1323.
13. Kouzarides T. Chromatin modifications and their function. *Cell.* 2007;128:693–705.
14. Hashimshony T, Zhang J, Keshet I, Bustin M, Cedar H. The role of DNA methylation in setting up chromatin structure during development. *Nat Genet.* 2003; 34:187–192.
15. Razin A. CpG methylation, chromatin structure and gene silencing—a three-way connection. *EMBO J.* 1998;17:4905–4908.
16. Aravin A, et al. A novel class of small RNAs bind to MILI protein in mouse testes. *Nature.* 2006;442:203–207.
17. Nagano T, et al. The Air noncoding RNA epigenetically silences transcription by targeting G9a to chromatin. *Science.* 2008;322:1717–1720.
18. Redrup L, et al. The long noncoding RNA Kcnq1ot1 organises a lineage-specific nuclear domain for epigenetic gene silencing. *Development.* 2009;136:525–530.
19. Pombo A, Dillon N. Three-dimensional genome architecture: players and mechanisms. *Nat Rev Mol Cell Biol.* 2015;16:245–257.
20. Lupiáñez DG, et al. Disruptions of topological chromatin domains cause pathogenic rewiring of gene-enhancer interactions. *Cell.* 2015;161:1012–1025.
21. Pombo A, Gilbert DM. Nucleus and gene expression: the structure and function conundrum. *Curr Opin Cell Biol.* 2010;22:269–270.
22. Gilbert N, Gilchrist S, Bickmore WA. Chromatin organization in the mammalian nucleus. *Int Rev Cytol.* 2005;242:283–336.
23. Dekker J, et al. The 4D nucleome project. *Nature.* 2017;549:219–226.
24. Lister R, et al. Human DNA methylomes at base resolution show widespread epigenomic differences. *Nature.* 2009;462:315–322.
25. Tomizawa S, et al. Dynamic stage-specific changes in imprinted differentially methylated regions during early mammalian development and prevalence of non-CpG methylation in oocytes. *Development.* 2011;138:811–820.
26. Ehrlich M, et al. Amount and distribution of 5-methylcytosine in human DNA from different types of tissues of cells. *Nucleic Acids Res.* 1982;10:2709–2721.
27. Bestor TH, Tycko B. Creation of genomic methylation patterns. *Nat Genet.* 1996;12:363–367.
28. Pradhan S, Bacolla A, Wells RD, Roberts RJ. Recombinant human DNA (cytosine-5) methyltransferase. I. Expression, purification, and comparison of de novo and maintenance methylation. *J Biol Chem.* 1999;274:33002–33010.
29. Law JA, Jacobsen SE. Establishing, maintaining and modifying DNA methylation patterns in plants and animals. *Nat Rev Genet.* 2010;11:204–220.
30. Ooi SK, O'Donnell AH, Bestor TH. Mammalian cytosine methylation at a glance. *J Cell Sci.* 2009;122:2787–2791.
31. Ooi SK, et al. DNMT3L connects unmethylated lysine 4 of histone H3 to de novo methylation of DNA. *Nature.* 2007;448:714–717.
32. Otani J, et al. Structural basis for recognition of H3K4 methylation status by the DNA methyltransferase 3A ATRX-DNMT3-DNMT3L domain. *EMBO Rep.* 2009;10:1235–1241.
33. Ball MP, et al. Targeted and genome-scale strategies reveal gene-body methylation signatures in human cells. *Nat Biotechnol.* 2009;27:361–368.
34. Ernst J, Kellis M. Discovery and characterization of chromatin states for systematic annotation of the human genome. *Nat Biotechnol.* 2010;28:817–825.
35. Meissner A, et al. Genome-scale DNA methylation maps of pluripotent and differentiated cells. *Nature.* 2008;454:766–770.
36. Weber M, et al. Distribution, silencing potential and evolutionary impact of promoter DNA methylation in the human genome. *Nat Genet.* 2007;39:457–466.

37. Ciccone DN, et al. KDM1B is a histone H3K4 demethylase required to establish maternal genomic imprints. *Nature.* 2009;461:415–418.
38. Epsztejn-Litman S, et al. De novo DNA methylation promoted by G9a prevents reprogramming of embryonically silenced genes. *Nat Struct Mol Biol.* 2008;15:1176–1183.
39. Feldman N, et al. G9a-mediated irreversible epigenetic inactivation of Oct-3/4 during early embryogenesis. *Nat Cell Biol.* 2006;8:188–194.
40. Neumann B, Barlow DP. Multiple roles for DNA methylation in gametic imprinting. *Curr Opin Genet Dev.* 1996;6:159–163.
41. Youngson NA, Kocialkowski S, Peel N, Ferguson-Smith AC. A small family of sushi-class retrotransposon-derived genes in mammals and their relation to genomic imprinting. *J Mol Evol.* 2005;61:481–490.
42. Schübeler D. Function and information content of DNA methylation. *Nature.* 2015;517:321–326.
43. Maunakea AK, et al. Conserved role of intragenic DNA methylation in regulating alternative promoters. *Nature.* 2010;466:253–257.
44. Laurent L, et al. Dynamic changes in the human methylome during differentiation. *Genome Res.* 2010;20:320–331.
45. Suzuki MM, Bird A. DNA methylation landscapes: provocative insights from epigenomics. *Nat Rev Genet.* 2008;9:465–476.
46. Weber M, et al. Chromosome-wide and promoter-specific analyses identify sites of differential DNA methylation in normal and transformed human cells. *Nat Genet.* 2005;37:853–862.
47. Hellman A, Chess A. Gene body-specific methylation on the active X chromosome. *Science.* 2007;315:1141–1143.
48. Wu H, et al. Dnmt3a-dependent nonpromoter DNA methylation facilitates transcription of neurogenic genes. *Science.* 2010;329:444–448.
49. Shen W, et al. The spectrum of DNMT3A variants in Tatton–Brown–Rahman syndrome overlaps with that in hematologic malignancies. *Am J Med Genet A.* 2017;173:3022–3028.
50. Amir RE, et al. Rett syndrome is caused by mutations in X-linked MECP2, encoding methyl-CpG-binding protein 2. *Nat Genet.* 1999;23:185–188.
51. Luikenhuis S, Giacometti E, Beard CF, Jaenisch R. Expression of MeCP2 in postmitotic neurons rescues Rett syndrome in mice. *Proc Natl Acad Sci USA.* 2004;101:6033–6038.
52. Jones PL, et al. Methylated DNA and MeCP2 recruit histone deacetylase to repress transcription. *Nat Genet.* 1998;19:187–191.
53. Nan X, et al. Transcriptional repression by the methyl-CpG-binding protein MeCP2 involves a histone deacetylase complex. *Nature.* 1998;393:386–389.
54. Skene PJ, et al. Neuronal MeCP2 is expressed at near histone-octamer levels and globally alters the chromatin state. *Mol Cell.* 2010;37:457–468.
55. Mellén M, Ayata P, Heintz N. 5-hydroxymethylcytosine accumulation in postmitotic neurons results in functional demethylation of expressed genes. *Proc Natl Acad Sci USA.* 2017;114:E7812–E7821. https://doi.org/10.1073/pnas.1708044114.
56. Mellén M, Ayata P, Dewell S, Kriaucionis S, Heintz N. MeCP2 binds to 5hmC enriched within active genes and accessible chromatin in the nervous system. *Cell.* 2012;151:1417–1430.
57. Ito S, et al. Role of Tet proteins in 5mC to 5hmC conversion, ES-cell self-renewal and inner cell mass specification. *Nature.* 2010;466:1129–1133.
58. Tahiliani M, et al. Conversion of 5-methylcytosine to 5-hydroxymethylcytosine in mammalian DNA by MLL partner TET1. *Science.* 2009;324:930–935.
59. Ito S, et al. Tet proteins can convert 5-methylcytosine to 5-formylcytosine and 5-carboxylcytosine. *Science.* 2011;333:1300–1303.

60. Inoue A, Shen L, Dai Q, He C, Zhang Y. Generation and replication-dependent dilution of 5fC and 5caC during mouse preimplantation development. *Cell Res.* 2011;21:1670–1676.
61. Iqbal K, Jin SG, Pfeifer GP, Szabo PE. Reprogramming of the paternal genome upon fertilization involves genome-wide oxidation of 5-methylcytosine. *Proc Natl Acad Sci USA.* 2011;108:3642–3647.
62. Kriaucionis S, Heintz N. The nuclear DNA base 5-hydroxymethylcytosine is present in Purkinje neurons and the brain. *Science.* 2009;324:929–930.
63. Penn NW. Modification of brain deoxyribonucleic acid base content with maturation in normal and malnourished rats. *Biochem J.* 1976;155:709–712.
64. Ficz G, et al. Dynamic regulation of 5-hydroxymethylcytosine in mouse ES cells and during differentiation. *Nature.* 2011;473:398–402.
65. Olins DE, Olins AL. Chromatin history: our view from the bridge. *Nat Rev Mol Cell Biol.* 2003;4:809–814.
66. Li B, Carey M, Workman JL. The role of chromatin during transcription. *Cell.* 2007;128:707–719.
67. Vakoc CR, Mandat SA, Olenchock BA, Blobel GA. Histone H3 lysine 9 methylation and HP1gamma are associated with transcription elongation through mammalian chromatin. *Mol Cell.* 2005;19:381–391.
68. Barski A, et al. High-resolution profiling of histone methylations in the human genome. *Cell.* 2007;129:823–837.
69. Bernstein BE, et al. A bivalent chromatin structure marks key developmental genes in embryonic stem cells. *Cell.* 2006;125:315–326.
70. Heintzman ND, et al. Distinct and predictive chromatin signatures of transcriptional promoters and enhancers in the human genome. *Nat Genet.* 2007;39:311–318.
71. Mikkelsen TS, et al. Genome-wide maps of chromatin state in pluripotent and lineage-committed cells. *Nature.* 2007;448:553–560.
72. Thomson JP, et al. CpG islands influence chromatin structure via the CpG-binding protein Cfp1. *Nature.* 2010;464:1082–1086.
73. Calo E, Wysocka J. Modification of enhancer chromatin: what, how, and why? *Mol Cell.* 2013;49:825–837.
74. Roadmap Epigenomics Consortium, et al. Integrative analysis of 111 reference human epigenomes. *Nature.* 2015;518:317–329.
75. Hon GC, Hawkins RD, Ren B. Predictive chromatin signatures in the mammalian genome. *Hum Mol Genet.* 2009;18:R195–201.
76. ENCODE Project Consortium, et al. An integrated encyclopedia of DNA elements in the human genome. *Nature.* 2012;489:57–74.
77. Bjornsson HT. The Mendelian disorders of the epigenetic machinery. *Genome Res.* 2015;25:1473–1481.
78. Martienssen RA, Zaratiegui M, Goto DB. RNA interference and heterochromatin in the fission yeast Schizosaccharomyces pombe. *Trends Genet.* 2005;21:450–456.
79. Kloc A, Zaratiegui M, Nora E, Martienssen R. RNA interference guides histone modification during the S phase of chromosomal replication. *Curr Biol.* 2008;18:490–495.
80. Aravin AA, Sachidanandam R, Girard A, Fejes-Toth K, Hannon GJ. Developmentally regulated piRNA clusters implicate MILI in transposon control. *Science.* 2007;316:744–747.
81. Kuramochi-Miyagawa S, et al. DNA methylation of retrotransposon genes is regulated by Piwi family members MILI and MIWI2 in murine fetal testes. *Genes Dev.* 2008;22:908–917.
82. Watanabe T, et al. Identification and characterization of two novel classes of small RNAs in the mouse germline: retrotransposon-derived siRNAs in oocytes and germline small RNAs in testes. *Genes Dev.* 2006;20:1732–1743.

83. Watanabe T, et al. Endogenous siRNAs from naturally formed dsRNAs regulate transcripts in mouse oocytes. *Nature*. 2008;453:539–543.
84. Aravin AA, Bourc'his D. Small RNA guides for de novo DNA methylation in mammalian germ cells. *Genes Dev*. 2008;22:970–975.
85. Barau J, et al. The DNA methyltransferase DNMT3C protects male germ cells from transposon activity. *Science*. 2016;354:909–912.
86. Brosnan CA, Voinnet O. The long and the short of noncoding RNAs. *Curr Opin Cell Biol*. 2009;21:416–425.
87. Lin SP, et al. Asymmetric regulation of imprinting on the maternal and paternal chromosomes at the Dlk1-Gtl2 imprinted cluster on mouse chromosome 12. *Nat Genet*. 2003;35:97–102.
88. Horsthemke B, Wagstaff J. Mechanisms of imprinting of the Prader-Willi/Angelman region. *Am J Med Genet A*. 2008;146A:2041–2052.
89. Pandey RR, et al. Kcnq1ot1 antisense noncoding RNA mediates lineage-specific transcriptional silencing through chromatin-level regulation. *Mol Cell*. 2008;32:232–246.
90. Peters J, Williamson CM. Control of imprinting at the Gnas cluster. *Epigenetics*. 2007;2:207–213.
91. Williamson CM, et al. Uncoupling antisense-mediated silencing and DNA methylation in the imprinted gnas cluster. *PLoS Genet*. 2011;7:e1001347.
92. Gong C, Maquat LE. lncRNAs transactivate STAU1-mediated mRNA decay by duplexing with 3' UTRs via Alu elements. *Nature*. 2011;470:284–288.
93. Edwards CA, Ferguson-Smith AC. Mechanisms regulating imprinted genes in clusters. *Curr Opin Cell Biol*. 2007;19:281–289.
94. McEwen KR, Ferguson-Smith AC. Distinguishing epigenetic marks of developmental and imprinting regulation. *Epigenetics Chromatin*. 2010;3:2.
95. Ferrón SR, et al. Differential genomic imprinting regulates paracrine and autocrine roles of IGF2 in mouse adult neurogenesis. *Nat Commun*. 2015;6:8265. https://doi.org/10.1038/ncomms9265.
96. Ferron SR, et al. Postnatal loss of Dlk1 imprinting in stem cells and niche astrocytes regulates neurogenesis. *Nature*. 2011;475:381–385.
97. Hirasawa R, Feil R. Genomic imprinting and human disease. *Essays Biochem*. 2010;48:187–200.
98. Hajkova P, et al. Chromatin dynamics during epigenetic reprogramming in the mouse germ line. *Nature*. 2008;452:877–881.
99. Morgan HD, Santos F, Green K, Dean W, Reik W. Epigenetic reprogramming in mammals. *Hum Mol Genet*. 2005;14:R47–58. Spec No.
100. Brykczynska U, et al. Repressive and active histone methylation mark distinct promoters in human and mouse spermatozoa. *Nat Struct Mol Biol*. 2010;17:679–687.
101. Hammoud SS, et al. Distinctive chromatin in human sperm packages genes for embryo development. *Nature*. 2009;460:473–478.
102. Sasaki H, Matsui Y. Epigenetic events in mammalian germ-cell development: reprogramming and beyond. *Nat Rev Genet*. 2008;9:129–140.
103. Seisenberger S, et al. The dynamics of genome-wide DNA methylation reprogramming in mouse primordial germ cells. *Mol Cell*. 2012;48:849–862.
104. Hajkova P, et al. Epigenetic reprogramming in mouse primordial germ cells. *Mech Dev*. 2002;117:15–23.
105. Lane N, et al. Resistance of IAPs to methylation reprogramming may provide a mechanism for epigenetic inheritance in the mouse. *Genesis*. 2003;35:88–93.
106. Rakyan V, Whitelaw E. Transgenerational epigenetic inheritance. *Curr Biol*. 2003;13:R6.
107. Tang WW, et al. A unique gene regulatory network resets the human germline epigenome for development. *Cell*. 2015;161:1453–1467.

108. Gkountela S, et al. DNA demethylation dynamics in the human prenatal germline. *Cell.* 2015;161:1425–1436.
109. Guo F, et al. The transcriptome and DNA methylome landscapes of human primordial germ cells. *Cell.* 2015;161:1437–1452.
110. Mayer W, Niveleau A, Walter J, Fundele R, Haaf T. Demethylation of the zygotic paternal genome. *Nature.* 2000;403:501–502.
111. Oswald J, et al. Active demethylation of the paternal genome in the mouse zygote. *Curr Biol.* 2000;10:475–478.
112. Zhu JK. Active DNA demethylation mediated by DNA glycosylases. *Annu Rev Genet.* 2009;43:143–166.
113. Koh KP, et al. Tet1 and Tet2 regulate 5-hydroxymethylcytosine production and cell lineage specification in mouse embryonic stem cells. *Cell Stem Cell.* 2011;8:200–213.
114. Hackett JA, et al. Germline DNA demethylation dynamics and imprint erasure through 5-hydroxymethylcytosine. *Science.* 2013;339:448–452.
115. Wossidlo M, et al. 5-Hydroxymethylcytosine in the mammalian zygote is linked with epigenetic reprogramming. *Nat Commun.* 2011;2:241.
116. Rougier N, et al. Chromosome methylation patterns during mammalian preimplantation development. *Genes Dev.* 1998;12:2108–2113.
117. Nakamura T, et al. PGC7/Stella protects against DNA demethylation in early embryogenesis. *Nat Cell Biol.* 2007;9:64–71.
118. Sales VM, Ferguson-Smith AC, Patti ME. Epigenetic mechanisms of transmission of metabolic disease across generations. *Cell Metab.* 2017;25:559–571.

CHAPTER FOUR

Neuroepigenetics of Aging and Age-Related Neurodegenerative Disorders

Roy Lardenoije*, Ehsan Pishva*,†, Katie Lunnon†, Daniel L. van den Hove*,1

*Department of Psychiatry and Neuropsychology, Maastricht University Medical Centre, Maastricht, The Netherlands
†University of Exeter Medical School, University of Exeter, Exeter, United Kingdom
[1]Corresponding author: e-mail address: d.vandenhove@maastrichtuniversity.nl

Contents

Abstract

Neurodegenerative diseases are complex, progressive disorders and affect millions of people worldwide, contributing significantly to the global burden of disease. In recent years, research has begun to investigate epigenetic mechanisms for a potential role in disease etiology. In this chapter, we describe the current state of play for epigenetic research into neurodegenerative disorders including Alzheimer's disease, Parkinson's disease and Huntington's disease. We focus on the recent evidence for a potential role of DNA modifications, histone modifications and non-coding RNA in the etiology of these disorders. Finally, we discuss how new technological and bioinformatics advances in the field of epigenetics could further progress our understanding about the underlying mechanisms of neurodegenerative diseases.

Progress in Molecular Biology and Translational Science, Volume 158
ISSN 1877-1173
https://doi.org/10.1016/bs.pmbts.2018.04.008

1. NEUROEPIGENETICS OF AGING

For many age-related neurodegenerative disorders, including Alzheimer's disease (AD), Parkinson's disease (PD) and Huntington's disease (HD), there is currently evidence for an etiopathogenic epigenetic component. In this context, however, it is important to consider the widespread epigenetic changes occurring with natural aging. Below, we provide an overview of findings from studies of age-related epigenetic modifications at the level of the DNA, chromatin and non-coding RNA.

1.1 Alterations in DNA Modifications With Age

Epigenetics refers to the reversible cell-type specific regulation of gene expression, independent of variations in DNA sequence. DNA modifications are the most well described epigenetic process in human brain studies.[1–3] DNA methylation, whereby a methyl group is added to the 5 position of a cytosine, forming 5-methylcytosine (5mC), is usually thought to bring about gene silencing. However, recent evidence suggests that this is dependent on the location of the CpG site, with gene body methylation associated sometimes with alternative splicing,[4] or increased gene expression. DNA methyltransferase 3A (DNMT3A), responsible for *de novo* DNA methylation, has been shown to be expressed in the human brain across all ages,[5] indicating that dynamic regulation of gene expression through DNA methylation is possible not only during development, but also during the later stages of life.

In mice, DNMT3A levels were observed to increase with age in the hippocampus,[6] along with increased DNA methylation and DNA hydroxymethylation levels.[7–9] Conversely, expression of a specific isoform of DNMT3A, DNMT3A2, was found to be decreased with age in the mouse hippocampus.[10] Expression of this isoform correlated with cognitive performance in mice, and restoring DNMT3A2 levels in the hippocampus of aged mice counteracted the age-related cognitive decline observed in these mice. Age-related increases in DNA methylation were also observed in the mouse cerebellum and rat cerebral cortex.[11,12] Despite these observations in the brain, it should be noted that studies in other murine and bovine tissues have shown age-related hypomethylation,[13,14] which has also been described during senescence of cultured human fibroblasts, an established model of cellular aging.[15] This age-related decline in global DNA methylation in fibroblasts was paired with a decline in maintenance DNMT activity, although there was an increase in *de novo* DNMT activity. Thus, whether global levels of DNA methylation increase or decrease with age remains to be firmly established and may be tissue-specific.[16] Discrepancies between studies may also be the result of methodological differences, as some rely on techniques detecting all methylated cytosines, while others focus on (selected) CpG sites or islands.

Site-specific analyses of CpGs showing age-related changes in DNA methylation do not show a general pattern, while the majority of sites show hypermethylation with age, some do show age-associated hypomethylation.[17,18] DNA methylation changes at specific CpG sites that show a linear correlation with aging have been used to determine the "DNA methylation age" of a sample, which is a fairly accurate predictor of chronological age.[18,19] The resulting differences in epigenetic aging rates between individuals may contribute to differences in age-related disease susceptibility. These epigenetic changes that are not genetically programmed, such as those in charge of development, may also result in the progressive epigenetic drift observed in monozygotic twins as they age and become phenotypically more distinct due to stochastic changes and environmental factors.[20] Interestingly, in line with the observation that women generally live longer than men,[21] men were found to have a higher epigenetic aging rate than women.[22] Genes with CpG sites predictive of age include genes associated with age-related diseases such as cancer and AD.[18,22]

1.2 Alterations in Chromatin Modifications During Aging

Covalent modifications of the histones amino terminal tails in nucleosomes have important roles in accessibility of DNA to transcription factors and subsequently gene expression. Numerous types of histone modifications have been shown to follow an age-dependent expression pattern. For example, H4 monoacetylation, was found to decrease with age in the cerebral cortex of rats, whilst other histone acetylation markers, such as H2A, H2B and H3 appear to remain stable.[23] Additionally, a decrease in monomethylated H4K20 with age was observed in brains of the senescence accelerated mouse prone 8 (SAMP8) model of aging.[24] Early *in vitro* work with human cells has also shown a decrease in the rate of histone acetylation with aging.[25] Age-related decreases in histone acetylation may be linked to observed age-related increases of histone deacetylase 2 (HDAC2) in the mouse hippocampus.[26] It is suggested that the age-related loss of monoacetylated H4 may contribute to the aging process by compromising the plasticity of chromatin.[23] In the hippocampus of aging mice, a decrease in H4K12 acetylation was linked to age-related memory deficits through the regulation of genes related to memory consolidation, the latter of which could be restored by reinstating H4K12 acetylation levels.[27] Age-related decreases in histone acetylation may be linked to observed age-related increases of neuron-specific HDAC2 in the mouse hippocampus.[26] Indeed, it was found in mice that specific overexpression of Hdac2 negatively impacted upon memory formation and synaptic plasticity.[28] These memory impairing effects of HDAC2 could be reversed through treatment with HDAC inhibitors and were also reduced in HDAC2-deficient mice. In contrast to HDAC2, the HDAC Sirtuin 1 (SIRT1) appears to decrease with age in the rat hippocampus and cerebral cortex,[29] and enhancing SIRT1 levels may enhance longevity.[30] Decreases in SIRT1 levels have also been observed during cell senescence in human fibroblasts.[31] However, other *in vitro* studies using human cells have shown that SIRT1 can also deacetylate the histone methyltransferase suppressor of variegation 3–9 homolog (SUV39H) 1, increasing its activity.[32] SUV39H1 is primarily responsible for H3K9 trimethylation, which accumulates in specific heterochromatin foci that are associated with senescence.[32,33] This would suggest the increased SIRT1 activity would induce senescence by activating SUV39H1. To complicate the role of SIRT1 in aging even further, its ortholog in *Caenorhabditis elegans*, sir-2.1, is thought to promote longevity through mitohormesis by the indirect generation of gen peroxide.[34,35] These latter findings stress that the non-epigenetic actions of enzymes involved in epigenetic regulation should not be overlooked. SIRT1 is also implicated in the life-prolonging effects of caloric restriction observed in yeast, invertebrates, and vertebrates,[36] which also appears to have a widespread effect on DNA methylation in mice.[6–8,11] Clearly, the exact role of histone-modifying enzymes in natural aging and associated brain functions requires further research.

1.3 Changes in Non-Coding RNAs

Micro-RNAs (miRNAs) are a group of small (~20 nucleotides), non-coding, single-stranded RNAs, which can regulate mRNA expression, with a single miRNA being able to regulate the expression of multiple mRNAs.[36a] In *C. elegans*, the miRNA lin-4 was found to be a pivotal regulator of development and aging, with overexpression leading to an expended lifespan, whereas reduced levels resulted in accelerated aging.[37] It was found that lin-4 and its target lin-14 likely regulate aging in *C. elegans* via the insulin/insulin-like growth factor (IGF)-1 pathway, a highly conserved nutrient sensing pathway that regulates longevity in multiple species.[38] Curiously, removal of the reproducing germ line cells in *C. elegans* leads to an extended lifespan, which is thought to be the result of a redistribution of resources from the gonads to bolster the immune system.[39] This increased longevity is mediated by the action of miR-71 in neurons and leading to an upregulation of the DAF-16 transcription factor in the intestine, which induces enhanced resistance to oxidative stress and microbial pathogens.[40] The regulation of oxidative resistance by miRNAs has also been observed in other species; for instance in brain and liver tissue of aged mice, several miRNAs regulating genes involved in resistance to oxidative stress and mitochondrial functioning were upregulated.[41,42] It was observed that apart from certain miRNAs that were specifically upregulated in either brain or liver tissue, a subset was also upregulated with aging in both tissues, suggesting both tissue-specific and whole organism regulatory mechanisms of aging.[42] Interestingly, those mice with a very long lifespan (33 months) showed less miRNA dysregulation than the group with a normal lifespan (24 months).

A study in human blood mononuclear cells found nine miRNAs (miR-24, miR-103, miR-107, miR-128, miR-130a, miR-155, miR-221, miR-496, and miR-1538) showing decreased levels with age. It still remains to be established whether these age-associated miRNAs contribute or are a consequence of the aging process, although miR-24 has been found to be involved in oxidative stress-induced cellular senescence through the regulation of DNA topoisomerase (TOP) 1 in human fibroblasts.[43] Another study found four additional miRNAs (miR-17, miR-19b, miR-20a, and miR-106a) to be downregulated in several models of senescence and human aging.[44] These miRNAs were suggested to affect longevity through the regulation of p21/CDKN1A, a cell cycle inhibitor involved in the stress response, that may limit the lifespan in mice and humans.[45,46] A small study looking at *post-mortem* brain tissue from chimpanzees, rhesus macaques, and humans found miR-144 levels to be consistently increased with aging in the cerebellum of all species.[47] Little is currently known about the regulatory function of miR-144 in aging, but it appears to be involved in regulating ataxin (ATXN) 1, the causative gene for spinocerebellar ataxia type 1,[47] and has a negative impact on insulin signaling by inhibiting the expression of insulin receptor substrate 1.[48]

2. NEUROEPIGENETICS OF AGE-RELATED NEURODEGENERATIVE DISORDERS

2.1 Epigenetic Alterations in Alzheimer's Disease

AD is a progressive neurodegenerative disorder and the most common form of dementia.[49] There are currently no therapies available to prevent, cure, or slow disease progression. AD is mainly characterized by cognitive decline, but can also involve other behavioral and physical impairments, including motor disabilities and mood instability.[50,51] Neurodegeneration in AD is thought to start in the brainstem, and mainly spreads to the frontal, temporal, and parietal cortex.[52,53] Primary hallmarks of neurodegeneration in AD are extracellular plaques, mainly consisting of aggregated amyloid-β (Aβ), and intracellular neurofibrillary tangles (NFTs), of which hyperphosphorylated tau is the primary constituent.[54] However, the exact etiology remains to be elucidated and is likely the result of an intricate interplay between genetic, epigenetic, and environmental factors.[53]

2.1.1 AD-Related DNA Modifications

In the earliest studies of epigenetic alterations in AD, S-adenosylmethionine (SAM), the methyl donor required by most methyltransferases, and S-adenosylhomocysteine (SAH), the demethylated product of SAM, were found to be markedly reduced in multiple brain regions of AD patients, including the frontal cortex, occipital cortex, temporal cortex, hippocampus and putamen.[55] However, this was later disputed by a study focusing on the prefrontal cortex of AD patients, which found SAM, but mainly SAH levels to be increased early in AD.[56] Kennedy et al. argue that not a lack of SAM, but a rise in SAH levels results in a decrease of methyltransferase activity in AD, as SAH is a potent inhibitor of most methyltransferases. SAH is normally converted into homocysteine (Hcy), but it is argued that the increased Hcy levels observed in AD reverse this reaction, resulting in Hcy-to-SAH conversion.[56–58] It is worth noting that while general methyltransferase inhibition by SAH may decrease DNMT activity and thus affect epigenetic regulation, it also has more widespread implications, as numerous processes affected in AD, such as neurotransmitter synthesis, and even dephosphorylation of tau, depend on methyltransferases.[59–62] Further highlighting the importance of disturbed DNA methylation due to high Hcy and SAH levels in AD, a study in human neuroblastoma cells indicated that a reduction of folate and vitamin B12 levels lead to gene promoter hypomethylation and increased expression of presenilin (PSEN) 1 and β-secretase (BACE), proteins that are critically involved in the production of Aβ.[63] The increase in PSEN1 and BACE levels were paired with an increase in Aβ levels. These findings were confirmed in a mouse model of AD, where a dietary deficiency of folate vitamin B12 and B6 led to elevated Hcy levels, increased SAH levels, and elevations in PSEN1 and BACE expression, as well as increased intraneuronal Aβ levels and extracellular Aβ deposition, and cognitive decline.[64] These detrimental effects induced by vitamin B deficiency, but also general AD-related neuropathology, could be counteracted through SAM supplementation.[65,66] Folate and vitamin B12 and B6 are important cofactors in Hcy metabolism, and the detection of low serum levels of these cofactors in AD makes them, together with SAM, attractive candidates for preventive treatment to counter cognitive decline in AD.[65,67] Two small pilot studies indeed show a beneficial effect of treatment with a formulation including folate, vitamin B12, and SAM in early and moderate to later-stage AD.[68,69]

Human neuroblastoma cells exposed to oligomeric Aβ showed reduced GTP-binding nuclear protein Ran expression, which is crucial for nucleocytoplasmic transport, and decreased nuclear localization of DNMT1 and RNA polymerase II.[70] This may offer a mechanism behind the global DNA hypomethylation associated with Aβ,[71] as decreased Ran mRNA levels and cytoplasmic DNMT1 sequestration was also observed in CA1 neurons of AD patients.

Inhibition of DNA methylation would suggest global DNA hypomethylation in AD, which has been reported in the entorhinal cortex and hippocampus of AD patients.[72,73] In contrast, it was found that LINE-1 repetitive elements, but not Alu or SAT-α elements, were hypermethylated in the blood of AD patients.[74] Curiously, within the AD group, higher LINE-1 methylation levels were associated with better cognitive performance. As the human genome consists for ~45% of repetitive elements, and ~15% specifically of LINE-1 repeats, the methylation status of these elements is thought to be a major contributor to global DNA methylation measurements.[74,75]

A gene-specific analysis found that the amyloid precursor protein (*APP*) promoter region becomes hypomethylated with age.[76] There is also a report of *APP* promoter hypomethylation in AD patients,[77] although this has not been successfully replicated.[78,79] Age-related changes in DNA methylation of transcription factor binding sites in the microtubule associated protein tau (*MAPT*) promoter suggest a downregulation of tau expression with age in human cerebral cortex.[80] Tau levels in the brain were indeed found to decrease with normal aging,[81] and were also found to be decreased in AD.[82] Whilst in non-demented brain samples decreased soluble tau levels do not correlate with neurofibrillary pathology, in AD it is thought that the decrease is due to an incorporation of tau into NFTs.[81,82] Looking at human brain tissue and lymphocytes from late-onset AD cases and controls, significant interindividual variability in the DNA methylation of methylenetetrahydrofolate reductase (*MTHFR*), *PSEN1*, *APOE*, and *DNMT1* was observed in AD cases.[83] This led to the proposition that age-related epigenetic drift may play a role in predisposing individuals to develop late-onset AD.[83] Sorbin and SH3 domain containing 3 (*SORBS3*) hypermethylation and S100 calcium binding protein A2 (*S100A*) hypomethylation have also been observed with aging, but this appears to be accelerated in AD patients.[5] *SORBS3* encodes a cell adhesion molecule that may be involved in synaptic functioning,[84] whereas the product of *S100A2* has been observed as part of *corpora amylacea*,[85] aggregates that are associated with aging and neurodegeneration and which may play a protective role by sequestering toxic molecules.[86]

With the rise of epigenome-wide studies (EWAS), numerous differentially methylated loci have been associated with AD pathology. One of the first AD EWAS highlighted neuropathology-associated differential methylation of ATP binding cassette subfamily A member 7 (*ABCA7*) and bridging integrator 1 (*BIN1*), which were previously identified as AD risk genes in genome-wide association studies (GWAS), and ankyrin 1 (*ANK1*), cadherin related 23 (*CDH23*), disco interacting protein 2 homolog A (*DIP2A*), rhomboid 5 homolog 2 (*RHBDF2*), ribosomal protein L13 (*RPL13*), ring finger protein 34 (*RNF34*), and serpin family F member 1 and 2 (*SERPINF1*, *SERPINF2*), which also showed altered gene expression in AD prefrontal cortex.[87,88] Another study also found *ANK1* to be hypermethylated in the entorhinal cortex, superior temporal gyrus and prefrontal cortex, but not in the cerebellum or blood.[89] A cell-type specific analysis showed that ANK1 expression was specifically upregulated in hippocampal microglia of AD patients, but not neurons and astrocytes.[90] Another study, focusing specifically on established AD susceptibility loci, found an association between methylation of *SORL1*, *ABCA7*, *HLA-DRB5*, *SLC24A4*, and *BIN1* and AD in the prefrontal cortex.[91] Of these genes, *SORL1* and *ABCA7* mRNA expression was associated with tangle density and BIN1 mRNA with Aβ levels.

A number of studies have demonstrated alterations in global DNA hydroxymethylation in AD. In the hippocampus of AD patients, a reduction in global levels of DNA hydroxymethylation has been observed, which showed a negative correlation with plaque load.[73] This was corroborated in the entorhinal cortex and cerebellum of AD patients by another study, although this study did not detect AD-related changes in 5mC, 5-formylcytosine (5fC), or 5-carboxylcytosine (5caC).[92] Conversely, it was reported that DNA methylation and hydroxymethylation increase in the middle frontal gyrus and middle temporal gyrus.[93] This study also established a correlation between

DNA methylation and hydroxymethylation and Aβ, tau, and ubiquitin levels and identified neurons, as opposed to astrocytes and microglia, as the main source of 5mC and 5hmC disease associated changes. Others have also reported increased levels of DNA methylation and hydroxymethylation in the hippocampus and parahippocampal gyrus of preclinical and late-state AD cases, paired with increased TET1 levels.[94] Additionally, this latter group found decreased 5fC and 5caC levels on both AD groups. These conflicting reports on the global status of epigenetic DNA modifications in AD thus call for larger, more conclusive studies, and may, apart from small sample sizes be attributed to differences in tissue processing and detection methods.[93,95]

Although the work on the one carbon metabolism and observed increases of Hcy in AD would point toward global DNA hypomethylation, which could lead to neurodegeneration in AD,[96] it has also been argued that it is DNA hypermethylation that promotes neuron loss in AD, by increasing the DNA mutation rate.[93] There are two published studies that quantified AD-associated epigenome-wide 5hmC at single base level.[97,98] However, more EWAS measuring 5hmC are expected in the field with larger sample sizes and higher sequencing resolution to conclusively differentiate 5mC from 5hmC.

2.1.2 AD-Related Chromatin Modifications

Histone acetylation has been associated with memory performance, and has been found to be decreased in animal models of AD,[99] and H3 acetylation has been shown to be decreased in the temporal lobe of AD patients.[100] Another study showed H4 acetylation was decreased in the hippocampus of an animal model of AD, and treatment with the HDAC inhibitor (HDACi) trichostatin A was able to rescue hippocampal long-term potentiation (LTP) and contextual memory.[101] These findings were corroborated with other HDACis, sodium valproate, sodium butyrate, and vorinostat, and it was determined these mainly acted on class I HDACs, including HDAC2.[102]

In an inducible transgenic mouse model overexpressing p25 in the forebrain, thereby inducing AD-related pathology, it was found that HDAC2 levels were elevated specifically in neurons of the CA1 hippocampal subregion and the prefrontal cortex, while other class I HDACs, i.e., HDAC1 and HDAC3, remained unchanged.[103] In contrast to these brain areas affected by neurodegeneration, the amygdala showed no changes in HDAC2 levels. The findings regarding HDAC2 overexpression, without changes in HDAC1 levels, were confirmed in another mouse model of AD, overexpressing human mutant *APP* and *PSEN1*. Mouse primary hippocampal neurons exposed to Aβ and H_2O_2 also expressed increased levels of HDAC2 mRNA, likely through the phosphorylation of glucocorticoid receptor (GR) 1 and subsequent activation of the HDAC2 gene promoter. It was found that HDAC2, but not HDAC1 and HDAC3, binding was enriched at several genes associated with learning and memory, paired with reduced levels of H2BK5, H3K14, H4K5, and H4K12 acetylation at these genes, but no overall changes in protein acetylation levels. Promoter hypoacetylation was associated with reduced RNA polymerase II binding and gene expression. Moreover, knockdown of HDAC2 resulted in elevated H4K12 acetylation and expression of neuroplasticity genes, and restoration of synaptic plasticity and cognitive performance, but did not improve neuronal survival. Importantly, HDAC2, but not HDAC1 and HDAC3, levels were determined to be increased in the CA1 area of the hippocampus and the entorhinal cortex, starting already at Braak stage I/II.

Although alterations in HDAC1 have not been observed in the brain of AD patients, a mouse model of AD overexpressing human mutant *APP* and *PSEN1* orally treated with MS-275, an HDACi favoring HDAC1, presented with decreased microglial activation and amyloid plaque formation, and better nesting behavior.[104] Of note, HDAC1 mRNA was actually found to be decreased in peripheral lymphocytes of AD patients.[105] Other HDACs have been implicated in AD, such as HDAC6 and SIRT1, although their role in AD pathogenesis is thought to be mainly due to their interaction with tau and not their involvement in epigenetic regulation.[106–108]

In contrast, *in vitro* work, using neurons from a triple transgenic mouse model of AD (3xTg-AD; PSEN1 (M146V), APPSwe (KM670/671NL), and tau (P301L)) at different ages, detected increased levels of H3 and H4 acetylation, starting at 4 months, which is around the age this model starts to develop cognitive impairments.[109] Exposing neurons from non-transgenic mice to Aβ also raised acetylation levels. Additionally, H3 lysine 9 (H3K9) methylation, which in contrast to histone acetylation is a marker generally associated with lower expression, was found to increase with age in neurons from both transgenic and non-transgenic mice, but more so in neurons from the AD model. A decrease in brain-derived neurotrophic factor (*Bdnf*) gene expression suggests that the repressive effect of H3K9 methylation prevailed over the enhancing effects of H3 and H4 acetylation. In another mouse model, only overexpressing human mutant APP, increased H3 acetylation and methylation was observed in the frontal cortex, while H3 methylation was decreased in the striatum and H4 acetylation was increased in the hippocampus.[110] In line with the findings regarding H3 methylation, in a monozygotic twin discordant for AD, H3K9 trimethylation was increased in the temporal cortex and hippocampus of the twin with AD.[111] Increased nuclear translocation of EP300 interacting inhibitor of differentiation 1 (EID1) in cortical neurons of AD patients has been reported.[112] EID1 is an inhibitor of the histone acetyltransferases (HATs) p300 and cAMP-response element binding protein (CREB)-binding protein (CBP), which are important for long-term memory formation.[113] Overexpression of EID1 led to impaired hippocampal LTP and spatial learning in mice,[112] which fits the observations regarding histone hypoacetylation in relation to AD. In the 3xTg-AD mouse model, enhancing CBP levels was able to restore learning and memory performance, although it was argued this was due to the interaction between transcription factor CREB and CBP, leading to elevated levels of BDNF, a pivotal player in memory formation.[114] Interestingly, mice not expressing KAT2B, another HAT, showed learning deficits, but appeared to be resistant to Aβ toxicity, leading to the suggestion that KAT2B regulates genes involved in Aβ metabolism.[115] A *BACE1*-specific investigation detected increased H3 acetylation levels in the *BACE1* promoter and increased mRNA levels in a mouse model of AD and blood cells from AD patients,[116] which is in line with the AD-associated DNA hypomethylation.[117] Globally elevated H3 phosphorylation was observed in the frontal cortex and hippocampus of AD patients and a mouse model of AD overexpressing human mutant *APP* (Tg2576).[110,118,119] H3 phosphorylation is associated with chromatin compaction during cell cycle activation, however it was observed in hippocampal neurons of AD patients that this marker was mainly located in the cytoplasm, suggesting dysregulated mitotic control may play a role in neurodegeneration.[119] The mitogen-activated protein kinase pathway is thought to be responsible for H3 phosphorylation, and kinases of this pathway were found to be upregulated in vulnerable neurons in AD.[119–121] H2AX phosphorylation, a marker of DNA damage, was also found to be increased in hippocampal astrocytes of AD patients.[122] Poly[ADP]-ribose polymerase (PARP) I has a strong association with AD and is suspected to play a role in inducing cell death after exposure to Aβ and oxidative stress, but is also responsible for H1 ADP-ribosylation involved in memory consolidation,[123,124] warranting the investigation of H1 ADP-ribosylation in AD. There is no EWAS publication yet on histone modifications in human AD.

2.1.3 Alterations in Non-Coding RNAs in AD

In *C. elegans* the APP homolog APP-like-1 (apl-1) was observed to be regulated by let-7 miRNAs, important for regulating gene expression in a time-dependent fashion during development.[125] SAMP8 mice develop cognitive deficits with age and show elevated hippocampal levels of APP, making these mice arguably not just a model of accelerated aging, but also of early AD pathology.[126] Importantly, these mice were shown to display low expression levels of miR-16, thought to repress *APP* expression and modulate tau phosphorylation.[126,127] In addition, with age, SAMP8 mice expressed decreased levels of miR-195, which targets β-site APP cleaving enzyme 1 (*BACE1*), an important player in Aβ generation. *In vitro* lowering miR-195 levels increases Aβ production, whilst increasing miR-195 levels reduces its production.[128]

It was found that the 3′UTR of *APP* has several miRNA binding sites, including some in, or close to, *APP* mutations associated with familial AD.[129] Several of these were shown to be able to regulate *APP in vitro*, and of these the binding of miR-147 and miR-20a was inhibited and enhanced, by the *APP* variants T117C and A454G, respectively. Others found that miR-101, miR-106a, miR-153, and miR-520c also bind and reduce APP protein expression in human cells.[130–132] In the cerebral cortex of the APP/PS1dE9 mouse model of AD, miR-153 has been shown to be decreased even in the early stages of pathology,[133] while miR-29c was observed to be overexpressed.[134] *APP* and amyloid precursor-like protein (*APLP*) *2*, a homolog of *APP*, both have binding sites for miR-153 and it was shown that miR-153 is able to suppress *APP* and *APLP2* expression *in vivo*. A subset of frontal cortex samples from AD patients with moderate AD-related pathology also exhibited reduced miR-153 levels, paired with increased *APP* levels.[132] *In vitro* and *in silico* work indicates that miR-29c actually represses *BACE1* expression,[134] suggesting its increased expression in the APP/PS1dE9 model may be a compensatory mechanism. In the anterior temporal cortex of a subset of sporadic AD patients, expression of the miR-29a/b-1 cluster and miR-124 was found to be lowered.[135,136] miR-124 and miR-29a/b-1 were also found to suppress *BACE1* expression in *in vitro* models of AD.[137] Additionally, *in vivo* and *in vitro* experiments indicated that miR-124 may regulate alternative splicing of *APP*, specifically the inclusion of exons 7 and 8, through the regulation of polypyrimidine tract binding protein (PTBP) 1.[135] *APP* including exons 7 and 8 is usually expressed in non-neuronal tissues, however, an increase in *APP* with these exons has been observed in the brains of AD patients. miR-107 is another miRNA thought to target *BACE1*, and which shows decreased levels in the temporal cortex even in the early stages of AD.[138] miR-107 also regulates *GRN*, a gene involved in inflammation that has been associated with AD.[139]

AD-associated cytoskeletal pathology has been linked to elevated levels of actin-binding protein cofilin, which is another target of miR-107, but also miR-103, which were both found to be decreased in Tg19959 mouse brains.[140]

The previously mentioned miR-29a/b-1 cluster that was found to be lowered in a subset of sporadic AD cases had also been implicated in the regulation of serine palmitoyltransferase (SPT), together with miR-9, miR-137, and miR-181c.[141] Lower levels of these miRNAs result in higher SPT levels, which in turn is associated with higher Aβ levels. Additionally, miR-9 and miR-181c were found to repress transforming growth factor-beta-induced protein ig-h3 (TGFBI), tripartite motif-containing protein 2 (TRIM2), NAD-dependent protein deacetylase SIRT1, and BTB/POZ domain-containing protein 3 (BRBD3).[142] Proliferation, differentiation, and migration of neuronal progenitors has further been linked to miR-9.[143,144] Note that others have detected an abundance of miR-9, as well as miR-125b, miR-128, and miR-146a in the temporal lobe of AD patients.[145,146]

In vitro exposure to Aβ42 oligomers was shown to induce and later reduce miR-106b levels, while TGF-beta receptor type-2 (TGFBR2) levels, a target of miR-106b, remained decreased after exposure.[147] A similar pattern was observed in APP/PS1dE9 mice, showing increased miR-106b levels at first, decreased levels at more advanced ages, and reduced TGFBR2 levels from 3 months onward. A miR-106b and TGFBR2 deficiency has also been detected in sporadic AD patients and the decrease in TGFBR2 was associated with AD pathology.[148,149] Apart from TGFBR2 regulation, miR-106b has also been associated with adult neurogenesis.[150]

Tau hyperphosphorylation, thought to be a major contributor to AD pathology, involves extracellular signal-regulated kinase 1 (ERK1) activity.[120] ERK1 is targeted by multiple miRNAs, including miR-15a/b, which were found to be decreased in AD patients.[127,151] miR-15, as well as previously mentioned miR-16, target proapoptotic protein Bcl-2, the levels of which were found to be increased in AD.[152,153] Glycogen synthase kinase 3 (GSK3) accumulation may also contribute to tau hyperphosphorylation in AD. *GSK3* is targeted by miR-26a, which showed altered expression in AD.[151,154,155] Contributing to tau pathology, may be an impairment of BAG family molecular chaperone regulator 2 (BAG2)-mediated tau clearance due to the abundance of BAG2-repressing miR-128a in AD.[156]

In addition to their involvement in tau and Aβ pathophysiology, miRNAs may contribute to a sustained inflammatory reaction in AD, such as through the repression of interleukin-1 receptor-associated kinase (IRAK) 1 by miR-146a, and activation of T cells by miR-181a and miR-142.[151,157] The abundance of miR-146a in the AD-affected brain has further been linked to a decrease in an important negative regulator of the brain's inflammatory response, i.e., complement factor H.[158] As an indication of the systemic dysregulation in AD, miRNAs have been studied in blood mononuclear cells of AD patients. In doing so miR-34a and -181b have been reported to be upregulated in AD patients.[159]

A microarray study charting miRNA expression in the temporal cortex of AD patients, distinguishing between white and gray matter, found most miRNAs to be downregulated in AD, although some were upregulated.[155] Among the dysregulated miRNAs were the previously mentioned miR-9, miR-15a/b, miR-16, miR-20a, miR-26a, miR-29a/b/c, miR-34a, miR-101, miR-106a, miR-106b, miR-107, miR-124, miR-125b, miR-128, miR-181b, miR-195, and 7 let-7 miRNAs. Some of these miRNAs were additionally detected in a brain region-specific analysis in AD patients, including miR-9 and miR-29a/b.[151] Multiple miRNAs were found to be altered in the cerebrospinal fluid of AD patients, such as miR-15b, mirR-142, miR-146b, miR-181a/c, miR-195, and let-7f.

While research into ncRNAs is heavily focused on miRNAs, some other species have been implicated in AD. Brain cytoplasmic RNA 200 (BC200) is highly expressed specifically in the brain and may play a role in long-term synaptic plasticity.[160] Initially, a marked decrease in BC200 levels was observed in the temporal neocortex of AD patients,[161] while a later study found BC200 to be mislocalized and overexpressed in brain regions affected by AD pathology.[162] Another ncRNA, neuroblastoma differentiation marker 29 (NDM29) was also found to be upregulated in the cortex of AD patients.[163] NDM29 induces *APP* expression and subsequent Aβ secretion and its expression may be induced in a pro-inflammatory environment, as seen in the AD-stricken brain. Similarly, the ncRNA 17A can enhance Aβ secretion, is elevated in response to inflammatory stimuli, and was enhanced in the brains of AD patients.[164] BACE1 was found to have a naturally expressed antisense transcript (BACE1-AS) which enhances BACE1 mRNA stability and translation, leading to downstream increases in Aβ42 generation.[165] Curiously, Aβ42 exposure can further induce BACE1-AS expression. BACE1-AS appeared to be overexpressed in AD patients and transgenic mice expressing human APP. Furthermore, it was found that BACE1-AS blocks the binding site of miR-485-5p, a miRNA that represses BACE1 and that showed decreased levels in AD patients.[166]

2.2 Epigenetic Alterations in Parkinson's Disease

Parkinson's disease is, after AD, the second most common neurodegenerative disorder in the elderly.[167] PD primarily affects the dopaminergic neurons of the substantia nigra in the midbrain, resulting in the progressive deterioration of motor control. Symptoms of PD include tremor, rigidity, bradykinesia, and gait disturbances, but cognitive and psychiatric disturbances can also occur in PD patients.[168,169] The main neuropathological hallmark of PD is the presence of cytoplasmic Lewy bodies, mainly consisting of α-synuclein, parkin, and ubiquitin.[170] These Lewy bodies not only occur in the midbrain, but may spread throughout the cortex, which is associated with the development of cognitive impairments.[171] As with AD, the etiopathogenesis of idiopathic PD remains unclear and is also likely to involve an interplay between genetic, epigenetic, and environmental factors.[1]

2.2.1 Alterations in DNA Modifications in PD

PD is also associated with elevated Hcy levels and a decreased methylation potential (SAM/SAH ratio).[172] Generally, a higher methylation potential has been associated with better cognitive performance. A lower methylation potential in PD is in line with observations of global DNA hypomethylation in the frontal cortex of PD patients, as well as in dementia with Lewy bodies (DLB).[173] Contributing to this global DNA hypomethylation may be an interaction between α-synuclein and DNMT1, causing DNMT1 to be trapped in the cytoplasm. Complementing this, it was observed that *SNCA*, the gene encoding α-synuclein, was hypomethylated in PD and DLB patients.[173] Protein kinase CAMP-dependent type II regulatory subunit alpha (*PRKAR2A*) was additionally found to be hypomethylated in PD cases only, whereas selenoprotein W (*SEPW1*) showed hypomethylation only in DLB cases. Apart from the frontal cortex, *SNCA* was found to be hypomethylated in the substantia nigra and putamen of sporadic PD patients, and it was shown that *SNCA* hypomethylation increases gene expression.[174,175] A study focusing on Lewy body disease pathology in general found no significant differences in *SNCA* methylation between cases and controls, although methylation levels were slightly higher in patients.[176] Apart from the case–control comparison, this study did find a minor increase in *SNCA* methylation levels with age in Lewy body disease cases. A larger study, combining genomic, methylomic, and transcriptomic data identified associations with PD on all three levels for *PARK16*, *GPNMB*, and *STX1B*.[177] *PARK16* is a known PD susceptibility gene,[178,179] while recently it was found that the PD-associated *GPNMB* locus may increase *GPNMB* expression,[180] and *STX1B* has also been associated with PD.[181] An EWAS in blood found *FANCC* and *TNKS2* to be hypermethylated in PD patients.[182] *FANCC*, encoding Fanconi anemia group C protein (FANCC) that is involved in DNA repair, functionally interacts with protein kinase regulated by RNA (PKR), which is thought to be involved in neuronal cell death in PD, but also AD and HD.[183,184] Tankyrase 2 (TANK2), encoded by *TNKS2*, is a positive regulator of telomere length, while its overexpression leads to necrosis.[185] Another EWAS identified multiple loci with differential PD-associated DNA methylation in both the brain and blood, including loci in *HLA-DQA1*, *GFPT2*, *MAPT*, and *MIR886*.[186]

2.2.2 Alterations in Chromatin Modifications in PD

H3 hypoacetylation has been observed in *Drosophila* and human neuroblastoma models of PD overexpressing α-synuclein, and contributed to increased nuclear α-synuclein levels.[187] By binding directly to histones, α-synuclein obstructs H3 acetylation. A loss of H3 acetylation and neurotoxicity could be prevented through the administration of the HDACis sodium butyrate or suberoylanilide hydroxamic acid.[188,189] Specific inhibition of SIRT2 in human neuroglioma cells overexpressing α-synuclein was also able to counter neurotoxicity, although due to the higher affinity for α-tubulin, the beneficial effects of SIRT2 inhibition may not be due to effects on histone acetylation.[188] Among the regions bound by α-synuclein was the peroxisome proliferator-activated receptor-gamma coactivator (PGC) 1α gene promoter, a transcriptional coactivator regulating mitochondrial gene expression, leading to lower expression levels and impaired mitochondrial functioning. Binding of nuclear α-synuclein to the PGC1α promoter has also been confirmed in tissue from sporadic PD patients, and PGC1α levels were found to be reduced in substantia nigra neurons from DLB patients with pathology similar to early PD.[190,191]

Contributing to histone hypoacetylation in PD may be the potential role of the HAT p300 in aggresome formation due to its prion-like domains, binding and aggregating toxic proteins as a protective mechanism.[192] This is in line with the observation of p300 in α-synuclein-positive inclusions and Lewy bodies in PD patients. Interestingly, a decrease in p300 levels and acetylation activity by α-synuclein in rat dopaminergic neurons is suggested to be neuroprotective by inhibiting expression of the proapoptotic protein kinase C-δ (PKCδ), possibly involving reduced NFκB acetylation.[193] Several mutations of the *SNCA* gene have been implicated in PD, and work in blood cells from an early-onset PD patient and a lymphoblastoid cell line with heterozygous p.Ala53Thr *SNCA* mutations indicate that the mutated *SNCA* allele is primarily silenced via histone deacetylation, but not DNA methylation, and that the other *SNCA* allele becomes hyperactive.[194]

In contrast to evidence of histone hypoacetylation related to α-synuclein, work on neurotoxic exposures related to the development of PD pathology points toward hyperacetylation of histones. In a mouse model of PD, based on neurotoxic 1-methyl-4-phenyl-1,2,3,6-tetrahydropyridine (MPTP) exposure, H3 acetylation levels were found to be increased in the striatum, while H4 acetylation and H3 trimethylation were reduced.[195] However, in a primate MPTP-based model of PD, H3 methylation was also reduced, while H3 acetylation remained unchanged, and H4 acetylation increased. H3 phosphorylation did not change in both models. Exposure to pesticides and herbicides has also been implicated in the development of PD. Specifically, the pesticide dieldrin, which can induce dopaminergic cell death, increased H3 and H4 acetylation in rat mesencephalic neuronal cultures and the striatum and substantia nigra of mice.[196] The proposed mechanism underlying the hyperacetylation revolves around an accumulation of HATs due to dieldrin-induced proteasomal dysfunction. Similarly, exposure to the herbicide paraquat, also linked to the development of PD, elevated H3 acetylation levels, but not H4 acetylation levels in rat dopaminergic cells.[197] Global HDAC activity was decreased, as well as HDAC4 and HDAC7 expression levels, and inhibition of HAT activity could prevent paraquat-induced cell death.

2.2.3 Alterations in Non-Coding RNAs in PD

On a translational level, α-synuclein is suppressed by brain-enriched miR-7 and miR-153, and miR-7 is reduced after MPTP exposure *in vitro* and in mice.[198,199] A single nucleotide polymorphism (SNP) in *FGF20* has been associated with an increased risk of developing PD in Caucasians.[200] Interestingly, this SNP was found to interfere with miR-433 binding, leading to increased expression of fibroblast growth factor (FGF) 20, which in turn is correlated with enhanced α-synuclein expression.[201]

In mice expressing human mutant *SNCA*, miR-10a/b, miR-132, miR-212, and miR-495 expression was found to be decreased in the brainstem.[202] Of these, miR-132 especially has been implicated in neuronal functioning.[203] Transgenic *C. elegans* models of PD exhibit decreases in miR-64 and miR-65, as well as let-7 expression, although it should be noted that the former two miRNAs do not have clear human orthologs.[204]

Leucine-rich repeat kinase 2 (LRRK2), another major player in PD pathogenesis that exacerbates α-synuclein pathology was found to be regulated by miR-205.[205] While miR-205 levels have been shown to be lowered in the brains of sporadic PD patients, LRRK2 was upregulated. Work in *Drosophila* indicates that the pathogenic effects of LRRK2 gain-of-function mutations leading to the development of PD may involve dysregulation of the miRNA pathway.[206] LRRK2 phosphorylates eukaryotic translation initiation factor 4E-binding protein (EIF4EBP) 1, which interacts with the Argonaute (Ago) 2 component of the RNA-induced silencing complex (RISC), hampering miRNA processing.[206,207] This was found to reduce let-7 and miR-184* levels, leading to an increase in their targets transcription factor E2F1 and differentiation regulated transcription factor protein (DP). E2F1 and DP are regulators of cell survival and proliferation and their upregulation may contribute to cell death in PD.[208]

In the brains of PD patients, levels of miR-34b/c were found to be reduced, already at pre-motor stages.[209] Curiously, reduced miR-34b/c levels were associated with decreased expression of protein/nucleic acid deglycase DJ-1 and the E3 ubiquitin-protein ligase parkin, which may disrupt mitochondrial functioning and promote oxidative stress. An investigation into early peripheral miRNA biomarkers for PD has identified miR-4639 as a promising candidate.[210] This miRNA was observed to be upregulated in early, untreated, PD patients, and it was shown to target DJ-1, resulting in oxidative stress and neuronal death. Interestingly, a dopaminergic neuron-specific miRNA, miR-133b, has been identified and appeared to be lowered in the midbrain of PD patients.[211] In the blood of PD patients miR-133b, as well as miR-433 and miR-29a/c were found to be reduced, indicating they may be used as biomarkers.[212,213]

Drug therapy appears to have a major effect on peripheral miRNA levels in PD patients; a comparison between treated and untreated PD patients found miR-7, miR-9-3p, miR-9-5p, miR-129, and miR-132 levels to be significantly increased in treated PD patients.[214] Another study found that while blood levels of miR-1, miR-22*, and miR-29a could be used to distinguish non-treated PD patients and healthy controls, miR-16-2*, miR-26a2* and miR30a showed altered levels between treated and untreated PD patients.[215] One of the targets of miR-1, which was downregulated in PD patients, is tubulin polymerization-promoting protein (TPPP), and upregulated TPPP has been associated with α-synuclein aggregation. Additionally, miR-1 has been linked to microglial activation and dopamine transport and may also target BDNF. miR-22* potentially targets tumor suppressor p53-binding protein 2 (TP53BP2) and the glutamate receptor 1 precursor (GRIA1). miR-29a, which was previously also mentioned in relation to AD, targets cell division control protein 42 homolog (CDC42). miR-1 and miR-29a also target IGF-1, important for normal functioning of the nervous system.

2.3 Epigenetic Alterations in Huntington's Disease

HD is a lethal neurodegenerative affliction that generally progresses fast once symptoms occur, without any treatment strategies currently available.[216] While the sporadic variants of AD and PD are more common than their familial counterparts, the autosomal-dominant form of HD is more prevalent than the sporadic form. HD is caused by an extended cytosine–adenine–guanine (CAG) repeat region of the huntingtin (HTT) gene, leading to the destruction of the neostriatal nuclei, caudate nucleus and putamen.[217–219] Due to the crucial role of these brain areas in movement, HD is associated with progressive motor impairments, particularly chorea, but can also involve cognitive and psychiatric symptoms.[220] Although the cause of HD is known, exactly how mutant HTT leads to neurodegeneration remains unclear, although it has been suggested that this could be partially driven by widespread epigenetic dysregulation induced by mutant HTT.[221,222]

2.3.1 Changes in DNA Modifications in HD

Even though HD has a clear genetic origin, it remains to be elucidated how the expression of mutant HTT leads to neurodegeneration. The effect of mutant HTT expression on DNA methylation was investigated in striatal cells from a mouse model of HD.[223] Most of the investigated sites showed hypomethylation, and genes with methylation changes were associated with processes such as neurogenesis, neuron differentiation, and signaling. Impaired hippocampal neurogenesis is thought to contribute to the cognitive decline seen in HD.[221] Interestingly, it appeared that mainly CpG-poor regions were affected by mutant HTT expression, and that the changes in DNA methylation mainly affected the binding of AP-1 and SOX2 transcription factors.[223] The epigenetic regulation of adenosine A2A receptor (A2AR) expression, which is greatly decreased in HD, has also been investigated.[224] In the striatum of a mouse model of HD it was observed that lowered A2AR expression was associated with hypohydroxymethylation of the 5′UTR of *ADORA2A*, which encodes A2AR. A similar pattern was found in the putamen of HD patients, which showed hypermethylation and hypohydroxymethylation of the 5′UTR of *ADORA2A*. Next to 5mC, 7mG, an epigenetic modification of both DNA and RNA, was also found to be disturbed in HD.[225] Brain tissue from an HD mouse model (CAG140 KI) indicated reduced nuclear 7mG levels, while 7mG levels were increased in the cytoplasm, reflecting global 7mG in RNA. This was confirmed in the motor cortex of HD patients; decreased 7mG levels in the nuclear fraction and increased 7mG levels in the cytoplasmic fraction. These differences were not observed in the primary visual cortex or the cingulate gyrus. The exact regulatory role of 7mG remains to be elucidated, although 7mG modifications may interfere with other epigenetic modifications such as 5mC and 8hG, or lead to spontaneous depurination and mutation events.[225] Note that as of yet direct interactions between mutant HTT and the enzymes involved in epigenetic DNA modifications have not been observed, and therefore the HD-associated alterations discussed here may be occurring downstream from the interactions between HTT and the chromatin-modifying apparatus discussed later.

2.3.2 Changes in Chromatin Modifications in HD

Due to a prohibitive direct interaction between mutant HTT and the transcriptional regulator and HAT CBP, HD is associated with histone hypoacetylation and chromatin condensation.[221] In the striatum of a mouse model of HD, H3 and H4 acetylation was decreased, while the repressive H3K9 methylation marker was increased.[226] Administration of the HDACi phenylbutyrate could attenuate these alterations, and restore transcriptional regulation of some of the genes associated with cell death in HD. Other work showed that administration of anthracyclines could reverse HD-associated H3K9 hypoacetylation and hypermethylation, restore gene expression of some of the downregulated genes, and improve behavior in R6/2 HD mouse models.[227] In a similar study, using the R6/2 mouse model of HD, no global changes in H3 acetylation were found, but an H3-specific hypoacetylation was detected in promoter regions of genes downregulated in these mice and two murine *in vitro* models of HD.[228] HDACi treatment could counter this transcriptional dysregulation, using phenylbutyrate in the mouse model, and sodium butyrate in the *in vitro* models. A more in-depth analysis of H3K9 and H3K14 acetylation in the striatum of a transgenic HD mouse model (R6/2) revealed a general decrease in acetylated sites at the gene level and fewer genes with acetylated H3 sites.[229] Coding regions were mainly affected by the loss of H3 acetylation, and the presence of H3 acetylation in a gene correlated with gene expression.

Similarly, in a R6/2 HD transgenic mouse model, increased levels of H3K9 methylation were observed, which could be reversed with mithramycin, an anthracycline that binds GC-rich DNA.[230] It was later determined that the H3K9 hypermethylation observed in these mice, but also in HD patients, is likely due to increased expression of histone-lysine *N*-methyltransferase SETDB1.[231] Increased SETDB1 expression is again linked to the sequestration of CBP by mutant HTT via the Ets-2 transcription factor, which promotes SETDB1 expression, but is inhibited by CBP.[232] Among the genes repressed by H3K9 hypermethylation is the gene encoding the muscarinic acetylcholine receptor M1 (CHRM1), lower levels of which were observed to hamper Ca^{2+}-dependent neuronal signaling in a murine *in vitro* model of HD.[233] Cholinergic signaling is pivotal for striatal functioning, and decreased levels of muscarinic acetylcholine receptors have been observed in the R6/2 HD transgenic mouse model and HD patients.[233–235] Of note, aberrant cytoplasmic localization of histones in neurons of HD patients has been reported, hinting at hampered nuclear transport in HD.[236]

2.3.3 Changes in Non-Coding RNAs in HD

Normally, HTT is thought to interact with repressor element 1 silencing transcription factor (REST), limiting its nuclear translocation in neurons.[237] Mutant HTT, however, fails to do so, leading to elevated nuclear REST levels. As REST is involved in the regulation of numerous genes, including some miRNAs, elevated nuclear REST levels lead to a widespread dysregulation of gene expression. Of the miRNAs targeted by REST, it appears that particularly the neuronal-specific miR-132 was reduced in HD patients, while miR-29a and miR-330 expression was increased. miR-132, which has also been implicated in PD, targets rho GTPase-activating protein 32, a neurite growth inhibitor.[238] Interestingly, miR-29a has been linked to AD and PD in addition to HD, as previously discussed. Work in two mouse models of HD (R6/2 and YAC128 mice) also detected a downregulation of miR-132, as well as miR-22, miR-29c, miR-128, miR-138, miR-218, miR-222, miR-344, and miR-674*.[239] Additionally, there were signs of increased miRNA synthesis at the very early stages of disease progression in these mouse models, whilst it was decreased at later stages.

In vitro work in STHdh^{Q111}/Hdh^{Q111} cells showed that mutant HTT can reduce miR-125b, miR-146a, and miR-150 levels, but that miR-146a repression is the result of increased cellular tumor antigen p53 levels, a target of miR-125b and miR-150.[240] Increased p53 levels, enhance suppression of transcription factor p65 and subsequently lowers miR-146a expression. These findings were confirmed in R6/2 mice. Due to multiple p53 response elements in the HTT gene, it has been suggested enhanced p53 levels may contribute to increased HTT levels, mitochondrial dysfunction and ultimately neurodegeneration.[241]

3. CONCLUDING REMARKS

This chapter provides an overview of epigenetic research in relation to aging and neurodegeneration. While epigenetics is a burgeoning field in neuroscience, it faces some important challenges that merit attention.[1,95] The question of cause and consequence may be of particular importance in neuroscience due to inaccessibility of the brain in living subjects.[242] Many of the studies on the epigenetics of neurodegeneration rely on *post-mortem* brain tissue and therefore it cannot be determined whether epigenetic alterations detected in such a setting are contributing to the pathology, or are a consequence of other pathological processes. Some studies thus focus on more accessible tissues, such as peripheral blood, but besides the identification of disease biomarkers, it is often unclear how epigenetic changes detected in systemic tissues reflect pathological processes in the brain. As epigenetic dysregulation can be highly regional and cell-type specific, studying tissues other than those primarily affected by disease pathology may thus be of limited use to gaining an insight into the role of epigenetic mechanisms in disease etiopathogenesis.[89,95] To tackle this causality conundrum, mechanistic *in vivo* and *in vitro* studies will be required, using appropriate models (e.g., induced pluripotent stem cell-derived human neurons) and techniques (e.g., epigenetic editing).[243] Such mechanistic validation, as well as replication of EWAS findings is also important in the face of the limited sample sizes many studies cope with.[95] Finally, it should be kept in mind that epigenetic regulation of gene expression encompasses a vast and complex molecular apparatus that operates at the DNA, chromatin, and ncRNA levels, and that these levels do not function independent of each other.[244] For instance, changes in DNA modifications are often paired with changes in histone modifications, and ncRNAs can regulate proteins responsible for adding or removing said modifications, while ncRNAs in turn can also be regulated by DNA and histone modifications. Thus, building integrated molecular networks of neurodegenerative diseases by combining genetic, epigenetic and transcriptomic information will allow researchers to better understand the complex nature of such diseases.

REFERENCES

1. Lardenoije R, Iatrou A, Kenis G, et al. The epigenetics of aging and neurodegeneration. *Prog Neurobiol.* 2015;131:21–64 [Epub ahead of print 2015].
2. Pishva E, Rutten BPF, van den Hove D. DNA methylation in major depressive disorder. *Adv Exp Med Biol.* 2017;978:185–196.

3. Dempster E, Viana J, Pidsley R, et al. Epigenetic studies of schizophrenia: progress, predicaments, and promises for the future. *Schizophr Bull.* 2013;39:11–16.
4. Anastasiadou C, Malousi A, Maglaveras N, et al. Human epigenome data reveal increased CpG methylation in alternatively spliced sites and putative exonic splicing enhancers. *DNA Cell Biol.* 2011;30:267–275.
5. Siegmund KD, Connor CM, Campan M, et al. DNA methylation in the human cerebral cortex is dynamically regulated throughout the life span and involves differentiated neurons. *PLoS One.* 2007;2:e895.
6. Chouliaras L, van den Hove DLA, Kenis G, et al. Caloric restriction attenuates age-related changes of DNA methyltransferase 3a in mouse hippocampus. *Brain Behav Immun.* 2011;25:616–623.
7. Chouliaras L, van den Hove DLA, Kenis G, et al. Prevention of age-related changes in hippocampal levels of 5-methylcytidine by caloric restriction. *Neurobiol Aging.* 2012;33:1672–1681.
8. Chouliaras L, van den Hove DLA, Kenis G, et al. Age-related increase in levels of 5-hydroxymethylcytosine in mouse hippocampus is prevented by caloric restriction. *Curr Alzheimer Res.* 2012;9:536–544.
9. Münzel M, Globisch D, Brückl T, et al. Quantification of the sixth DNA base hydroxymethylcytosine in the brain. *Angew Chem Int Ed Engl.* 2010;49:5375–5377.
10. Oliveira AMM, Hemstedt TJ, Bading H. Rescue of aging-associated decline in Dnmt3a2 expression restores cognitive abilities. *Nat Neurosci.* 2012;15: 1111–1113.
11. Lardenoije R, van den Hove DA, Vaessen TSJ, et al. Epigenetic modifications in mouse cerebellar Purkinje cells: effects of aging, caloric restriction, and overexpression of superoxide dismutase 1 on 5-methylcytosine and 5-hydroxymethylcytosine. *Neurobiol Aging.* 2015;36:3079–3089.
12. Thakur MK, Kanungo MS. Methylation of chromosomal proteins and DNA of rat brain and its modulation by estradiol and calcium during aging. *Exp Gerontol.* 1981;16:331–336.
13. Singhal RP, Mays-Hoopes LL, Eichhorn GL. DNA methylation in aging of mice. *Mech Ageing Dev.* 1987;41:199–210.
14. Hornsby PJ, Yang L, Gunter LE. Demethylation of satellite I DNA during senescence of bovine adrenocortical cells in culture. *Mutat Res.* 1992;275:13–19.
15. Lopatina N, Haskell JF, Andrews LG, et al. Differential maintenance and de novo methylating activity by three DNA methyltransferases in aging and immortalized fibroblasts. *J Cell Biochem.* 2002;84:324–334.
16. Thompson RF, Atzmon G, Gheorghe C, et al. Tissue-specific dysregulation of DNA methylation in aging. *Aging Cell.* 2010;9:506–518.
17. Hernandez DG, Nalls MA, Gibbs JR, et al. Distinct DNA methylation changes highly correlated with chronological age in the human brain. *Hum Mol Genet.* 2011;20: 1164–1172.
18. Horvath S. DNA methylation age of human tissues and cell types. *Genome Biol.* 2013;14:R115.
19. Bocklandt S, Lin W, Sehl ME, et al. Epigenetic predictor of age. *PLoS One.* 2011;6: e14821.
20. Fraga MF, Ballestar E, Paz MF, et al. Epigenetic differences arise during the lifetime of monozygotic twins. *Proc Natl Acad Sci USA.* 2005;102:10604–10609.
21. Austad SN. Why women live longer than men: sex differences in longevity. *Gend Med.* 2006;3:79–92.
22. Hannum G, Guinney J, Zhao L, et al. Genome-wide methylation profiles reveal quantitative views of human aging rates. *Mol Cell.* 2013;49:359–367.
23. Pina B, Martinez P, Suau P. Differential acetylation of core histones in rat cerebral cortex neurons during development and aging. *Eur J Biochem.* 1988;174:311–315.

24. Wang CM, Tsai SN, Yew TW, et al. Identification of histone methylation multiplicities patterns in the brain of senescence-accelerated prone mouse 8. *Biogerontology.* 2010;11:87–102.
25. Ryan JM, Cristofalo VJ. Histone acetylation during aging of human cells in culture. *Biochem Biophys Res Commun.* 1972;48:735–742.
26. Chouliaras L, van den Hove DLA, Kenis G, et al. Histone deacetylase 2 in the mouse hippocampus: attenuation of age-related increase by caloric restriction. *Curr Alzheimer Res.* 2013;10:868–876.
27. Peleg S, Sananbenesi F, Zovoilis A, et al. Altered histone acetylation is associated with age-dependent memory impairment in mice. *Science.* 2010;328:753–756.
28. Guan J-S, Haggarty SJ, Giacometti E, et al. HDAC2 negatively regulates memory formation and synaptic plasticity. *Nature.* 2009;459:55–60.
29. Quintas A, de Solís AJ, Díez-Guerra FJ, et al. Age-associated decrease of SIRT1 expression in rat hippocampus: prevention by late onset caloric restriction. *Exp Gerontol.* 2012;47:198–201 [Epub ahead of print].
30. Sommer M, Poliak N, Upadhyay S, et al. ΔNp63alpha overexpression induces downregulation of Sirt1 and an accelerated aging phenotype in the mouse. *Cell Cycle.* 2006;5:2005–2011.
31. Sasaki T, Maier B, Bartke A, et al. Progressive loss of SIRT1 with cell cycle withdrawal. *Aging Cell.* 2006;5:413–422.
32. Vaquero A, Scher M, Erdjument-Bromage H, et al. SIRT1 regulates the histone methyl-transferase SUV39H1 during heterochromatin formation. *Nature.* 2007;450:440–444.
33. Narita M, Nũnez S, Heard E, et al. Rb-mediated heterochromatin formation and silencing of E2F target genes during cellular senescence. *Cell.* 2003;113:703–716.
34. Schmeisser K, Mansfeld J, Kuhlow D, et al. Role of sirtuins in lifespan regulation is linked to methylation of nicotinamide. *Nat Chem Biol.* 2013;9:693–700.
35. Ristow M, Zarse K. How increased oxidative stress promotes longevity and metabolic health: the concept of mitochondrial hormesis (mitohormesis). *Exp Gerontol.* 2010;45:410–418.
36. Guarente L, Picard F, Ramsey CS, et al. Calorie restriction—the SIR2 connection. *Cell.* 2005;120:473–482. (a) Pasquinelli AE. MicroRNAs and their targets: recognition, regulation and an emerging reciprocal relationship. *Nat Rev Genet.* 2012;13:271–282.
37. Boehm M, Slack F. A developmental timing microRNA and its target regulate life span in C. elegans. *Science.* 2005;310:1954–1957.
38. Mathew R, Pal Bhadra M, Bhadra U. Insulin/insulin-like growth factor-1 signalling (IIS) based regulation of lifespan across species. *Biogerontology.* 2017;18:35–53.
39. Yunger E, Safra M, Levi-Ferber M, et al. Innate immunity mediated longevity and longevity induced by germ cell removal converge on the C-type lectin domain protein IRG-7. *PLoS Genet.* 2017;13:e1006577.
40. Boulias K, Horvitz HR, Suchanek-Kavipurapu M, et al. The C. elegans microRNA mir-71 acts in neurons to promote germline-mediated longevity through regulation of DAF-16/FOXO. *Cell Metab.* 2012;15:439–450.
41. Maes OC, An J, Sarojini H, et al. Murine microRNAs implicated in liver functions and aging process. *Mech Ageing Dev.* 2008;129:534–541.
42. Li N, Bates DJ, An J, et al. Up-regulation of key microRNAs, and inverse down-regulation of their predicted oxidative phosphorylation target genes, during aging in mouse brain. *Neurobiol Aging.* 2011;32:944–955.
43. Bu H, Baraldo G, Lepperdinger G, et al. mir-24 activity propagates stress-induced senescence by down regulating DNA topoisomerase 1. *Exp Gerontol.* 2016;75:48–52.
44. Hackl M, Brunner S, Fortschegger K, et al. miR-17, miR-19b, miR-20a, and miR-106a are down-regulated in human aging. *Aging Cell.* 2010;9:291–296.

45. Choudhury AR, Ju Z, Djojosubroto MW, et al. Cdkn1a deletion improves stem cell function and lifespan of mice with dysfunctional telomeres without accelerating cancer formation. *Nat Genet.* 2007;39:99–105.
46. Gravina S, Lescai F, Hurteau G, et al. Identification of single nucleotide polymorphisms in the p21 (CDKN1A) gene and correlations with longevity in the Italian population. *Aging (Albany, NY).* 2009;1:470–480.
47. Persengiev S, Kondova I, Otting N, et al. Genome-wide analysis of miRNA expression reveals a potential role for miR-144 in brain aging and spinocerebellar ataxia pathogenesis. *Neurobiol Aging.* 2011;32: 2316: e17–2316: e27.
48. Karolina DS, Armugam A, Tavintharan S, et al. MicroRNA 144 impairs insulin signaling by inhibiting the expression of insulin receptor substrate 1 in type 2 diabetes mellitus. *PLoS One.* 2011;6:e22839.
49. Selkoe DJ. Preventing Alzheimer's disease. *Science.* 2012;337:1488–1492.
50. Bediou B, Ryff I, Mercier B, et al. Impaired social cognition in mild Alzheimer disease. *J Geriatr Psychiatry Neurol.* 2009;22:130–140.
51. Budson AE, Solomon PR. New criteria for Alzheimer disease and mild cognitive impairment: implications for the practicing clinician. *Neurologist.* 2012;18:356–363.
52. Wenk GL. Neuropathologic changes in Alzheimer's disease. *J Clin Psychiatry.* 2003; 64(suppl 9):7–10.
53. Iatrou A, Kenis G, Rutten BPF, et al. Epigenetic dysregulation of brainstem nuclei in the pathogenesis of Alzheimer's disease: looking in the correct place at the right time? *Cell Mol Life Sci.* 2016;74:1–15.
54. Alzheimer's Association. 2015 Alzheimer's disease facts and figures. *Alzheimers Dement.* 2015;11:332–384.
55. Morrison LD, Smith DD, Kish SJ. Brain S-adenosylmethionine levels are severely decreased in Alzheimer's disease. *J Neurochem.* 1996;67:1328–1331.
56. Kennedy BP, Bottiglieri T, Arning E, et al. Elevated S-adenosylhomocysteine in Alzheimer brain: influence on methyltransferases and cognitive function. *J Neural Transm.* 2004;111:547–567.
57. Hoffman DR, Cornatzer WE, Duerre JA. Relationship between tissue levels of S-adenosylmethionine, S-adenylhomocysteine, and transmethylation reactions. *Can J Biochem.* 1979;57:56–65.
58. Morris MS. Homocysteine and Alzheimer's disease. *Lancet Neurol.* 2003;2:425–428.
59. Zhu BT. On the mechanism of homocysteine pathophysiology and pathogenesis: a unifying hypothesis. *Histol Histopathol.* 2002;17:1283–1291.
60. Burke WJ, Chung HD, Nakra BRS, et al. Phenylethanolamine N-methyltransferase activity is decreased in Alzheimer's disease brains. *Ann Neurol.* 1987;22:278–280.
61. Sontag E, Nunbhakdi-Craig V, Sontag J-M, et al. Protein phosphatase 2A methyltransferase links homocysteine metabolism with tau and amyloid precursor protein regulation. *J Neurosci.* 2007;27:2751–2759.
62. Zhou X-W, Gustafsson J-Å, Tanila H, et al. Tau hyperphosphorylation correlates with reduced methylation of protein phosphatase 2A. *Neurobiol Dis.* 2008;31:386–394.
63. Fuso A, Seminara L, Cavallaro RA, et al. S-adenosylmethionine/homocysteine cycle alterations modify DNA methylation status with consequent deregulation of PS1 and BACE and beta-amyloid production. *Mol Cell Neurosci.* 2005;28:195–204.
64. Fuso A, Nicolia V, Cavallaro RA, et al. B-vitamin deprivation induces hyperhomocysteinemia and brain S-adenosylhomocysteine, depletes brain S-adenosylmethionine, and enhances PS1 and BACE expression and amyloid-beta deposition in mice. *Mol Cell Neurosci.* 2008;37:731–746.
65. Fuso A, Nicolia V, Ricceri L, et al. S-adenosylmethionine reduces the progress of the Alzheimer-like features induced by B-vitamin deficiency in mice. *Neurobiol Aging.* 2012;33: 1482: e1–16.

66. Lee S, Lemere CA, Frost JL, et al. Dietary supplementation with S-adenosyl methionine delayed amyloid-β and tau pathology in 3xTg-AD mice. *J Alzheimers Dis.* 2012;28:423–431.
67. Cacciapuoti F. Lowering homocysteine levels with folic acid and B-vitamins do not reduce early atherosclerosis, but could interfere with cognitive decline and Alzheimer's disease. *J Thromb Thrombolysis.* 2013;36:258–262.
68. Chan A, Paskavitz J, Remington R, et al. Efficacy of a vitamin/nutriceutical formulation for early-stage Alzheimer's disease: a 1-year, open-label pilot study with an 16-month caregiver extension. *Am J Alzheimers Dis Other Demen.* 2008;23:571–585.
69. Remington R, Chan A, Paskavitz J, et al. Efficacy of a vitamin/nutriceutical formulation for moderate-stage to later-stage Alzheimer's disease: a placebo-controlled pilot study. *Am J Alzheimers Dis Other Demen.* 2009;24:27–33.
70. Mastroeni D, Chouliaras L, Grover A, et al. Reduced RAN expression and disrupted transport between cytoplasm and nucleus; a key event in Alzheimer's disease pathophysiology. *PLoS One.* 2013;8:e53349 [Epub ahead of print 2013].
71. Chen K-L, Wang SS-S, Yang Y-Y, et al. The epigenetic effects of amyloid-beta(1-40) on global DNA and neprilysin genes in murine cerebral endothelial cells. *Biochem Biophys Res Commun.* 2009;378:57–61.
72. Mastroeni D, Grover A, Delvaux E, et al. Epigenetic changes in Alzheimer's disease: decrements in DNA methylation. *Neurobiol Aging.* 2010;31:2025–2037.
73. Chouliaras L, Mastroeni D, Delvaux E, et al. Consistent decrease in global DNA methylation and hydroxymethylation in the hippocampus of Alzheimer's disease patients. *Neurobiol Aging.* 2013;34:2091–2099.
74. Bollati V, Galimberti D, Pergoli L, et al. DNA methylation in repetitive elements and Alzheimer disease. *Brain Behav Immun.* 2011;25:1078–1083.
75. Ehrlich M. DNA methylation in cancer: too much, but also too little. *Oncogene.* 2002;21:5400–5413.
76. Tohgi H, Utsugisawa K, Nagane Y, et al. Reduction with age in methylcytosine in the promoter region -224 approximately -101 of the amyloid precursor protein gene in autopsy human cortex. *Brain Res Mol Brain Res.* 1999;70:288–292.
77. West RL, Lee JM, Maroun LE. Hypomethylation of the amyloid precursor protein gene in the brain of an Alzheimer's disease patient. *J Mol Neurosci.* 1995;6:141–146.
78. Barrachina M, Ferrer I. DNA methylation of Alzheimer disease and tauopathy-related genes in postmortem brain. *J Neuropathol Exp Neurol.* 2009;68:880–891.
79. Brohede J, Rinde M, Winblad B, et al. A DNA methylation study of the amyloid precursor protein gene in several brain regions from patients with familial Alzheimer disease. *J Neurogenet.* 2010;24:179–181.
80. Tohgi H, Utsugisawa K, Nagane Y, et al. The methylation status of cytosines in a τ gene promoter region alters with age to downregulate transcriptional activity in human cerebral cortex. *Neurosci Lett.* 1999;275:89–92 [Epub ahead of print].
81. Mukaetova-Ladinska EB, Harrington CR, Roth M, et al. Alterations in tau protein metabolism during normal aging. *Dement Geriatr Cogn Disord.* 1996;7:95–103.
82. Mukaetova-Ladinska EB, Harrington CR, Roth M, et al. Biochemical and anatomical redistribution of tau protein in Alzheimer's disease. *Am J Pathol.* 1993;143:565–578.
83. Wang S-C, Oelze B, Schumacher A, et al. Age-specific epigenetic drift in late-onset Alzheimer's disease. *PLoS One.* 2008;3:e2698.
84. Urdinguio RG, Sanchez-Mut JV, Esteller M, et al. Epigenetic mechanisms in neurological diseases: genes, syndromes, and therapies. *Lancet Neurol.* 2009;8:1056–1072.
85. Hoyaux D, Decaestecker C, Heizmann CW, et al. S100 proteins in Corpora amylacea from normal human brain. *Brain Res.* 2000;867:280–288.
86. Rohn TT. Corpora amylacea in neurodegenerative diseases: cause or effect? *Int J Neurol Neurother.* 2015;2:031.

87. De Jager PL, Srivastava G, Lunnon K, et al. Alzheimer's disease: early alterations in brain DNA methylation at ANK1, BIN1, RHBDF2 and other loci. *Nat Neurosci.* 2014;17:1156–1163.
88. Lambert J-C, Ibrahim-Verbaas CA, Harold D, et al. Meta-analysis of 74,046 individuals identifies 11 new susceptibility loci for Alzheimer's disease. *Nat Genet.* 2013;45:1452–1458.
89. Lunnon K, Smith R, Hannon E, et al. Methylomic profiling implicates cortical deregulation of ANK1 in Alzheimer's disease. *Nat Neurosci.* 2014;17:1164–1170.
90. Mastroeni D, Sekar S, Nolz J, et al. ANK1 is up-regulated in laser captured microglia in Alzheimer's brain; the importance of addressing cellular heterogeneity. *PLoS One.* 2017;12:e0177814.
91. Yu L, Chibnik LB, Srivastava GP, et al. Association of brain DNA methylation in *SORL1, ABCA7, HLA-DRB5, SLC24A4,* and *BIN1* with pathological diagnosis of Alzheimer disease. *JAMA Neurol.* 2015;72:15.
92. Condliffe D, Wong A, Troakes C, et al. Cross-region reduction in 5-hydroxymethylcytosine in Alzheimer's disease brain. *Neurobiol Aging.* 2014;35: 1850–1854.
93. Coppieters N, Dieriks BV, Lill C, et al. Global changes in DNA methylation and hydroxymethylation in Alzheimer's disease human brain. *Neurobiol Aging.* 2014;35: 1334–1344.
94. Bradley-Whitman MA, Lovell MA. Epigenetic changes in the progression of Alzheimer's disease. *Mech Ageing Dev.* 2013;134:486–495.
95. Lunnon K, Mill J. Epigenetic studies in Alzheimer's disease: current findings, caveats, and considerations for future studies. *Am J Med Genet B Neuropsychiatr Genet.* 2013;162B:789–799.
96. Hutnick LK, Golshani P, Namihira M, et al. DNA hypomethylation restricted to the murine forebrain induces cortical degeneration and impairs postnatal neuronal maturation. *Hum Mol Genet.* 2009;18:2875–2888.
97. Ellison EM, Bradley-Whitman MA, Lovell MA. Single-base resolution mapping of 5-hydroxymethylcytosine modifications in hippocampus of Alzheimer's disease subjects. *J Mol Neurosci.* 2017;63:185–197.
98. Zhao J, Zhu Y, Yang J, et al. A genome-wide profiling of brain DNA hydroxymethylation in Alzheimer's disease. *Alzheimers Dement.* 2017;13:674–688 [Epub ahead of print].
99. Gräff J, Kim D, Dobbin MM, et al. Epigenetic regulation of gene expression in physiological and pathological brain processes. *Physiol Rev.* 2011;91:603–649.
100. Zhang K, Schrag M, Crofton A, et al. Targeted proteomics for quantification of histone acetylation in Alzheimer's disease. *Proteomics.* 2012;12:1261–1268.
101. Francis YI, Fà M, Ashraf H, et al. Dysregulation of histone acetylation in the APP/PS1 mouse model of Alzheimer's disease. *J Alzheimers Dis.* 2009;18:131–139.
102. Kilgore M, Miller CA, Fass DM, et al. Inhibitors of class 1 histone deacetylases reverse contextual memory deficits in a mouse model of Alzheimer's disease. *Neuropsychopharmacology.* 2010;35:870–880.
103. Gräff J, Rei D, Guan J-S, et al. An epigenetic blockade of cognitive functions in the neurodegenerating brain. *Nature.* 2012;483:222–226.
104. Zhang Z-Y, Schluesener HJ. Oral administration of histone deacetylase inhibitor MS-275 ameliorates neuroinflammation and cerebral amyloidosis and improves behavior in a mouse model. *J Neuropathol Exp Neurol.* 2013;72:178–185.
105. Bayram A, Akkafa F, Ozer A, et al. Decreased HDAC1 gene expression in patients with Alzheimer's disease. *Int J Hum Genet.* 2014;14:177–182.
106. Ding H, Dolan PJ, Johnson GVW. Histone deacetylase 6 interacts with the microtubule-associated protein tau. *J Neurochem.* 2008;106:2119–2130.

107. Julien C, Tremblay C, Émond V, et al. Sirtuin 1 reduction parallels the accumulation of tau in Alzheimer disease. *J Neuropathol Exp Neurol*. 2009;68:48–58.
108. Min S-W, Cho S-H, Zhou Y, et al. Acetylation of tau inhibits its degradation and contributes to tauopathy. *Neuron*. 2010;67:953–966.
109. Walker MP, LaFerla FM, Oddo SS, et al. Reversible epigenetic histone modifications and Bdnf expression in neurons with aging and from a mouse model of Alzheimer's disease. *Age (Dordr)*. 2013;35:519–531.
110. Lithner CU, Hernandez CM, Nordberg A, et al. Epigenetic changes related to beta-amyloid-implications for Alzheimer's disease. *Alzheimer's Dement J Alzheimer's Assoc*. 2009;5:P304.
111. Ryu H, Barrup M, Kowall NW, et al. P3-260: epigenetic modification in a monozygotic twin with Alzheimer's disease. *Alzheimer's Dement*. 2008;4:T598.
112. Liu R, Lei JX, Luo C, et al. Increased EID1 nuclear translocation impairs synaptic plasticity and memory function associated with pathogenesis of Alzheimer's disease. *Neurobiol Dis*. 2012;45:902–912.
113. Oliveira AMM, Wood MA, McDonough CB, et al. Transgenic mice expressing an inhibitory truncated form of p300 exhibit long-term memory deficits. *Learn Mem*. 2007;14:564–572.
114. Caccamo A, Maldonado MA, Bokov AF, et al. CBP gene transfer increases BDNF levels and ameliorates learning and memory deficits in a mouse model of Alzheimer's disease. *Proc Natl Acad Sci USA*. 2010;107:22687–22692.
115. Duclot F, Meffre J, Jacquet C, et al. Mice knock out for the histone acetyltransferase p300/CREB binding protein-associated factor develop a resistance to amyloid toxicity. *Neuroscience*. 2010;167:850–863.
116. Marques SCF, Lemos R, Ferreiro E, et al. Epigenetic regulation of BACE1 in Alzheimer's disease patients and in transgenic mice. *Neuroscience*. 2012;220:256–266.
117. Do Carmo S, Hanzel CE, Jacobs ML, et al. Rescue of early bace-1 and global DNA demethylation by S-adenosylmethionine reduces amyloid pathology and improves cognition in an Alzheimer's model. *Sci Rep*. 2016;6:34051.
118. Rao JS, Keleshian VL, Klein S, et al. Epigenetic modifications in frontal cortex from Alzheimer's disease and bipolar disorder patients. *Transl Psychiatry*. 2012;2:e132.
119. Ogawa O, Zhu X, Lee H-G, et al. Ectopic localization of phosphorylated histone H3 in Alzheimer's disease: a mitotic catastrophe? *Acta Neuropathol*. 2003;105:524–528.
120. Perry G, Roder H, Nunomura A, et al. Activation of neuronal extracellular receptor kinase (ERK) in Alzheimer disease links oxidative stress to abnormal phosphorylation. *Neuroreport*. 1999;10:2411–2415.
121. Zhu X, Castellani RJ, Takeda A, et al. Differential activation of neuronal ERK, JNK/SAPK and p38 in Alzheimer disease: the 'two hit' hypothesis. *Mech Ageing Dev*. 2001;123:39–46.
122. Myung N-H, Zhu X, Kruman II, et al. Evidence of DNA damage in Alzheimer disease: phosphorylation of histone H2AX in astrocytes. *Age (Dordr)*. 2008;30:209–215.
123. Strosznajder JB, Czapski GA, Adamczyk A, et al. Poly(ADP-ribose) polymerase-1 in amyloid beta toxicity and Alzheimer's disease. *Mol Neurobiol*. 2012;46:78–84.
124. Fontán-Lozano Á, Suárez-Pereira I, Horrillo A, et al. Histone H1 poly[ADP]-ribosylation regulates the chromatin alterations required for learning consolidation. *J Neurosci*. 2010;30:13305–13313.
125. Niwa R, Zhou F, Li C, et al. The expression of the Alzheimer's amyloid precursor protein-like gene is regulated by developmental timing microRNAs and their targets in Caenorhabditis elegans. *Dev Biol*. 2008;315:418–425.
126. Liu W, Liu C, Zhu J, et al. MicroRNA-16 targets amyloid precursor protein to potentially modulate Alzheimer's-associated pathogenesis in SAMP8 mice. *Neurobiol Aging*. 2012;33:522–534.

127. Hébert SS, Sergeant N, Buée L. MicroRNAs and the regulation of tau metabolism. *Int J Alzheimers Dis*. 2012;2012:1–6.
128. Zhu H-C, Wang L-M, Wang M, et al. MicroRNA-195 downregulates Alzheimer's disease amyloid-β production by targeting BACE1. *Brain Res Bull*. 2012;88:596–601.
129. Delay C, Calon F, Mathews P, et al. Alzheimer-specific variants in the 3'UTR of amyloid precursor protein affect microRNA function. *Mol Neurodegener*. 2011;6:70.
130. Long JM, Lahiri DK. MicroRNA-101 downregulates Alzheimer's amyloid-β precursor protein levels in human cell cultures and is differentially expressed. *Biochem Biophys Res Commun*. 2011;404:889–895 [Epub ahead of print].
131. Patel N, Hoang D, Miller N, et al. MicroRNAs can regulate human APP levels. *Mol Neurodegener*. 2008;3:10.
132. Long JM, Ray B, Lahiri DK. MicroRNA-153 physiologically inhibits expression of amyloid-β precursor protein in cultured human fetal brain cells and is dysregulated in a subset of Alzheimer disease patients. *J Biol Chem*. 2012;287:31298–31310.
133. Liang C, Zhu H, Xu Y, et al. MicroRNA-153 negatively regulates the expression of amyloid precursor protein and amyloid precursor-like protein 2. *Brain Res*. 2012;1455:103–113.
134. Zong Y, Wang H, Dong W, et al. miR-29c regulates BACE1 protein expression. *Brain Res*. 2011;1395:108–115.
135. Smith P, Al Hashimi A, Girard J, et al. In vivo regulation of amyloid precursor protein neuronal splicing by microRNAs. *J Neurochem*. 2011;116:240–247.
136. Hébert SS, Horré K, Nicolaï L, et al. Loss of microRNA cluster miR-29a/b-1 in sporadic Alzheimer's disease correlates with increased BACE1/beta-secretase expression. *Proc Natl Acad Sci USA*. 2008;105:6415–6420.
137. Fang M, Wang J, Zhang X, et al. The miR-124 regulates the expression of BACE1/β-secretase correlated with cell death in Alzheimer's disease. *Toxicol Lett*. 2012;209:94–105.
138. Wang W-X, Rajeev BW, Stromberg AJ, et al. The expression of microRNA miR-107 decreases early in Alzheimer's disease and may accelerate disease progression through regulation of β-site amyloid precursor protein-cleaving enzyme 1. *J Neurosci*. 2008;28:1213–1223.
139. Wang W-X, Wilfred BR, Madathil SK, et al. miR-107 regulates granulin/progranulin with implications for traumatic brain injury and neurodegenerative disease. *Am J Pathol*. 2010;177:334–345.
140. Yao J, Hennessey T, Flynt A, et al. MicroRNA-related cofilin abnormality in Alzheimer's disease. *PLoS One*. 2010;5:e15546.
141. Geekiyanage H, Chan C. MicroRNA-137/181c regulates serine palmitoyltransferase and in turn amyloid β, novel targets in sporadic Alzheimer's disease. *J Neurosci*. 2011;31:14820–14830.
142. Schonrock N, Humphreys DT, Preiss T, et al. Target gene repression mediated by miRNAs miR-181c and miR-9 both of which are down-regulated by amyloid-β. *J Mol Neurosci*. 2012;46:324–335.
143. Delaloy C, Liu L, Lee J-A, et al. MicroRNA-9 coordinates proliferation and migration of human embryonic stem cell-derived neural progenitors. *Cell Stem Cell*. 2010;6:323–335.
144. Zhao C, Sun G, Li S, et al. A feedback regulatory loop involving microRNA-9 and nuclear receptor TLX in neural stem cell fate determination. *Nat Struct Mol Biol*. 2009;16:365–371.
145. Lukiw WJ. Micro-RNA speciation in fetal, adult and Alzheimer's disease hippocampus. *Neuroreport*. 2007;18:297–300.
146. Sethi P, Lukiw WJ. Micro-RNA abundance and stability in human brain: specific alterations in Alzheimer's disease temporal lobe neocortex. *Neurosci Lett*. 2009;459:100–104 [Epub ahead of print].

147. Wang H, Liu J, Zong Y, et al. miR-106b aberrantly expressed in a double transgenic mouse model for Alzheimer's disease targets TGF-β type II receptor. *Brain Res.* 2010;1357:166–174.
148. Tesseur I, Zou K, Esposito L, et al. Deficiency in neuronal TGF-β signaling promotes neurodegeneration and Alzheimer's pathology. *J Clin Invest.* 2006;116:3060–3069.
149. Hébert SS, Horré K, Nicolaï L, et al. MicroRNA regulation of Alzheimer's amyloid precursor protein expression. *Neurobiol Dis.* 2009;33:422–428.
150. Brett JO, Renault VM, Rafalski VA, et al. The microRNA cluster miR-106b∼25 regulates adult neural stem/progenitor cell proliferation and neuronal differentiation. *Aging (Albany NY).* 2011;3:108–124.
151. Cogswell JP, Ward J, Taylor IA, et al. Identification of miRNA changes in Alzheimer's disease brain and CSF yields putative biomarkers and insights into disease pathways. *J Alzheimers Dis.* 2008;14:27–41.
152. Cimmino A, Calin GA, Fabbri M, et al. miR-15 and miR-16 induce apoptosis by targeting BCL2. *Proc Natl Acad Sci USA.* 2005;102:13944–13949.
153. Lu G, Kwong WH, Li Q, et al. bcl2, bax, and nestin in the brains of patients with neurodegeneration and those of normal aging. *J Mol Neurosci.* 2005;27:167–174.
154. Mohamed JS, Lopez MA, Boriek AM. Mechanical stretch up-regulates microRNA-26a and induces human airway smooth muscle hypertrophy by suppressing glycogen synthase kinase-3β. *J Biol Chem.* 2010;285:29336–29347.
155. Wang W-X, Huang Q, Hu Y, et al. Patterns of microRNA expression in normal and early Alzheimer's disease human temporal cortex: white matter versus gray matter. *Acta Neuropathol.* 2011;121:193–205.
156. Carrettiero DC, Hernandez I, Neveu P, et al. The cochaperone BAG2 sweeps paired helical filament- insoluble tau from the microtubule. *J Neurosci.* 2009;29: 2151–2161.
157. Cui JG, Li YY, Zhao Y, et al. Differential regulation of interleukin-1 receptor-associated kinase-1 (IRAK-1) and IRAK-2 by microRNA-146a and NF-kappaB in stressed human astroglial cells and in Alzheimer disease. *J Biol Chem.* 2010;285: 38951–38960.
158. Lukiw WJ, Zhao Y, Cui JG. An NF-kappaB-sensitive micro RNA-146a-mediated inflammatory circuit in Alzheimer disease and in stressed human brain cells. *J Biol Chem* 2008;283:31315–31322.
159. Schipper HM, Maes OC, Chertkow HM, et al. MicroRNA expression in Alzheimer blood mononuclear cells. *Gene Regul Syst Bio.* 2007;1:263–274.
160. Smalheiser NR. The RNA-centred view of the synapse: non-coding RNAs and synaptic plasticity. *Philos Trans R Soc Lond B Biol Sci.* 2014;369:20130504 [Epub ahead of print 26 September 2014].
161. Lukiw WJ, Handley P, Wong L, et al. BC200 RNA in normal human neocortex, non-Alzheimer dementia (NAD), and senile dementia of the Alzheimer type (AD). *Neurochem Res.* 1992;17:591–597.
162. Mus E, Hof PR, Tiedge H. Dendritic BC200 RNA in aging and in Alzheimer's disease. *Proc Natl Acad Sci USA.* 2007;104:10679–10684.
163. Massone S, Ciarlo E, Vella S, et al. NDM29, a RNA polymerase III-dependent non coding RNA, promotes amyloidogenic processing of APP and amyloid β secretion. *Biochim Biophys Acta.* 2012;1823:1170–1177.
164. Massone S, Vassallo I, Fiorino G, et al. 17A, a novel non-coding RNA, regulates GABA B alternative splicing and signaling in response to inflammatory stimuli and in Alzheimer disease. *Neurobiol Dis.* 2011;41:308–317.
165. Faghihi MA, Modarresi F, Khalil AM, et al. Expression of a noncoding RNA is elevated in Alzheimer's disease and drives rapid feed-forward regulation of β-secretase. *Nat Med.* 2008;14:723–730.

166. Faghihi MA, Zhang M, Huang J, et al. Evidence for natural antisense transcript-mediated inhibition of microRNA function. *Genome Biol.* 2010;11:R56.
167. Jankovic J. Parkinson's disease: clinical features and diagnosis. *J Neurol Neurosurg Psychiatry.* 2008;79:368–376.
168. Naimark D, Jackson E, Rockwell E, et al. Psychotic symptoms in Parkinson's disease patients with dementia. *J Am Geriatr Soc.* 1996;44:296–299.
169. Riedel O, Klotsche J, Spottke A, et al. Cognitive impairment in 873 patients with idiopathic Parkinson's disease. *J Neurol.* 2008;255:255–264.
170. Jellinger KA. Formation and development of Lewy pathology: a critical update. *J Neurol.* 2009;256:270–279.
171. van de Berg WDJ, Hepp DH, Dijkstra AA, et al. Patterns of α-synuclein pathology in incidental cases and clinical subtypes of Parkinson's disease. *Parkinsonism Relat Disord.* 2012;18(suppl 1):S28–S30.
172. Obeid R, Schadt A, Dillmann U, et al. Methylation status and neurodegenerative markers in Parkinson disease. *Clin Chem.* 2009;55:1852–1860.
173. Desplats P, Spencer B, Coffee E, et al. Alpha-synuclein sequesters Dnmt1 from the nucleus: a novel mechanism for epigenetic alterations in Lewy body diseases. *J Biol Chem.* 2011;286:9031–9037.
174. Jowaed A, Schmitt I, Kaut O, et al. Methylation regulates alpha-synuclein expression and is decreased in Parkinson's disease patients' brains. *J Neurosci.* 2010;30:6355–6359.
175. Matsumoto L, Takuma H, Tamaoka A, et al. CpG demethylation enhances alpha-synuclein expression and affects the pathogenesis of Parkinson's disease. *PLoS One.* 2010;5:e15522.
176. de Boni L, Tierling S, Roeber S, et al. Next-generation sequencing reveals regional differences of the α-synuclein methylation state independent of Lewy body disease. *Neuromolecular Med.* 2011;13:310–320.
177. International Parkinson's Disease Genomics Consortium (IPDGC), Wellcome Trust Case Control Consortium 2 (WTCCC2). A two-stage meta-analysis identifies several new loci for Parkinson's disease. *PLoS Genet.* 2011;7:e1002142.
178. Simón-Sánchez J, Schulte C, Bras JM, et al. Genome-wide association study reveals genetic risk underlying Parkinson's disease. *Nat Genet.* 2009;41:1308–1312.
179. Satake W, Nakabayashi Y, Mizuta I, et al. Genome-wide association study identifies common variants at four loci as genetic risk factors for Parkinson's disease. *Nat Genet.* 2009;41:1303–1307.
180. Murthy MN, Blauwendraat C, Guelfi S, et al. Increased brain expression of GPNMB is associated with genome wide significant risk for Parkinson's disease on chromosome 7p15.3. *Neurogenetics.* 2017;18:121–133.
181. Wang J-Y, Gong M-Y, Ye Y-L, et al. The RIT2 and STX1B polymorphisms are associated with Parkinson's disease. *Parkinsonism Relat Disord.* 2015;21:300–302.
182. Moore K, McKnight AJ, Craig D, et al. Epigenome-wide association study for Parkinson's disease. *Neuromolecular Med.* 2014;16:845–855.
183. Bando Y, Onuki R, Katayama T, et al. Double-strand RNA dependent protein kinase (PKR) is involved in the extrastriatal degeneration in Parkinson's disease and Huntington's disease. *Neurochem Int.* 2005;46:11–18.
184. Zhang X, Li J, Sejas DP, et al. The fanconi anemia proteins functionally interact with the protein kinase regulated by RNA (PKR). *J Biol Chem.* 2004;279:43910–43919.
185. Kaminker PG, Kim SH, Taylor RD, et al. TANK2, a new TRF1-associated poly(ADP-ribose) polymerase, causes rapid induction of cell death upon overexpression. *J Biol Chem.* 2001;276:35891–35899.
186. Masliah E, Dumaop W, Galasko D, et al. Distinctive patterns of DNA methylation associated with Parkinson disease: identification of concordant epigenetic changes in brain and peripheral blood leukocytes. *Epigenetics.* 2013;8:1030–1038.

187. Kontopoulos E, Parvin JD, Feany MB. Alpha-synuclein acts in the nucleus to inhibit histone acetylation and promote neurotoxicity. *Hum Mol Genet.* 2006;15:3012–3023.
188. Outeiro TF, Kontopoulos E, Altmann SM, et al. Sirtuin 2 inhibitors rescue α-synuclein-mediated toxicity in models of Parkinson's disease. *Science.* 2007;317:516–519.
189. St Laurent R, O'Brien LM, Ahmad ST. Sodium butyrate improves locomotor impairment and early mortality in a rotenone-induced Drosophila model of Parkinson's disease. *Neuroscience.* 2013;246:382–390.
190. Zheng B, Liao Z, Locascio JJ, et al. PGC-1, a potential therapeutic target for early intervention in Parkinson's disease. *Sci Transl Med.* 2010;2:52ra73.
191. Siddiqui A, Chinta SJ, Mallajosyula JK, et al. Selective binding of nuclear alpha-synuclein to the PGC1alpha promoter under conditions of oxidative stress may contribute to losses in mitochondrial function: implications for Parkinson's disease. *Free Radic Biol Med.* 2012;53:993–1003.
192. Kirilyuk A, Shimoji M, Catania J, et al. An intrinsically disordered region of the acetyltransferase p300 with similarity to prion-like domains plays a role in aggregation. *PLoS One.* 2012;7:e48243.
193. Jin H, Kanthasamy A, Ghosh A, et al. α-synuclein negatively regulates protein kinase Cδ expression to suppress apoptosis in dopaminergic neurons by reducing p300 histone acetyltransferase activity. *J Neurosci.* 2011;31:2035–2051.
194. Voutsinas GE, Stavrou EF, Karousos G, et al. Allelic imbalance of expression and epigenetic regulation within the alpha-synuclein wild-type and p.Ala53Thr alleles in Parkinson disease. *Hum Mutat.* 2010;31:685–691.
195. Nicholas AP, Lubin FD, Hallett PJ, et al. Striatal histone modifications in models of levodopa-induced dyskinesia. *J Neurochem.* 2008;106:486–494.
196. Song C, Kanthasamy A, Anantharam V, et al. Environmental neurotoxic pesticide increases histone acetylation to promote apoptosis in dopaminergic neuronal cells: relevance to epigenetic mechanisms of neurodegeneration. *Mol Pharmacol.* 2010;77:621–632.
197. Song C, Kanthasamy A, Jin H, et al. Paraquat induces epigenetic changes by promoting histone acetylation in cell culture models of dopaminergic degeneration. *Neurotoxicology.* 2011;32:586–595.
198. Junn E, Lee K-W, Jeong BS, et al. Repression of alpha-synuclein expression and toxicity by microRNA-7. *Proc Natl Acad Sci USA.* 2009;106:13052–13057.
199. Doxakis E. Post-transcriptional regulation of alpha-synuclein expression by mir-7 and mir-153. *J Biol Chem.* 2010;285:12726–12734.
200. Wang X, Sun X, Zhang X, et al. Quantitative assessment of the effect of FGF20 rs12720208 variant on the risk of Parkinson's disease: a meta-analysis. *Neurol Res.* 2017;39:374–380.
201. Wang G, van der Walt JM, Mayhew G, et al. Variation in the miRNA-433 binding site of FGF20 confers risk for Parkinson disease by overexpression of alpha-synuclein. *Am J Hum Genet.* 2008;82:283–289.
202. Gillardon F, Mack M, Rist W, et al. MicroRNA and proteome expression profiling in early-symptomatic α-synuclein(A30P)-transgenic mice. *Proteomics Clin Appl.* 2008;2: 697–705.
203. Vo N, Klein ME, Varlamova O, et al. From the cover: a cAMP-response element binding protein-induced microRNA regulates neuronal morphogenesis. *Proc Natl Acad Sci USA.* 2005;102:16426–16431.
204. Asikainen S, Rudgalvyte M, Heikkinen L, et al. Global microRNA expression profiling of Caenorhabditis elegans Parkinson's disease models. *J Mol Neurosci.* 2010;41:210–218.
205. Cho HJ, Liu G, Jin SM, et al. MicroRNA-205 regulates the expression of Parkinson's disease-related leucine-rich repeat kinase 2 protein. *Hum Mol Genet.* 2013;22:608–620.
206. Gehrke S, Imai Y, Sokol N, et al. Pathogenic LRRK2 negatively regulates microRNA-mediated translational repression. *Nature.* 2010;466:637–641.

207. Imai Y, Gehrke S, Wang H-Q, et al. Phosphorylation of 4E-BP by LRRK2 affects the maintenance of dopaminergic neurons in Drosophila. *EMBO J*. 2008;27:2432–2443.
208. Hoglinger GU, Breunig JJ, Depboylu C, et al. The pRb/E2F cell-cycle pathway mediates cell death in Parkinson's disease. *Proc Natl Acad Sci USA*. 2007;104:3585–3590.
209. Minones-Moyano E, Porta S, Escaramis G, et al. MicroRNA profiling of Parkinson's disease brains identifies early downregulation of miR-34b/c which modulate mitochondrial function. *Hum Mol Genet*. 2011;20:3067–3078.
210. Chen Y, Gao C, Sun Q, et al. MicroRNA-4639 is a regulator of DJ-1 expression and a potential early diagnostic marker for Parkinson's disease. *Front Aging Neurosci*. 2017;9:232.
211. Kim J, Inoue K, Ishii J, et al. A microRNA feedback circuit in midbrain dopamine neurons. *Science*. 2007;317:1220–1224.
212. Zhang X, Yang R, Hu B-L, et al. Reduced circulating levels of miR-433 and miR-133b are potential biomarkers for Parkinson's disease. *Front Cell Neurosci*. 2017;11:170.
213. Bai X, Tang Y, Yu M, et al. Downregulation of blood serum microRNA 29 family in patients with Parkinson's disease. *Sci Rep*. 2017;7:5411.
214. Alieva AK, Filatova EV, Karabanov AV, et al. miRNA expression is highly sensitive to a drug therapy in Parkinson's disease. *Parkinsonism Relat Disord*. 2015;21:72–74.
215. Margis R, Margis R, Rieder CRM. Identification of blood microRNAs associated to Parkinson's disease. *J Biotechnol*. 2011;152:96–101.
216. Ross CA, Tabrizi SJ. Huntington's disease: from molecular pathogenesis to clinical treatment. *Lancet Neurol*. 2011;10:83–98.
217. A novel gene containing a trinucleotide repeat that is expanded and unstable on Huntington's disease chromosomes. The Huntington's Disease Collaborative Research Group. *Cell*. 1993;72:971–983.
218. Myers RH, MacDonald ME, Koroshetz WJ, et al. De novo expansion of a (CAG)n repeat in sporadic Huntington's disease. *Nat Genet*. 1993;5:168–173.
219. Kowall NW, Ferrante RJ, Martin JB. Patterns of cell loss in Huntington's disease. *Trends Neurosci*. 1987;10:24–29.
220. Ryu H, Rosas HD, Hersch SM, et al. The therapeutic role of creatine in Huntington's disease. *Pharmacol Ther*. 2005;108:193–207.
221. Lee J, Hwang YJ, Kim KY, et al. Epigenetic mechanisms of neurodegeneration in Huntington's disease. *Neurotherapeutics*. 2013;10:664–676.
222. Bassi S, Tripathi T, Monziani A, et al. Epigenetics of Huntington's disease. *Adv Exp Med Biol*. 2017;978:277–299.
223. Ng CW, Yildirim F, Yap YS, et al. Extensive changes in DNA methylation are associated with expression of mutant huntingtin. *Proc Natl Acad Sci USA*. 2013;110:2354–2359.
224. Villar-Menéndez I, Blanch M, Tyebji S, et al. Increased 5-methylcytosine and decreased 5-hydroxymethylcytosine levels are associated with reduced striatal A2AR levels in Huntington's disease. *Neuromolecular Med*. 2013;15:295–309.
225. Thomas B, Matson S, Chopra V, et al. A novel method for detecting 7-methyl guanine reveals aberrant methylation levels in Huntington disease. *Anal Biochem*. 2013;436:112–120.
226. Gardian G, Browne SE, Choi D-K, et al. Neuroprotective effects of phenylbutyrate in the N171-82Q transgenic mouse model of Huntington's disease. *J Biol Chem*. 2005;280:556–563.
227. Stack EC, Del Signore SJ, Luthi-Carter R, et al. Modulation of nucleosome dynamics in Huntington's disease. *Hum Mol Genet*. 2007;16:1164–1175.
228. Sadri-Vakili G, Bouzou B, Benn CL, et al. Histones associated with downregulated genes are hypo-acetylated in Huntington's disease models. *Hum Mol Genet*. 2007;16:1293–1306.

229. McFarland KN, Das S, Sun TT, et al. Genome-wide histone acetylation is altered in a transgenic mouse model of Huntington's disease. *PLoS One.* 2012;7:e41423.
230. Ferrante RJ, Ryu H, Kubilus JK, et al. Chemotherapy for the brain: the antitumor antibiotic mithramycin prolongs survival in a mouse model of Huntington's disease. *J Neurosci.* 2004;24:10335–10342.
231. Ryu H, Lee J, Hagerty SW, et al. ESET/SETDB1 gene expression and histone H3 (K9) trimethylation in Huntington's disease. *Proc Natl Acad Sci USA.* 2006;103: 19176–19181.
232. Lee J, Hagerty S, Cormier KA, et al. Monoallele deletion of CBP leads to pericentromeric heterochromatin condensation through ESET expression and histone H3 (K9) methylation. *Hum Mol Genet.* 2008;17:1774–1782.
233. Lee J, Hwang YJ, Shin J-Y, et al. Epigenetic regulation of cholinergic receptor M1 (CHRM1) by histone H3K9me3 impairs Ca^{2+} signaling in Huntington's disease. *Acta Neuropathol.* 2013;125:727–739.
234. Cha JH, Kosinski CM, Kerner JA, et al. Altered brain neurotransmitter receptors in transgenic mice expressing a portion of an abnormal human huntington disease gene. *Proc Natl Acad Sci USA.* 1998;95:6480–6485.
235. Calabresi P, Centonze D, Gubellini P, et al. Acetylcholine-mediated modulation of striatal function. *Trends Neurosci.* 2000;23:120–126.
236. Iqbal K, Tellez-Nagel I, Grunke-Iqbal I. Protein abnormalities in Huntington's chorea. *Brain Res.* 1974;76:178–184 [Epub ahead of print].
237. Johnson R, Zuccato C, Belyaev ND, et al. A microRNA-based gene dysregulation pathway in Huntington's disease. *Neurobiol Dis.* 2008;29:438–445.
238. Vo N, Klein ME, Varlamova O, et al. A cAMP-response element binding protein-induced microRNA regulates neuronal morphogenesis. *Proc Natl Acad Sci USA.* 2005;102:16426–16431.
239. Lee S-T, Chu K, Im W-S, et al. Altered microRNA regulation in Huntington's disease models. *Exp Neurol.* 2011;227:172–179.
240. Ghose J, Sinha M, Das E, et al. Regulation of miR-146a by RelA/NFkB and p53 in STHdhQ111/HdhQ111 cells, a cell model of Huntington's disease. *PLoS One.* 2011;6:e23837.
241. Bae B-I, Xu H, Igarashi S, et al. p53 mediates cellular dysfunction and behavioral abnormalities in Huntington's disease. *Neuron.* 2005;47:29–41.
242. van den Hove DLA, Kenis G, Rutten BP. Epigenetic dysregulation in Alzheimer's disease: cause or consequence? *Epigenomics.* 2014;6:9–11.
243. Imm J, Kerrigan TL, Jeffries A, et al. Using induced pluripotent stem cells to explore genetic and epigenetic variation associated with Alzheimer's disease. *Epigenomics.* 2017;9:1455–1468.
244. Bianchi M, Renzini A, Adamo S, et al. Coordinated actions of microRNAs with other epigenetic factors regulate skeletal muscle development and adaptation. *Int J Mol Sci.* 2017;18:840.

CHAPTER FIVE

Neuroepigenetics of Prenatal Psychological Stress

Stephan Claes[1]
Department of Neuroscience, KU Leuven, Leuven, Belgium
[1]Corresponding author: e-mail address: stephan.claes@uzleuven.be

Contents

Abstract

Life does not start at birth but at conception. What a person experiences during the first 9 months of life in the intrauterine environment will have lasting effects on health and disease later in life. Psychological stress in pregnant women has a number of potential negative consequences for the physical and psychological function of their children. The mechanisms driving this association are, among others, epigenetic modifications at specific loci in the infant DNA.

In the chapter, the relevant animal and human studies underlying these assumptions are reviewed, and it is argued that they are convincingly supported by the evidence. However, a number of methodological issues are discussed, and it is shown that the actual search for the genetic loci that are epigenetically modified by prenatal psychological stress is still in its early phases. A new exciting field of research has opened up that is particularly relevant for our understanding of how early life experiences shape our biology and psychology.

Progress in Molecular Biology and Translational Science, Volume 158
ISSN 1877-1173
https://doi.org/10.1016/bs.pmbts.2018.04.007

1. INTRODUCTION

1.1 The DOHaD Hypothesis

The observation that the emotional state of a pregnant woman can have profound and lasting effects on the offspring has been repeatedly made throughout history. Hippocrates (460–377 BC) remarked that what a mother is thinking can transmute the fetus in different ways, and also made a link between emotional distress in the mother and the risk for spontaneous abortion. Leonardo Da Vinci (1452–1519) remarked in the early 16th century in his Quaderni: "... one fear that the mother has, or mental pain, has more power over the child than over the mother, since frequently the child loses its life thereby."[1] In the same era, Paracelsus (1493–1541) remarked: "an infant in the mother's womb is as much in the hands and under the will of the mother as clay in the hands of the potter, who from it makes whatever pleases him. Any strong desire, appetite or inclination can be impressed upon the fetus."

In modern times, the hypothesis that events early in life, including in the prenatal period, can shape the biology of an individual and increase or lower the risk for a number of health related outcomes, is often referred to as the "Developmental Origins of Health and Disease" (DOHaD) hypothesis. The DOHaD concept was formed by three influential papers in The Lancet,[2–4] in which Barker and colleagues demonstrated how fetal undernutrition and birth weight were associated to the risk of coronary heart disease 50 years later.

Since it origin, the DOHaD hypothesis has been the field of an impressive amount of empirical studies, and the awareness has grown that the fetus is able to adapt its biology to a number of environmental cues and triggers. One the most famous examples is the Dutch famine study, examining children who were exposed in utero to severe undernutrition at the final stages of the World War II. Children exposed to famine late in gestation had an increased tendency to develop glucose intolerance and diabetes later in life, whereas undernutrition in early gestation led to a more atherogenic lipid profile and an increased risk for coronary artery disease.[5] The Southampton Women's Survey is a large prospective evaluation of women before, during and after pregnancy. Among other results, studies on this cohort showed that maternal diet and central adiposity influenced ductus venosus shunting and blood flow to the liver in the fetus,[6] while another study demonstrated that the relative fat mass in then newborn was dependent on a number of factors related to maternal life style.[7]

1.2 Maternal Psychological Stress

This chapter is devoted to the neuroepigenetics of prenatal psychological stress. Defining what the term "psychological stress" means precisely is not straightforward; "Stress" is a broad term that can cover a number of different concepts. It can refer to a stressor, a threatening or negative life event that the mother is confronted with. In the case of prenatal stress research, a broad range of stressors such as food deprivation, natural calamities, war situations and psychosocial adversity have been examined regarding the consequences for pregnant women and their unborn offspring. While some of these stressors seem more psychological in nature (e.g., psychosocial adversity) and other more material (famine), in practice all these stressors will have important psychological consequences. For example, in the case the Dutch hunger winter study cited above, it is impossible to distinguish the effects of the malnutrition on the fetus from the effects induced by maternal anxiety, which beyond doubt was present in many of these women exposed to the atrocities of war while carrying a baby. For that reason, we will include all kinds of stressors as maternal psychological stress.

The term stress can also refer to the emotional situation of the mother, e.g., a state of anxiety and depression. The relation between stress as a negative life event and stress as the emotional state of the mother is not straightforward. The degree to which an adverse condition such as being exposed to a natural disaster provokes anxiety in pregnant women will not only be related to the severity of the stressor, but also very much to the liability of the individual person exposed to it and to contextual factors. On the other hand, negative emotional states, such as depression during pregnancy, can also occur in women who have not been exposed to identifiable major threatening or negative life events. We believe that a negative emotional status of the mother, even unrelated to specific events, qualifies as maternal psychological stress, and studies on the effects on the fetus will be included in this review chapter.

2. THE CONSEQUENCES OF PRENATAL PSYCHOLOGICAL STRESS

It seems intuitively logical to assume that children of mothers who are highly anxious or depressed during pregnancy, will be more anxious or depressed themselves. This assumption however is not so easy to validate scientifically. First, long-term prospective studies are tedious and costly.

Second, in human studies, it is very hard to distinguish the effects from prenatal stress from other factors. Mothers who are highly anxious during pregnancy might be very well genetically prone to such an emotional status, and could transmit this genetic liability to their offspring. Further, these mothers will in many cases also show increased negative emotions postnatally, which again might confound the effect of prenatal stress on the children. In animal studies, these confounding factors can be controlled for. We will discuss both animal and human studies below.

2.1 Animal Studies

Most animal studies examining the effects of prenatal stress were performed in rodents. The data are extensively reviewed in a recent publication.[8] A range of strategies have been applied to subject pregnant rats and mice to prenatal stress, such as intermittent electric shocks, restraint, exposure to bright light, cat meowing, food deprivation, sleep prevention, overcrowded housing. These conditions have been applied as a single stressor, or in combination schemes. Whether these models are really imposing significant stress upon the pregnant animals can be deduced from parameters such as the rise of corticosterone (CORT) in response to the stressor, and the evolution of the body weight. Most stressors do indeed lead to rise in CORT, but after applying the same stressor a number of days at the same time of the day, the effect seems to disappear,[9] indicating that the animal has adapted to the aversive context. In a number of studies, stress exposure leads to weight loss in the pregnant dam,[10–12] but in other studies this was not the case. The duration of the stressor, rather than its nature, seems to be more relevant to eventual weight loss.[8]

Taken together, the available studies confirm that pups from dams exposed to prenatal stress show increased levels of anxiety, as evaluated with classical tests such as the elevated plus maze.[13–15] Also, offspring of prenatally stressed rats shows an increase in depressive behavior when measuring "learned helplessness" behavior in response to a forced swim test, a frequently used paradigm with a substantial construct and predictive validity for depression.[16] The effect of prenatal stress on this depression model in the offspring of rats seems to be critically dependent on the timing of the stressor. Only when applied during the last week of gestation, the effect seems robust.[17,18]

Further, studies show that offspring of prenatally stressed dams also shows cognitive deficits.[8] Spatial learning, as assessed through the water maze test, is slower, especially in male offspring.[19]

2.2 Human Studies

Most studies in humans are prospective longitudinal birth cohort studies. The length of the follow-up is very variable, from a few months to several years. Several very large studies have been published, such as the Raine and the Mater-University of Queensland Study of Pregnancy (MUSP) studies from Australia, the Avon Longitudinal Study of Parents and Children (ALSPAC) study (UK), Generation R and ABCD studies (The Netherlands) and the Early Developmental Stages of Psychopathology (EDSP) study (Germany).

These studies used a variety of questionnaires to assess prenatal anxiety and depression. Several studies used the State Trait Anxiety Questionnaire (STAI) (e.g., Ref. 20) and the Pregnancy Related Anxiety Scale.[21,22] Maternal depression was assessed in most studies by means of the Edinburgh Postnatal Depression Scale (EDPS) (e.g., Ref. 23).

Most studies also assessed the occurrence of negative life events during pregnancy, such as divorce, financial problems, and loss of a beloved person.

The literature seems to confirm that several kinds of prenatal stress, whether in the form of negative life events or of prenatal maternal anxiety, are associated with increased psychopathology in the children.

2.2.1 Prenatal Stress and Internalizing Psychopathology

Internalizing psychopathology refers to mental problems with prominent negative emotionality, such as symptoms of depression and anxiety. The MUSP study[24] showed that adolescents with anxiety disorders were more likely, in comparison with healthy controls, to stem from pregnancies with increased prenatal stress, whereas depression in adolescents was predicted more by early postnatal childhood variables.

Other studies did show a link between prenatal maternal stress and the development of depression in children. The ALSPAC study showed that prenatal maternal depression increases the risk for offspring to develop depression at the age of 18 by 1.28 times, independent of postnatal maternal depression, whereas paternal prenatal depression did not seem to increase the depression risk.[25] Similar findings were reported in a smaller population study by Pawlby et al.,[26] who prospectively followed up 151 mother–child dyads starting from pregnancy until the age of 16 years. The risk of depression for the 16-year-olds exposed to antenatal depression was 4.7 times greater than for offspring not so exposed.

As stated before, the field is confronted with many methodological issues, in the first place with correcting properly for genetic and postnatal effects. While both the ALSPAC and Generation R studies reported a link between maternal depressive symptoms during pregnancy and child attention problems, these associations were considerably attenuated when controlling properly for maternal symptoms 3 years after childbirth.[27] The same thing is true when examining the link between antenatal maternal depression and anxiety symptoms and anxiety and other emotional problems in children, but some of the associations remained significant, such as the link between prenatal maternal anxiety (but not depression) and childhood emotional problems in the Generation R study, and prenatal maternal depression (but not anxiety) and childhood emotional problems in ALSPAC.[27] Taken together, the evidence that prenatal psychological enhances negative emotionality in children is convincing, but the size of the effect is probably overestimated in a number of studies that have not sufficiently corrected for confounding pre- and postnatal factors.

2.2.2 Prenatal Stress and Externalizing Psychopathology

Symptoms of depression and anxiety in children are mostly referred to as internalizing pathology, while attention deficit and hyperactivity disorder (ADHD) and conduct disorders are often taken together under the label "externalizing disorders." According to a longitudinal cohort study including almost 1000 German subjects,[28] children of mothers with high levels of self-perceived distress during pregnancy, showed higher levels of ADHD, conduct disorder and separation anxiety disorders when followed up until early adulthood. A smaller but very long-term prospective follow-up study in 192 adolescents found a link between prenatal maternal depressive symptoms and externalizing problems during adolescence, while in this study postnatal maternal depression in the first months after birth was associated with more internalizing pathology.[29]

Recently, the data on externalizing disorders collected in children from the ALSPAC study were published.[30] This very large analysis included over 10,000 mother–offspring pairs, and followed up the children at 2–3 year intervals until the age of 16. Children exposed to the highest quartile of prenatal stress had a higher probability to belong to the group with a high symptom trajectory for hyperactivity and conduct disorder. The analyses were adjusted for a large number of pre- and postnatal environmental and parental variables, again indicating that the link between prenatal maternal stress and postnatal psychopathology statistically survives proper statistical correction for confounders.

2.2.3 Prenatal Stress and Other Mental Disorders

A link between prenatal stress and autism has been established in some studies,[31,32] but a very large ($n = 37{,}275$) prospective cohort study in Denmark found no evidence for this association.[33] The research on the link between prenatal stress and schizophrenia is scarce, but at least two papers show that severe form of stress, particularly in the first trimester of pregnancy, do lead to an increase in the risk for the disorder in the offspring.[32,34]

2.2.4 Underlying Endophenotypes and Neurobiological System Changes

The effects of prenatal stress cannot only be related to psychopathology, but to underlying brain functions that can be assessed in infants, and that are indirectly supposed to be increasing the risk for psychopathology later in life. The large Generation R study showed that infant neuromotor development at the age of 3 months was negatively influenced by maternal prenatal anxiety.[35] In a study using event-related potentials from EEG measurements in 9-month old babies, Otte et al.[36] concluded that prenatal maternal anxiety is related to a more extensive brain processing of fear-related stimuli in the infants.

The neurobiological system changes that underlie the link between prenatal maternal stress and child health outcome remain largely unknown, but a lot of focus has been placed on the hypothalamic–pituitary–adrenal (HPA) axis. The HPA axis is intrinsically involved in the proper adaptation of the organism to different demands from the environment, so, from an evolutionary perspective, it seems plausible to assume that prenatal stress can modulate the set point of the system at in early life. In order to test this hypothesis, studies have looked at various HPA axis functional measures, such as diurnal patterns of cortisol secretion and cortisol secretion in reaction to specific stressors. Some studies have confirmed that baseline cortisol secretion is related to behavioral problems in maltreated children.[37] Exaggerated reactive cortisol secretion in response to specific stressors has been related to internalizing problems such as anxiety and depression.[38] Very different prenatal variables, such as prenatal maternal substance use and stress, have been linked to changes in the intrauterine development of the HPA axis.[39] The mechanism between parental stress and altered HPA axis activity in the child is postulated to be increased exposure of the fetus to maternal cortisol.[40] This hypothesis is supported by some animal studies, showing for example that treating pregnant rats with glucocorticoids altered the reactivity of the HPA axis in the offspring.[41]

Evidently, the neurobiology of the link between parental stress and child outcomes cannot simply be reduced to maternal cortisol impacting the offspring HPA axis. As demonstrated in a recent review,[42] prenatal maternal psychosocial stress affects the fetus through a network of different pathways and mechanisms, such as a compromised functioning of the placenta, increased secretion of catecholamines, immune changes and alterations in the intestinal microbiome.

3. PRENATAL PSYCHOLOGICAL STRESS AND EPIGENETIC MODIFICATIONS

As discussed, prenatal psychological maternal stress leads to lasting changes in physiological functions in the offspring, which can only be explained by enduring modifications in gene expression. Three major mechanisms regulating gene expression are DNA methylation (binding of methyl groups to cytosine), histone modification (several processes influencing the structure and function of histones, the molecules around which the DNA strand is wrapped) and microRNA function. In the frame of prenatal stress, the focus has been on DNA methylation changes. Studies are providing evidence that environmental and social cues, pre-and postnatally, can alter gene expression profiles through specific alterations in DNA methylation. For obvious reasons, a lot of attention has been given to genes involved in the HPA axis function, such as NR3C1, the gene encoding the glucocorticoid receptor (GR), and FKBP5, a GR co-chaperone protein. While promising data have been found regarding these genes, it is clear now that many more genes are differentially methylated in function of prenatal stress, thereby modulation the expression of numerous proteins. We will subsequently describe the major animal and human studies in the field. Not only prenatal psychosocial stress, but also prenatal alcohol use, seems to decrease GR expression in the offspring, at least in rodents.[43]

3.1 Animal Studies

In a seminal paper, Weaver et al.[44] reported how mothering behavior in rats in the early postnatal period had a lasting influence on methylation patterns at the promoter site of the NR3C1 gene, resulting in lifelong changes in GR expression and HPA axis responses to stress. These effects were reversible when pups exposed to poor mothering behavior were cross-fostered to dams with strong mothering, indicating how the epigenetic machinery flexibly adjusts itself to changing circumstances. The Weaver study refers to postnatal stress, but other studies demonstrated comparable effects of prenatal stress. In a study in mice, Mueller and Bale[45] demonstrated that male offspring of mice exposed to stress in early gestation suffered from anhedonia and a maladaptive responses to stress, had a higher expression of corticotrophin releasing factor (the hormone driving the HPA axis), higher corticosterone secretion in response to stress, and a lower expression of GR in the hippocampus. Exposed pups had an increased methylation of the 1F promotor of the GR gene, which supposedly drives the decreased GR expression.

Several other genes have been found to be differentially methylated in function of prenatal stress in animals. In a rat study, Palacios-García et al.[46] used a restraint stress model to examine the behavioral and neurobiological consequences of prenatal stress, focusing on Reelin, a protein known to exert a critical role on cortical lamination and synapse formation. Prenatal stress induced changes in Reelin expression and in DNA methylation levels of the Reelin gene promoter. Also applying a prenatal restraint stress model in rats, Van den Hove et al.[47] found changes in in gene expression in several pathways, including glutamatergic and GABAergic neurotransmission, with sex-specific profiles. Schraut et al.[48] showed that prenatal restriction stress in serotonin transporter (5-HTT) deficient transgenic mice interacted with the 5-HTT genotype in causing specific epigenetic and expression changes, for example for the gene encoding myelin basic protein.

Mychasiuk et al.[49] evaluated the effect of prenatal maternal stress at a genome-wide level, applying a bystander stress paradigm in rats (housing the pregnant dam with another female that underwent elevated platform stress). Pups from these dams performed differently on a number of behavioral tasks. Both at the frontal cortex and the hippocampus, global DNA methylation levels were higher in these animals, compared to no-stress controls. Gene expression profiling indicated a total of 558 genes from multiple biological processes and pathways being differentially expressed between both groups.

3.2 Human Studies

3.2.1 The Glucocorticoid Receptor Gene (NR3C1)

In the light of the putative crucial function of the HPA axis in mediating the effects of prenatal stress, and encouraged by the animal findings, several groups set out to examine methylation changes at the level of the GR gene in response to prenatal stress.

Oberlander et al.[50] studied 82 mother–child pairs, and discovered that exposure of the fetus to depression or anxiety of the mother during the third trimester of pregnancy led to increased methylation at exon 1F of the NR3C1 gene in the newborn, which was associated to an increased salivary cortisol response at the age of 3 months. Our group published a study looking at NR3C1 promoter methylation in cord blood from 83 infants, relating these data to prenatal anxiety, depression and cortisol secretion in the mothers, stemming from our Prenatal and Early Life Stress (PELS) sample.[51] This study looked at three alternative NR3C1 promoters: 1B, 1D and 1F, the latter being identical to the region explored by Oberlander.[50] We found that prenatal maternal anxiety specifically induced methylation changes at the 1F promoter site of NR3C1. However, there was no link between methylation patterns at birth, and infant parameters such as intrauterine growth, birth weight, of cortisol secretion at the ages of 3, 6 and 12 months. In an independent large study, Conradt et al.[52] showed that infants whose mothers reported depression during pregnancy showed greater methylation of a specific CpG within the NR3C1 1F promoter region, and had poorer self-regulation and more lethergia and hypotonia. The effect of maternal prenatal anxiety and depression might be gender specific. One study found increased NR3C1 1F methylation specifically in males stemming from mothers with depression during pregnancy,[53] while another project reported that prenatal stressful events modulated methylation at the same locus, but now in female offspring.[54]

While these studies focused on the emotional status of the mother, others have investigated the effect of negative or threatening life events on infant DNA methylation. Mothers exposed to intimate partner violence had children with increased NR3C1 promoter methylation in a German study.[55] These children were examined at the age of 10–19 years, which might indicate stability of the stress-induced methylation changes, although examination at that age makes it difficult to distinguish pre- from postnatal stress effects. A study in Congolese women exposed to situations again indicated an effect of such an extreme stressor on infant NR3C1 methylation and on birth weight.[56] Similar findings were reported in other-child dyads exposed to the cruelties of the Tutsi genocide.[57]

While these data seem to yield convincing evidence for a role of the GR gene methylation and expression in effect of prenatal stress, some cautionary remarks are appropriate. First, while all these studies reported a positive finding with the 1F promoter region, the actual CpGs that were differentially methylated are not the same. In a recent meta-analysis, Palma-Gudiel et al.[58]

combined data from 7 studies, totaling 977 study subjects. They found that the published studies showed overlap across 5 consecutive CpG sites. Only methylation levels at CpG site 36 displayed a significant correlation to prenatal stress ($r = 0.14$, 95% CI: 0.05–0.23, $P = 0.002$) when all 7 studies were combined. CpG36 is human homologue of one of the CpG sites that were originally reported by Weaver et al.[44] to be associated with prenatal stress in rats.

Second, an important remark should be made about the tissue under study. While animal studies can examine methylation levels in relevant brain regions such as the hippocampus, most human studies will rely on DNA extracted from whole blood or saliva. It is clear that DNA methylation and gene expression patterns are highly tissue specific. A study characterizing intra- and inter-individual methylomic variation across whole blood and multiple regions of the brain found that while the variation in methylation between tissues greatly exceeds the variation between individuals for a given tissue, variation between individuals was reflected in both brain and in blood, indicating that peripheral tissues could be useful for studying complex neurobiological phenotypes.[59] Nevertheless caution is necessary. A comparative study suggests that methylations patterns of DNA isolated from saliva may be more similar to brain tissue than DNA from blood.[60]

Third, the differences in methylation percentage at the NR3C1 promoter region induced by prenatal stress are in most studies very small, in the order of 1%–2%. While such minimal differences are statistically significant, the functional relevance needs to be established further.

3.2.2 Other Candidate Genes

What follows is not a comprehensive overview of all studies that have been published in this rapidly evolving field, but rather a selection of a few examples in which convergent evidence seems to indicate a role of these genes in the epigenetic machinery linking prenatal stress to health outcomes in the offspring.

FKBP5 is a co-chaperone of the glucocorticoid receptor, and FKBP5 genetic polymorphisms have been implicated in the neurobiology of depression,[61] so the gene is also an evident candidate for studies regarding prenatal stress. HSD11B2 (11β-hydroxysteroid dehydrogenase type 2) is a barrier enzyme that inactivates maternal glucocorticoids passing through the placenta, and therefore also an important regulator of the link between prenatal maternal stress and fetal physiology. In a recent study in 61 women, perceived stress during pregnancy was associated with modestly elevated methylation of FKBP5 and of HSD11B2.[62] A potential role of HSD11B2 expression and methylation as mediator of the effects of prenatal maternal adversities has been demonstrated as well by Appleton et al.[63]

BDNF (Brain derived Neurotropic Factor) plays an important role in the growth, maturation and maintenance of brain cells, and has since long been suspected to play an important role in the neurobiology of depression. Studies in mice[64] and in rats[65] showed that prenatal stress led to BDNF underexpression and higher methylation at the BDNF gene promoter in crucial brain regions. Studies in humans indicate that both maternal depressive symptoms and maternal exposure to war trauma were associated with increased methylation at the BDNF gene promoter.[66] The latter study has the advantage of finding a robust effect in three tissues (cord blood, placental tissue, maternal venous blood), although the sample size was quite limited ($n = 24$).

Imprinted genes are also of particular interest for epigenetic studies, as the mechanism of imprinting is methylation of CpG sites. The interest for imprinted genes in the study of prenatal stress was further enhanced by the study of the Dutch famine, which examines the link between the undernutrition of pregnant women in The Netherlands at the end of World War II, and its consequences regarding the physical and mental health of children born form these pregnancies in the course of 1945. These data are summarized by Lumey et al.[67] together with other studies on the consequences of prenatal famine. Among the negative health outcomes were increased body weight and BMI at ages 18 and 50, decreased glucose tolerance at age 50, a higher prevalence of type 2 diabetes at age 59, and increased levels of the ration low-density lipoproteins/high-density lipoproteins at ages 50 and 59. Regarding mental disorders, a twofold increase in the risk for schizophrenia was found. Heijmans et al.[68] showed that individuals exposed in utero to the Dutch war famine had lower levels of DNA methylation at the imprinted Insulin-like Growth Factor (IGF2) gene at the age of 59 years. The methylation difference between exposed ($n = 60$) and control ($n = 60$) individuals was highly significant over the 5 CpGs that were assessed. The association was specific for early, periconceptional exposure; the effect was absent in individuals exposed to famine in late in the pregnancy. Recent studies suggest that IGF2 methylation at birth are not only modulated by prenatal famine, but also by maternal emotional factors during pregnancy. In a study on 80 mother–child dyads from the PELS sample, our research group found a negative correlation between anxiety during pregnancy and methylation at the same locus that was reported by Heijmans et al. in the Dutch hunger winter population.[69] This finding is replicated by an independent larger study from Mansell et al.,[70] who found that higher prenatal maternal anxiety led to lower methylation at the IGF2-H19 imprinting control region. A weaker association was observed with prenatal maternal depression. In a large study (n =922) focusing on prenatal maternal depression, birth weight and methylation of several imprinted genes (IGF2/H19, DLK1/MEG3, MEST, PEG3, PEG10/SGCE, NNAT and PLAGL1), depressed mood was associated with a more that threefold higher risk of low birth weight.[71] Compared with infants of women without depressed mood, infants born to women with severe depressed mood had a 2.4% higher methylation at the MEG3 DMR. Whereas low birth weight infants had 1.6% lower methylation at the IGF2 DMR, high birth weight (>4500 g) infants had 5.9% higher methylation at the PLAGL1 DMR compared with normal birth weight infants.

3.2.3 Epigenome-Wide Approaches

The research on the epigenetic marks of prenatal stress faces the same difficult challenges as the field of psychiatric genetics in general. One the one hand, the study of specific candidate epigenetic regions has the advantage of allowing a detailed analysis of the region under study, and sufficient power can be generated with a relatively moderate sample size. Major disadvantages of this approach are the arbitrary selection of candidate regions, and the increased probability of false positive findings and publication bias favoring positive replications. The more solid approach seems to be epigenome-wide, eliminating the problem of having to select a candidate region on hypothetical grounds. On the other hand, epigenome-wide approaches look simultaneously at between many thousands of CpGs. This brings about an evident power problem, because samples of thousands of subjects would be required to generate sufficient power. This is particularly challenging in epigenetic research regarding the consequences of prenatal stress, as these studies require a prospective long-term follow-up of mother–child dyads, which is labor-intensive and costly. The problem can be mitigated by not looking at individual CpGs when analyzing the data, but at "differentially methylated regions" (DMRs), that can be identified and ranked by statistical packages such as DMRcate,[72] but still then there are statistical challenges. The proper analysis of DNA methylome association studies will require the development of new analytical methods, together with integration of data from other studies and different species.[73]

In the Dutch hunger winter sample, Tobi et al.[74] generated DNA methylation data in whole blood on 1.2 million individual CpG dinucleotides in 24 individuals prenatally exposed to famine in early gestation, and 24 unexposed same-sex sibling as controls. They showed that prenatal famine induces methylation changes at regulatory regions that map to genes enriched for expression during early development. The DMRs identified study played a role in pathways relevant for growth and metabolism. The study further observed tentative associations of DNA methylation at the intragenic INSR and CPT1A P-DMRs with birth weight and lipid metabolism, respectively. Another study in the same cohort[75] looked at a larger sample, consisting of 422 individuals exposed to prenatal famine and 463 controls. The timing of the prenatal famine once again proved crucial, identifying early gestation as the most critical period to modulated methylation patterns. Of the four CpG dinucleotides showing the strongest evidence for association, CpG cg20823026 was located between FAM150B, involved in cell growth and differentiation, and TMEM18, a known obesity gene. Another significant CpG (cg10354880) was situated within a strong enhancer mapping to the SLC38A2 gene, whose function includes matching nutrient supply to fetal demand, together with IGF2, for which we previously discussed differential methylation in a subset of this cohort.[68]

In a study on 36 children at the age of 13 from mothers who were exposed to a natural disaster, the 1998 Quebec ice storm, Cao-Lei et al. analyzed the results of the Infinium Methylation 450 BeadChip array.[76] No statistically significant correlations were found between subjective prenatal stress (the emotional status of the mother) and methylation levels in the children. However, 1675 CpGs related to 957 genes were differentially methylated in function of what is referred to as "objective hardship," a measure of the severity of the stressor. This could lead to the conclusion that in this study, the objective severity of the stressor was more important in terms of epigenetic effects than the maternal emotional reaction to it. However, this conclusion should be handled with care, as the questionnaire measuring "objective hardship" also contained items on threat to the self, which is difficult to imagine without anxiety, while the questionnaire assessing "subjective stress" actually measured post-traumatic stress-like symptoms, which might be quite different from acute stress symptoms.

Non et al.[77] also used the methylation 450 BeadChip array to study 58 mother–child dyads from the Harvard Epigenetic Birth Cohort: 13 women exposed to non-medicated anxiety or depression during pregnancy, 22 exposed to SSRIs and 23 control dyads. Applying a relatively mild method of correction for multiple testing (false discovery rate [FDR]-adjusted $P < 0.1$), they found 42 DMRs when comparing children from the non-medicated anxious/depressed mothers to the control group. This number was reduced to 10 when applying a stricter approach (FDR-adjusted level of $P \leq 0.05$).

Our research group selected 22 versus 23 newborns from our Prenatal Early Life Stress (PELS) cohort, exposed to the lowest or highest degree of maternal pregnancy anxiety, respectively.[78] Cord blood genome-wide DNA methylation was assayed using the Human Methylation450 BeadChip. The top signals coming out of the genome-wide analysis were subsequently validated using a different method (EpiTYPER) in a larger sample ($n = 80$). We found 10 DMRs when comparing newborns exposed to low versus high pregnancy anxiety scores. We validated a top DMR in the GABA-B receptor gene (GABBR1) revealing the association with pregnancy anxiety particularly in male newborns. Cord blood GABBR1 methylation was associated with child cortisol levels in response to a stressor at 2 and 4 months and with cortisol awakening response at 12 months of age, indicating a role in GABBR1 expression in HPA axis function in the newborn.

In a study on 201 newborns from mothers suffering from mental illness during pregnancy, Schroeder et al.,[79] using the 27,578 CpG analysis chip, and found no significant changes in neonatal DNA methylation attributable to maternal psychiatric diagnosis or depressive symptoms during pregnancy.

By far the largest analysis of methylation changes induced by prenatal maternal stress was recently published by Rijlaarsdam et al.,[80] studying two large independent population-based studies, the Generation R ($n = 912$) and the ALSPAC ($n = 828$) studies with the Infinium Methylation 450 BeadChip array. At the individual CpG level, none of the associations between prenatal maternal stress and newborn DNA methylation survived proper correction for multiple testing. Interestingly, the 39,308 (out of approximately 450,000) probes with weak evidence of differential methylation in the EWAS meta-analysis showed a strong correlation across the two samples [$r(39{,}308) = 0.75$]. Of these 39,308 CpG correlations, 38,000 (97%) went in the same direction across Generation R and ALSPAC. This seems to indicate a high consistence of these weak signals across both samples. The pathway analysis pointed out that the methyltransferase pathway, involved in the methylation process, was enriched for CpGs modulated by prenatal stress, a finding that came out of several statistical approaches. Finally, the DMR analysis did not yield any DMRs, at least at a significance level robust enough to survive Bonferroni correction. In the Generation R sample three DMRs showed marginal associations, which could not be replicated in the ALSPAC sample.

It should be mentioned that none of the epigenome-wide approaches comes up with strong findings in the candidate genes that have been described higher, such as NR3C1, BDNF or IGF2. This is comparable for example to the fact that none of the GWAS studies in depression detected strong signals for a putative candidate gene for depression such as the serotonin transporter. This does not necessarily imply that the candidate gene findings are falsely positive. The methylome studies, even when analyzing 450,000 CpGs, in most cases only have a few CpGs in the candidate genes of interest, and can easily miss relevant signals. This means that the epigenome-wide studies do not contradict the candidate gene analyses, but rather constitute an alternative approach.

4. CONCLUSIONS

Research on the clinical, physiological and pathological consequences of prenatal psychological stress is developing rapidly. Taken together, the animal and human studies deliver convincing evidence that maternal stress—whether defined as the emotional status of the mother of as objective stressors to which she was exposed—during pregnancy can have lasting effects on neurobiology and disease vulnerability. Evidently, human studies trying to link prenatal stress to postnatal child outcomes are fraught with methodological difficulties, such as correcting properly for a potentially endless list of prenatal and postnatal confounders. Yet the main findings survive these corrective approaches, especially in the larger studies.

The reason why the prenatal environment influences postnatal (neuro-) biological functions is not hard to grasp from an evolutionary point of view. It makes sense that the biological systems of newborn systems who were confronted with a threatening environment in utero is tuned in order to be more alert and ready to react rapidly to threats.[81] Such an evolutionary adaptation makes sense in terms of reproductive success, even if it comes at the price of an increased vulnerability for somatic or mental disorders later in life. It is therefore safe to assume that the whole machinery of epigenetic changes in response to prenatal stress has developed over millions of years in a careful trade-off between enabling the individual to react adequately to a potentially threatening environment on the one hand, and minimizing the negative consequences of an overly alert stress system on the other hand.

From the same evolutionary perspective, it also makes sense that these epigenetic adaptations are not endlessly transmitted through the generations, but do persist, maybe in a mitigated way, for a few generations. Studying marmoset primates, Buchwald et al.[82] were able to prove that the epigenetic effects of administering glucocorticoids to pregnant animals could be detected over the next three generations. However, the data on transgenerational effects of prenatal factors in humans are scarce, and studies are fraught with great methodological differences.[83]

An important question is whether it is possible to interfere with these epigenetic changes in order to reverse (some of) the negative effects of prenatal psychological stress. The seminal study of Weaver et al.[44] already indicated that the NR3C1 methylation changes induced by poor mothering in rats could be prevented by administering a pharmacological agent (such as a histone deacetylase inhibitor) or reversed by changing the environment through a cross-fostering experiment. In a human study, Sharp et al.[84] found that infants of mothers who were depressed during pregnancy had a decreased physiological adaptability and increased negative emotionality, but only in the presence of low maternal postnatal stroking. If the mothers performed highly on maternal stroking in the first few weeks postpartum, the negative effects of prenatal maternal depression were neutralized. While evidently much more research is needed, such findings are encouraging toward the development of pharmacological and environmental strategies to prevent the negative health outcome related to prenatal maternal stress.

The search for the actual genes that are epigenetically modified by prenatal stress is ongoing, but far from easy. Candidate genes studies might be prone to type I error and publication bias, whereas genome-wide approaches are hampered by power problems, and are even with the newest 850,000 CpG chips only examining a fraction (<10%) of the epigenome. The most obvious way forward at this stage seems to validate candidate gene studies with functional data and animal studies, something that has been done with some success for NR3C1 methylation changes.

In general, the field of epigenetics in relation to prenatal stress had opened a new alley of research that is of particular interesting for psychologists and psychiatrists, as it enriches the established knowledge about the importance of the early environment for psychological development with insights about the underlying neurobiological mechanisms, potentially opening the road to early preventive interventions.

REFERENCES

1. Van den Hove DLA. *Prenatal Stress and Adult Psychopathology*. PhD thesis, Maastricht. 2006.
2. Barker DJ, Osmond C. Infant mortality, childhood nutrition, and ischaemic heart disease in England and Wales. *Lancet*. 1986;1(8489):1077–1081.
3. Barker DJ, Winter PD, Osmond C, Margetts B, Simmonds SJ. Weight in infancy and death from ischaemic heart disease. *Lancet*. 1989;2(8663):577–580.
4. Barker DJ, Gluckman PD, Godfrey KM, Harding JE, Owens JA, Robinson JS. Fetal nutrition and cardiovascular disease in adult life. *Lancet*. 1993;341(8850):938–941.
5. Roseboom TJ, van der Meulen JH, Ravelli AC, Osmond C, Barker DJ, Bleker OP. Effects of prenatal exposure to the Dutch famine on adult disease in later life: an overview. *Mol Cell Endocrinol*. 2001;185:93–98.
6. Haugen G, Hanson M, Kiserud T, Crozier S, Inskip H, Godfrey KM. Fetal liver-sparing cardiovascular adaptations linked to mother's slimness and diet. *Circ Res*. 2005;96:12–14.
7. Harvey NC, Poole JR, Javaid MK, et al. Parental determinants of neonatal body composition. *J Clin Endocrinol Metab*. 2007;92(2):523–526.
8. Weinstock M. Prenatal stressors in rodents: effects on behavior. *Neurobiol Stress*. 2016;6:3–13.
9. Melia KR, Ryabinin AE, Schroeder R, Bloom FE, Wilson MC. Induction and habituation of immediate early gene expression in rat brain by acute and repeated restraint stress. *J Neurosci*. 1994;14:5929–5938.
10. Van den Hove DL, Leibold NK, Strackx E, et al. Prenatal stress and subsequent exposure to chronic mild stress in rats; interdependent effects on emotional behavior and the serotonergic system. *Eur Neuropsychopharmacol*. 2014;24:595–607.
11. Fujita S, Ueki S, Miyoshi M, Watanabe T. "Green odor" inhalation by stressed rat dams reduces behavioral and neuroendocrine signs of prenatal stress in the offspring. *Horm Behav*. 2010;58:264–272.
12. Baker S, Chebli M, Rees S, et al. Effects of gestational stress: 1. Evaluation of maternal and juvenile offspring behavior. *Brain Res*. 2008;213:98–110.
13. Akatsu S, Ishikawa C, Takemura K, Ohtani A, Shiga T. Effects of prenatal stress and neonatal handling on anxiety, spatial learning and serotonergic system of male offspring mice. *Neurosci Res*. 2015;101:15–23.
14. Salomon S, Bejar C, Schorer-Apelbaum D, Weinstock M. Corticosterone mediates some but not other behavioural changes induced by prenatal stress in rats. *J Neuroendocrinol*. 2011;23:118–128.
15. Wang Y, Ma Y, Cheng W, et al. Sexual differences in long-term effects of prenatal chronic mild stress on anxiety-like behavior and stress-induced regional glutamate receptor expression in rat offspring. *Int J Dev Neurosci*. 2015;41:80–91.
16. Porsolt RD, Anton G, Blavet N, Jalfre M. Behavioural despair in rats: a new model sensitive to antidepressant treatments. *Eur J Pharmacol*. 1978;47:379–391.
17. Guan L, Jia N, Zhao X, et al. The involvement of ERK/CREB/Bcl-2 in depression-like behavior in prenatally stressed offspring rats. *Brain Res Bull*. 2013;99:1–8.
18. Sickmann HM, Arentzen TS, Dyrby TB, Plath N, Kristensen MP. Prenatal stress produces sex-specific changes in depression-like behavior in rats: implications for increased vulnerability in females. *J Dev Orig Health Dis*. 2015;6:462–474.
19. Barzegar M, Sajjadi FS, Talaei SA, Hamidi G, Salami M. Prenatal exposure to noise stress: anxiety, impaired spatial memory, and deteriorated hippocampal plasticity in postnatal life. *Hippocampus*. 2015;25:187–196.
20. Loomans EM, van der Stelt O, van Eijsden M, et al. Antenatal maternal anxiety is associated with problem behaviour at age five. *Early Hum Dev*. 2011;87:565–570.
21. Blair MM, Glynn LM, Sandman CA, Davis EP. Prenatal maternal anxiety and early childhood temperament. *Stress*. 2011;14:644–651.

22. Davis EP, Sandman CA. Prenatal psychobiological predictors of anxiety risk in preadolescent children. *Psychoneuroendocrinology*. 2012;37:1224–1233.
23. O'Connor TG, Heron J, Golding J, Glover V, ALSPAC Study Team. Maternal antenatal anxiety and behavioural/emotional problems in children: a test of a programming hypothesis. *J Child Psychol Psychiatry*. 2003;44:1025–1036.
24. Phillips NK, Hammen CL, Brennan PA, Najman JM, Bor W. Early adversity and the prospective prediction of depressive and anxiety disorders in adolescents. *J Abnorm Child Psychol*. 2005;33:13–24.
25. Pearson RM, Evans J, Kounali D, Lewis G, Heron J. Maternal depression during pregnancy and the postnatal period: risks and possible mechanisms for offspring depression at age 18 years. *JAMA Psychiatry*. 2013;70:1312–1319.
26. Pawlby S, Hay DF, Sharp D, Waters CS, O'Keane V. Antenatal depression predicts depression in adolescent offspring: prospective longitudinal community-based study. *J Affect Disord*. 2009;113:236–243.
27. Van Batenburg-Eddes T, Brion MJ, Henrichs J, et al. Parental depressive and anxiety symptoms during pregnancy and attention problems in children: a cross-cohort consistency study. *J Child Psychol Psychiatry*. 2013;54(5):591–600.
28. Martini J, Knappe S, Beesdo-Baum K, Lieb R, Wittchen HU. Anxiety disorders before birth and self-perceived distress during pregnancy: associations with maternal depression and obstetric, neonatal and early childhood outcomes. *Early Hum Dev*. 2010;86: 305–310.
29. Korhonen M, Luoma I, Salmelin R, Tamminen T. Maternal depressive symptoms: associations with adolescents' internalizing and externalizing problems and social competence. *Nord J Psychiatry*. 2014;68:323–332.
30. MacKinnon N, Kingsbury M, Mahedy L, Evans J, Colman I. The association between prenatal stress and externalizing symptoms in childhood: evidence from the Avon Longitudinal Study of Parents and Children. *Biol Psychiatry*. 2018;83:100–108.
31. Beversdorf DQ, Manning SE, Hillier A, et al. Timing of prenatal stressors and autism. *J Autism Dev Disord*. 2005;35:471–478.
32. Class QA, Abel KM, Khashan AS, et al. Offspring psychopathology following preconception, prenatal and postnatal maternal bereavement stress. *Psychol Med*. 2014;44: 71–84.
33. Li J, Vestergaard M, Obel C, et al. A nationwide study on the risk of autism after prenatal stress exposure to maternal bereavement. *Pediatrics*. 2009;123:1102–1107.
34. van Os J, Selten JP. Prenatal exposure to maternal stress and subsequent schizophrenia. The May 1940 invasion of The Netherlands. *Br J Psychiatry*. 1998;172:324–326.
35. van Batenburg-Eddes T, de Groot L, Huizink AC, et al. Maternal symptoms of anxiety during pregnancy affect infant neuromotor development: the generation R study. *Dev Neuropsychol*. 2009;34:476–493.
36. Otte RA, Donkers FC, Braeken MA, Van den Bergh BR. Multimodal processing of emotional information in 9-month-old infants II: prenatal exposure to maternal anxiety. *Brain Cogn*. 2015;95:107–117.
37. Cicchetti D, Rogosch FA. Diverse patterns of neuroendocrine activity in maltreated children. *Dev Psychopathol*. 2001;13:677–693.
38. Luby JL, Mrakotsky C, Heffelfinger A, Brown K, Spitznagel E. Characteristics of depressed preschoolers with and without anhedonia: evidence for a melancholic depressive subtype in young children. *Am J Psychiatry*. 2004;161:1998–2004.
39. Pearson J, Tarabulsy GM, Bussières EL. Foetal programming and cortisol secretion in early childhood: a meta-analysis of different programming variables. *Infant Behav Dev*. 2015;40:204–215.
40. Gitau R, Cameron A, Fisk NM, Glover V. Fetal exposure to maternal cortisol. *Lancet*. 1998;352:707–708.

41. Nyirenda MJ, Lindsay RS, Kenyon CJ, Burchell A, Seckl JR. Glucocorticoid exposure in late gestation permanently programs rat hepatic phosphoenolpyruvate carboxykinase and glucocorticoid receptor expression and causes glucose intolerance in adult offspring. *J Clin Invest*. 1998;101:2174–2181.
42. Beijers R, Buitelaar JK, de Weerth C. Mechanisms underlying the effects of prenatal psychosocial stress on child outcomes: beyond the HPA axis. *Eur Child Adolesc Psychiatry*. 2014;23:943–956.
43. Ngai YF, Sulistyoningrum DC, O'Neill R, Innis SM, Weinberg J, Devlin AM. Prenatal alcohol exposure alters methyl metabolism and programs serotonin transporter and glucocorticoid receptor expression in brain. *Am J Physiol Regul Integr Comp Physiol*. 2015;309:R613–R622.
44. Weaver IC, Cervoni N, Champagne FA, et al. Epigenetic programming by maternal behavior. *Nat Neurosci*. 2004;7:847–854.
45. Mueller BR, Bale TL. Sex-specific programming of offspring emotionality after stress early in pregnancy. *J Neurosci*. 2008;28:9055–9065.
46. Palacios-García I, Lara-Vásquez A, Montiel JF, et al. Prenatal stress down-regulates Reelin expression by methylation of its promoter and induces adult behavioral impairments in rats. *PLoS One*. 2015;10:e0117680.
47. Van den Hove DL, Kenis G, Brass A, et al. Vulnerability versus resilience to prenatal stress in male and female rats; implications from gene expression profiles in the hippocampus and frontal cortex. *Eur Neuropsychopharmacol*. 2013;23:1226–1246.
48. Schraut KG, Jakob SB, Weidner MT, et al. Prenatal stress-induced programming of genome-wide promoter DNA methylation in 5-HTT-deficient mice. *Transl Psychiatry*. 2014;4:e473.
49. Mychasiuk R, Schmold N, Ilnytskyy S, Kovalchuk O, Kolb B, Gibb R. Prenatal bystander stress alters brain, behavior, and the epigenome of developing rat offspring. *Dev Neurosci*. 2011;33:159–169.
50. Oberlander TF, Weinberg J, Papsdorf M, Grunau R, Misri S, Devlin AM. Prenatal exposure to maternal depression, neonatal methylation of human glucocorticoid receptor gene (NR3C1) and infant cortisol stress responses. *Epigenetics*. 2008;3:97–106.
51. Hompes T, Izzi B, Gellens E, et al. Investigating the influence of maternal cortisol and emotional state during pregnancy on the DNA methylation status of the glucocorticoid receptor gene (NR3C1) promoter region in cord blood. *J Psychiatr Res*. 2013;47: 880–891.
52. Conradt E, Lester BM, Appleton AA, Armstrong DA, Marsit CJ. The roles of DNA methylation of NR3C1 and 11β-HSD2 and exposure to maternal mood disorder in utero on newborn neurobehavior. *Epigenetics*. 2013;8:1321–1329.
53. Braithwaite EC, Kundakovic M, Ramchandani PG, Murphy SE, Champagne FA. Maternal prenatal depressive symptoms predict infant NR3C1 1F and BDNF IV DNA methylation. *Epigenetics*. 2015;10:408–417.
54. Ostlund BD, Conradt E, Crowell SE, Tyrka AR, Marsit CJ, Lester BM. Prenatal stress, fearfulness, and the epigenome: exploratory analysis of sex differences in DNA methylation of the glucocorticoid receptor gene. *Front Behav Neurosci*. 2016;10:147. https://doi.org/10.3389/fnbeh.2016.00147.
55. Radtke KM, Ruf M, Gunter HM, et al. Transgenerational impact of intimate partner violence on methylation in the promoter of the glucocorticoid receptor. *Transl Psychiatry*. 2011;1:e21.
56. Mulligan CJ, D'Errico NC, Stees J, Hughes DA. Methylation changes at NR3C1 in newborns associate with maternal prenatal stress exposure and newborn birth weight. *Epigenetics*. 2012;7:853–857.
57. Perroud N, Rutembesa E, Paoloni-Giacobino A, et al. The Tutsi genocide and transgenerational transmission of maternal stress: epigenetics and biology of the HPA axis. *World J Biol Psychiatry*. 2014;15:334–345.

58. Palma-Gudiel H, Córdova-Palomera A, Eixarch E, Deuschle M, Fañanás L. Maternal psychosocial stress during pregnancy alters the epigenetic signature of the glucocorticoid receptor gene promoter in their offspring: a meta-analysis. *Epigenetics*. 2015;10:893–902.
59. Davies MN, Volta M, Pidsley R, et al. Functional annotation of the human brain methylome identifies tissue-specific epigenetic variation across brain and blood. *Genome Biol.* 2012;13:R43. https://doi.org/10.1186/gb-2012-13-6-r43.
60. Smith AK, Kilaru V, Klengel T, et al. DNA extracted from saliva for methylation studies of psychiatric traits: evidence tissue specificity and relatedness to brain. *Am J Med Genet B Neuropsychiatr Genet.* 2015;168B:36–44.
61. Binder EB, Salyakina D, Lichtner P, et al. Polymorphisms in FKBP5 are associated with increased recurrence of depressive episodes and rapid response to antidepressant treatment. *Nat Genet.* 2004;36:1319–1325.
62. Monk C, Feng T, Lee S, Krupska I, Champagne FA, Tycko B. Distress during pregnancy: epigenetic regulation of placenta glucocorticoid-related genes and fetal neurobehavior. *Am J Psychiatry*. 2016;173:705–713.
63. Appleton AA, Armstrong DA, Lesseur C, et al. Patterning in placental 11-B hydroxysteroid dehydrogenase methylation according to prenatal socioeconomic adversity. *PLoS One*. 2013;8:e74691. https://doi.org/10.1371/journal.pone.0074691.
64. Dong E, Dzitoyeva SG, Matrisciano F, Tueting P, Grayson DR, Guidotti A. Brain-derived neurotrophic factor epigenetic modifications associated with schizophrenia-like phenotype induced by prenatal stress in mice. *Biol Psychiatry*. 2015;77:589–596.
65. Boersma GJ, Lee RS, Cordner ZA, et al. Prenatal stress decreases Bdnf expression and increases methylation of Bdnf exon IV in rats. *Epigenetics*. 2014;9:437–447.
66. Kertes DA, Bhatt SS, Kamin HS, Hughes DA, Rodney NC, Mulligan CJ. BNDF methylation in mothers and newborns is associated with maternal exposure to war trauma. *Clin Epigenetics*. 2017;9:68. https://doi.org/10.1186/s13148-017-0367-x.
67. Lumey LH, Stein AD, Susser E. Prenatal famine and adult health. *Annu Rev Public Health*. 2011;32:237–362.
68. Heijmans BT, Tobi EW, Stein AD, et al. Persistent epigenetic differences associated with prenatal exposure to famine in humans. *Proc Natl Acad Sci USA*. 2008;105:17046–17049.
69. Vangeel EB, Izzi B, Hompes T, et al. DNA methylation in imprinted genes IGF2 and GNASXL is associated with prenatal maternal stress. *Genes Brain Behav.* 2015;14(8):573–582.
70. Mansell T, Novakovic B, Meyer B, et al. The effects of maternal anxiety during pregnancy on IGF2/H19 methylation in cord blood. *Transl Psychiatry*. 2016;6:e765. https://doi.org/10.1038/tp.2016.32.
71. Liu Y, Murphy SK, Murtha AP, et al. Depression in pregnancy, infant birth weight and DNA methylation of imprint regulatory elements. *Epigenetics*. 2012;7:735–746.
72. DRMcate package: https://bioconductor.org/packages/release/bioc/html/DMRcate.html.
73. Lin X, Barton S, Holbrook JD. How to make DNA methylome wide association studies more powerful. *Epigenomics*. 2016;8:1117–1129.
74. Tobi EW, Goeman JJ, Monajemi R, et al. DNA methylation signatures link prenatal famine exposure to growth and metabolism. *Nat Commun*. 2014;5:5592.
75. Tobi EW, Slieker RC, Stein AD, et al. Early gestation as the critical time-window for changes in the prenatal environment to affect the adult human blood methylome. *Int J Epidemiol.* 2015;44:1211–1223.
76. Cao-Lei L, Massart R, Suderman MJ, et al. DNA methylation signatures triggered by prenatal maternal stress exposure to a natural disaster: project ice storm. *PLoS One*. 2014;9:e107653. https://doi.org/10.1371/journal.pone.0107653.
77. Non AL, Binder AM, Kubzansky LD, Michels KB. Genome-wide DNA methylation in neonates exposed to maternal depression, anxiety, or SSRI medication during pregnancy. *Epigenetics*. 2014;9:964–972.

78. Vangeel EB, Pishva E, Hompes T, et al. Newborn genome-wide DNA methylation in association with pregnancy anxiety reveals a potential role for GABBR1. *Clin Epigenetics*. 2017;9:107.
79. Schroeder JW, Smith AK, Brennan PA, et al. DNA methylation in neonates born to women receiving psychiatric care. *Epigenetics*. 2012;7:409–414.
80. Rijlaarsdam J, Pappa I, Walton E, et al. An epigenome-wide association meta-analysis of prenatal maternal stress in neonates: a model approach for replication. *Epigenetics*. 2016;11:140–149.
81. Glover V. Annual research review: prenatal stress and the origins of psychopathology: an evolutionary perspective. *J Child Psychol Psychiatry*. 2011;52:356–367.
82. Buchwald U, Teupser D, Kuehnel F, et al. Prenatal stress programs lipid metabolism enhancing cardiovascular risk in the female F1, F2, and F3 generation in the primate model common marmoset (Callithrix jacchus). *J Med Primatol*. 2012;41:231–240.
83. Aiken CE, Tarry-Adkins JL, Ozanne SE. Transgenerational effects of maternal diet on metabolic and reproductive ageing. *Mamm Genome*. 2016;27:430–439.
84. Sharp H, Pickles A, Meaney M, Marshall K, Tibu F, Hill J. Frequency of infant stroking reported by mothers moderates the effect of prenatal depression on infant behavioural and physiological outcomes. *PLoS One*. 2012;7:e45446.

CHAPTER SIX

Central Neuroepigenetic Regulation of the Hypothalamic–Pituitary–Adrenal Axis

Alec Dick*, Nadine Provencal[†,‡,1]
*Max Planck Institute of Psychiatry, Munich, Germany
[†]Simon Fraser University, Vancouver, Canada
[‡]BC Children's Hospital Research Institute, Vancouver, Canada
[1]Corresponding author: e-mail address: nadine_provencal@sfu.ca

Contents

Abstract

Dynamic adaptation to stressful life events requires the co-ordinated action of the central stress response, which is mediated by the hypothalamic–pituitary–adrenal (HPA) axis, to restore and maintain homeostasis. Excessive exposure to stress or traumatic life events, such as childhood maltreatment, has been linked to HPA axis dysfunction increasing the risk of developing stress-related psychopathologies such as major depressive disorder and post-traumatic-stress-disorder. Mounting evidence supports the notion that stressors throughout pre- and postnatal development as well as adulthood can induce neuroepigenetic regulation of gene expression within key nodes of the brain, which may in part mediate such HPA axis dysfunction. Neuroepigenetic mechanisms, particularly DNA methylation and small non-coding RNAs, are therefore considered to be molecular mechanisms by which stressful life events may perpetuate aberrant behavioral phenotypes associated with psychiatric disorders throughout one's life and even across generations. In this chapter we outline the progress made toward understanding the effects of stress-induced neuroepigenetic changes upon HPA axis function and highlight the need for novel research strategies to deepen our

Progress in Molecular Biology and Translational Science, Volume 158
ISSN 1877-1173
https://doi.org/10.1016/bs.pmbts.2018.04.006

understanding of the establishment, maintenance and reversibility of neuroepigenetic regulation following stress to enable realization of potential novel therapeutic and preventative strategies for stress-related psychiatric disorders.

1. INTRODUCTION

Throughout life, an organism is required to constantly adapt to ever changing environmental milieu for survival. A key component of this adaptation is the central stress system, which encompasses the co-ordinated action of endocrine, autonomic and behavioral responses to stress. This is mediated by the hypothalamic–pituitary–adrenal (HPA) axis. Upon stress, the neuropeptide corticotrophin releasing factor (CRF), expressed in and secreted from the parvocellular neurons of the paraventricular nucleus (PVN) of the hypothalamus, activates and modulates HPA axis activity. Specifically, CRF stimulates adrenocorticotropic hormone release from the anterior pituitary into the peripheral blood stream that in turn initiates the secretion of glucocorticoids (corticosterone [CORT] in rodents and cortisol in humans) from the adrenal cortex.[1] Glucocorticoids act via the glucocorticoid and mineralocorticoid receptors (GR and MR) in both peripheral tissues and the brain providing negative feedback throughout the HPA axis and also acting as adaptive transcriptional regulators in response to stressful life experiences.[2]

Excessive exposure to stress or traumatic life events, such as childhood maltreatment, has been associated with a wide range of health problems such as increased reactivity to stress, cognitive deficits, psychiatric and behavioral disorders.[3–9] Mounting evidence suggests that exposure to stressful life events also results in dysregulation of the HPA axis, which has been linked to stress-related psychopathologies.[10] Not only does the HPA axis mediate a co-ordinated response to a given stressor, negative feedback within this system also results in the restoration of homeostasis among several physiological processes including the immune system, energy storage, and expenditure, digestion, mood and emotional responsivity to stress.[11] This ensures that a given stress response is limited in duration and precipitated only upon stressful stimuli, which is a core aberrant process of several psychiatric disorders. For instance, major depressive disorder (MDD) patients commonly present with basal hyperactivation of the HPA axis and impaired negative feedback,[12] while PTSD is considered to be characterized by increased sensitivity of the GR, ultimately leading to decreased basal cortisol levels.[13,14] As core features of these disorders it begs the question as to the molecular underpinnings of this dysregulation.

Recent evidence suggests that stressful life events in humans, non-human primates and rodents can dynamically alter epigenetic mechanisms within peripheral tissues and the central nervous system. Epigenesis (or epigenetics) was first described by Conrad Waddington in the mid-20th century as a mechanism functioning "above the genome" by which differentiation of diverse cellular phenotypes could be derived from identical genomic information and be subsequently propagated through cellular replication. However, with time the definition of epigenetics has shifted such that it commonly refers to the complex regulation of chromosomal regions that alters transcription without alteration of the DNA sequence itself. Such epigenetic mechanisms include DNA methylation states, post-translational histone modifications as well as post-transcriptional mechanisms such as non-coding RNA signaling (e.g., microRNA [miRNAs], see Glossary) and RNA modifications (e.g., N^6-methyladenosine). Although epigenetic mechanisms are dynamically regulated throughout development it has only recently become appreciated that epigenetics is a dynamic process within the brain throughout life.[15] The discovery of such diverse epigenetic modifications within the brain, particularly within post-mitotic neurons, has led to the conceptualization of neuroepigenetics, which refers to epigenetic processes mainly within the central nervous system. Considering the post-mitotic state of the majority of neuronal populations within the brain, neuroepigenetic modulation following stressful life events not only enables dynamic cellular responses to environmental stimuli (e.g., stress) but also may maintain molecular changes within key nodes of the brain throughout life and even across generations. Thus, neuroepigenetics may be a molecular mechanism by which stressful life events perpetuate aberrant behavioral phenotypes associated with psychiatric disorder as well as risk factors for psychiatric disorders.

This chapter will focus on neuroepigenetic mechanisms, particularly DNA methylation and miRNAs, underpinning dysfunction of HPA axis following stressful life events. We must note that although we will focus on DNA methylation and miRNAs, neuroepigenetic processes encompass a co-ordinated regulation of chromatin state and post-transcriptional regulation of RNA in response to environmental stimuli. For stress-induced regulation of histone modifications the authors refer the readers to several recent reviews.[16,17]

2. NEUROEPIGENETIC MECHANISMS OF HPA AXIS DYSREGULATION

Exposure to psychosocial stressors throughout life can propagate long-term deficits in HPA axis function, albeit with differential periods of sensitivity within various developmental stages, which involves a complex interaction of the qualitative and quantitative nature of a given stressor, genetic background and the timing of this exposure. How is it that an exposure to a stressor in early life, such as maternal separation (MS) for example,[18] can result in a stably maintained alteration in HPA axis function? Mounting evidence suggests that epigenetic modifications of genes associated with the HPA axis in various organs may mediate stable changes in transcriptional programs that ultimately result in hyper or hypo HPA function or impaired negative feedback of this system.[19,20] Many studies to date have focused on models of prenatal or early postnatal stress, such as maternal stress during pregnancy and deficient maternal care, however it is now evident that dynamic alterations of the epigenome occur within the brain throughout life and such changes can alter HPA axis function even in the adult brain.[17,19] Although we acknowledge the effects of stress upon all components of the HPA axis, we will focus on neuroepigenetic mechanisms occurring within the central nervous system. CRF secretion within the PVN of the hypothalamus acts as an activating and co-ordinating factor of the HPA axis, which in turn is regulated by glucocorticoid-induced modulation of neurons within corticolimbic circuitry sending descending projections to the PVN including the hippocampus (HPC), prefrontal cortex (PFC), and the amygdala (AMG) via the bed nucleus of the stria terminalis (BNST).[21] These brain regions also mediate the autonomic and behavioral responses to stress.[22] In this chapter we will demonstrate that neuroepigenetic mechanisms within this network are critical factors underlying the propagation of HPA axis dysfunction following acute or repeated stressors. To date, most of the empirical evidence for neuroepigenetic modulation of the HPA axis has been conducted in preclinical studies and thus we will focus on this literature with mention of clinical data when available. Preclinical models of physical and psychosocial stress are very diverse in the severity, physicality and duration of a given stressor (see Glossary). Such variability, not unsurprisingly, has led to inconsistent findings with regards to neuroepigenetic changes following stress. Despite this, these studies evidently highlight the importance of epigenetic-mediated alteration of HPA axis genes and ultimately function and behavior and will be outlined in this chapter.

2.1 CRF Signaling

Signaling via the neuropeptide CRF, expressed and secreted from the parvocellular neurons of the PVN, mediates the activation and modulation of the HPA axis and is considered the final common pathway for the integration of the neuroendocrine stress response within the brain.[23] As such, neuroepigenetic regulation genes encoding CRF and CRF receptors has been studied in various models of early and adult stress exposure associated with HPA axis dysfunction, which are outlined in the following section.

2.1.1 Prenatal Stress

Prenatal and early life stressors have both been demonstrated to alter CRF signaling within the brain where epigenetic factors may underline the observed alterations in HPA axis function. For example, chronic variable mild stress (CVMS, see Glossary) of pregnant female mice early in gestation results in maladaptive responsivity to stress associated with hyperactivation of the HPA axis in the offspring.[24] The authors also demonstrated a stress-induced decrease in methylation of two CG dinucleotides or CpGs (see Glossary) within the *Crf* promoter, one within a glucocorticoid responsive element (GRE), which was associated with increased *Crf* mRNA expression within the central nucleus of the amygdala (CeA) in male offspring only. Interestingly, *Crf* overexpression (OE) within the CeA has been shown to increase *Crf* mRNA within the PVN and result in stress-induced hyperactivation of the HPA axis, a likely mechanism for HPA axis dysfunction observed in Mueller and Bale,[24] although this was not confirmed.[25] Mueller and Bale also demonstrated altered DNA methylation in the *Nr3c1* promoter, encoding GR, within the HPC emphasizing the multifaceted nature of epigenetic-mediated regulation of the HPA axis acting on multiple genes to co-ordinate the adaptive response.

2.1.2 Postnatal Stress

In contrast to prenatal stress, Chen et al.[26] demonstrated that maternal deprivation (MD, see Glossary), a model of early postnatal stress, from postnatal day (PND) 2–13 in both male and female Sprague-Dawley rats resulted in altered HPA axis function in adulthood, which was associated with differential transcriptional activation of *Crf* within the PVN but not CeA in response to acute restraint stress (see Glossary). Within the PVN, MD-induced demethylation of two CpG sites within the *Crf* promoter associated with a cyclic AMP-response element which was suggested to mediate the increase in transcriptional activity of *Crf* mRNA observed following MD.[26] Such epigenetic priming of HPA-associated transcripts without alteration of basal levels may significantly alter stress-induced transcription and ultimately stress responsivity throughout life. Importantly, epigenetic priming also emphasizes the need for subsequent acute stressful challenges in chronic stress experiments to investigate latent neuroepigenetic regulation of transcriptional activity not immediately evident at baseline following chronic stress.

2.1.3 Adult Stress

Over the past 10–15 years, mounting evidence has definitively identified that the neuroepigenome is both stably and dynamically regulated in response to neuronal activity and environmental stimuli (e.g., stress) not only during critical developmental periods but also in adulthood. A leading study by Elliott et al. demonstrated that chronic social defeat stress (CSDS, see Glossary) increases *Crf* mRNA within the PVN, which was accompanied by a decrease in methylation of several CpG sites within the *Crf* promoter specifically in mice susceptible to CSDS compared to resilient and control mice. Interestingly, this study demonstrated similar decreases in DNA methylation of one CpG also altered by prenatal stress in mice,[24] and one CpG within a CRE binding site orthologous to a CpG altered in rats following MD,[26] indicative of a degree of commonality for decreased *Crf* promoter methylation within the PVN and CeA across stress modalities in rodents. These changes were alleviated by chronic imipramine treatment, a tricyclic antidepressant, throughout the CSDS paradigm normalizing *Crf* mRNA and *Crf* promoter methylation.[27] CSDS also decreases *Dnmt3a* mRNA levels, which encodes the DNA methyltransferase DNMT3A (see Glossary), and decreases global DNA methylation within the mouse PFC.[28] Viral-mediated *Dnmt3a* OE in this region also rescues the CSDS-induced behavioral phenotypes, although rescue of global DNA methylation levels were not assessed following *Dnmt3a* OE.

CVMS in adult rats has also been demonstrated to induce brain region and sex-specific effects upon *Crf* promoter methylation and *Crf* mRNA expression in which only female rats demonstrated altered *Crf* promoter methylation within the PVN. However, both males and females displayed altered *Crf* promoter methylation within the BNST and CeA with increased and decreased DNA methylation observed at specific CpGs emphasizing the importance of single base resolution analysis of stimuli-induced changes in DNA methylation.[29] Although not the focus of this chapter, it should be noted that known sex-specific differences in HPA axis activity may mediate altered neuroepigenetic responses to stress within the brain as observed in

Sterrenburg et al.,[29] which requires further investigation. These studies indicate that prenatal, early postnatal and adult stress alters DNA methylation within the *Crf* promoter region within the PVN, CeA, and BNST with minimal overlap of regulated CpGs reported. Despite such discrepancies, stress throughout different developmental stages primarily leads to demethylation of the *Crf* promoter and is correlated with increased *Crf* mRNA and subsequent increases of HPA axis activity, although this may be sex and region-dependent and requires further investigation. Thus, further mechanistic understanding of the effects of stress-induced methylation changes at single and/or multiple CpGs throughout the *Crf* locus upon *Crf* transcription are required. Future studies should focus on single nucleotide resolution analysis throughout an expanded region of the *Crf* locus beyond the promoter region, which for example was recently demonstrated for the *fkbp5* locus (~10 kb) with PCR-based locus amplification followed by bisulfite sequencing.[30]

CRF elicits its effects via activation of the CRF receptor type 1 (CRFR1) and CRF receptor type 2 (CRFR2) of which CRFR1 is more widely expressed throughout the mammalian brain and is critical for HPA axis activation.[21,31–33] To date, miRNA-mediated regulation of the *Crfr1* transcript and CRFR1 protein levels has been demonstrated within the rodent brain adding to the mounting evidence of the importance of non-coding RNAs in the central stress response and regulation of the HPA axis. Following initial observations that ablation of Dicer, a key component of miRNA biogenesis (see Glossary), from the CeA of mice resulted in an anxiogenic phenotype, Haramati et al.[34] demonstrated that both acute and chronic stress alters miRNA expression within the amygdala. In particular, miR-34c was upregulated within the amygdala 90 min following acute restraint and 10 days following CSDS in this region and was shown to target the 3′UTR of the *Crfr1* transcript in vitro indicative of endogenous repression of *Crfr1* mRNA (see Glossary).[34] Utilizing the neuroblastoma N2A cell line endogenously expressing *Crfr1*, the authors demonstrated that overexpression of miR-34c as well as miR-34a attenuates the response of these cells to CRF, which was in line with the anxiolytic phenotype observed following viral-mediated OE of mir34c within the CeA. Recent evidence further suggest a role of miR-34 family in regulation of CRF signaling as miR-34b targets the 3′UTR of *Crfr1* mRNA in rat primary hypothalamic neurons.[35]

Collectively these studies demonstrate that CRF signaling within the PVN and other modulatory regions are regulated via stress-induced epigenetic modifications following various stressors in the early life as well as in adulthood. Such stress-induced epigenetic changes appear to perpetuate stable increases in CRF signaling altering HPA axis responsivity to stress. However, miRNA-mediated regulation of CRF receptors also appears to buffer the effects of stress upon the brain with the miR-34 family regulating CRFR1 levels within the amygdala and hypothalamus. Further work will be required to investigate the dynamic and complex interaction between these epigenetic changes, particularly within specific neuronal populations. For example, it was recently shown that PVN CRFR1 positive neurons are a distinct neuronal population within the mouse PVN that are recruited specifically following chronic stress and are important for regulating HPA axis activity and anxiety-like behavior.[36] One may posit that these neurons undergo stable epigenetic changes in response to chronic stress such that they are activated/recruited upon exposure to subsequent stressors contributing to altered HPA axis activity and behavioral deficits following chronic stress.

2.2 Glucocorticoid Signaling via the GR/MR

2.2.1 Postnatal Stress

Regulation of the central stress response requires orchestrated negative feedback via glucocorticoid-induced activation of both GR and MR in corticolimbic circuitry. As such, stable alteration of GR and MR expression and function has been implicated in mediating the long-term consequences of stress, particularly in early life. Initial neuroepigenetic studies focused on a model of low level maternal care in rats (licking and grooming model, see Glossary and Buschdorf and Meaney[37]), which resulted in stable decreases in *Nr3c1* transcription (gene coding for GR) within the HPC induced by increased methylation of a single CpG within the promoter region of this gene.[18,38] Throughout a series of studies, it was demonstrated that licking and grooming from the mother (but not tactile stimulation alone) stimulates a serotonergic response activating nerve growth factor-inducible protein A (NGFI-A) that binds specifically to a glucocorticoid response element within the *Nr3c1* exon 1_7 promoter. (The highly conserved *Nr3c1* gene has multiple first exons, each with its own promoter.[39]) This induces demethylation of a single 5′ CpG within this NGFI-A response element in the adult HPC associated with decreased *Nr3c1* mRNA expression within this region. The interaction of multiple epigenetic components was emphasized in these studies as this demethylation required the interaction of the histone acetyltransferase CREB binding protein in complex with NGFI-A. The binding of this complex to the exon 1_7 promoter region increases H3K9 acetylation inducing an active chromatin state and enables demethylation. This active demethylation process is absent within the HPC of pups exposed to low level maternal care (i.e., low licking and grooming), which results in decreased *Nr3c1* transcription and increased HPA axis responsivity.[18,38,40] In one of the few studies examining neuroepigenetic modifications in human post-mortem brain following stressful life experiences, Mcgowan et al.[41] suggested that similar mechanisms may be involved in the increased methylation levels of the orthologous human promoter of this loci (GR 1F) within the HPC of suicide victims exposed to child abuse.[41] A recent study also identified genome-wide alterations of intragenic DNA hydroxymethylation (see Glossary) within the PFC of suicide victims compared to psychiatrically healthy controls, although the aforementioned HPA axis associated genes were not identified in this study.[42]

2.2.2 Adult Stress

The recent re-discovery of DNA hydroxymethylation (5-hydroxymethylcytosine or 5hmC)[43–45] has prompted intense research into its function as an independent and functional epigenetic modification, particularly within the brain. This is due to the abundance of 5hmC within post-mitotic neurons, its dynamic regulation during development of the central nervous system, and its high inter-individual variability, which renders it a particularly interesting candidate for involvement in neurological and psychiatric disorders.[46] DNA hydroxymethylation is also dynamically regulated within the brain in response to neuronal activity and various environmental stimuli.[46–49] For example, acute restraint stress in late adolescent mice results in global increases in 5hmC levels within the HPC and a locus-specific increase of 5hmC at a single CpG within the 3′UTR of the *Nr3c1* gene, although the authors did not analyze *Nr3c1* mRNA or GR protein expression in this study to enable correlation of stress-induced 5hmC and GR expression in this context.[50] Using a different 5hmC profiling technique, follow up studies from the same lab failed to replicate altered 5hmC within the *Nr3c1* gene in the HPC of male mice despite identification of loci with differential stress-induced 5hmC levels, particularity within intragenic regions. Here, only a small subset of these loci were associated with differentially expressed genes known to be linked to stress-related functions.[51,52] Employing a similar chemical-based labeling approach for 5hmC profiling, Papale et al.[53] suggested that semi-chronic variable mild stress in the pre-pubertal stage (PND 12–18), but not prenatal stress, induces an anxiogenic behavioral phenotype in females later in life associated with differential 5hmC profiles within the hypothalamus. Comparison with hypothalamic RNA-seq data identified 118 differentially expressed genes associated with stress-induced 5hmC changes including the *Nr3c2* gene encoding MR. The functional consequences of this change upon HPA axis function were not however assessed in this recent study,[53] emphasizing the need for further investigation of stress-induced 5hmC alterations on HPA axis function. It must be noted that traditional bisulfite sequencing techniques, the most utilized technique for DNA methylation analysis, do not distinguish between 5mC and 5hmC such that future studies should employ targeted or genome-wide oxidative bisulfite (oxBS) sequencing[54] and/or tet-assisted bisulfite (TAB) sequencing[55] to analyze 5mC and 5hmC at single base resolution, respectively. Stimuli-induced changes of 5mC and 5hmC are also increasingly appreciated to occur predominantly at distal regulatory elements (e.g., enhancers) and within gene bodies emphasizing the need for broader analysis DNA modifications far beyond promoter regions. Although still cost prohibitive at a genome-wide level, in combination with ever expanding genomic capture techniques, oxBS and TAB-seq will prove to be invaluable methods for such analysis of stress-induced changes to the DNA methylome.

Several studies have also identified stress-induced miRNA expression as a potential mechanism of regulating GR and MR expression within the adult rodent brain. For example, repeated restraint stress increases miR-18a within the adult rat PVN and decreases GR protein but not mRNA expression within this region, which is likely mediated via direct binding of miR-18a to the 3′UTR of the *Nr3c1* gene.[56] In vitro studies using an immortalized neuronal-like cell line also identified mir-29b and mir-181a as direct repressors of *Nr3c1* mRNA and GR protein expression.[57] A link between stress-induced regulation of miR-124 and mir-135a has also been suggested for MR as mir-124 and miR-135a repress MR expression in vitro, and are negatively correlated with increased MR expression within the adult rat AMG following acute restraint stress.[58,59]

2.3 Indirect Regulation of GR Signaling

Indirect regulation of GR signaling is also subject to miRNA-mediated regulation within the adult mouse HPC as repeated restraint stress decreased expression levels of the miR-17-92 cluster (miR-17, miR-18, miR-19a, and miR-92a) and increased *Sgk1* mRNA levels, a downstream effector of GR signaling, which was validated as a target of miR-19a and miR-92a from this cluster.[60] Employing inducible KO and OE mouse lines of this miR-17-92 cluster within neuronal progenitors of the mouse HPC the authors demonstrated the significance of these miRNAs in regulating anxiety and depression-like behaviors and hippocampal neurogenesis under basal and stress conditions. The authors suggest that this miRNA cluster is critical for maintaining low levels of *Sgk1* mRNA and thus normal levels of GR signaling in the face of repeated stressful challenges buffering the effects of stress upon hippocampal function.[60]

Further evidence for indirect epigenetic regulation of GR signaling has emerged from an expanding area of research focusing on GR interacting proteins, such as the FK506 binding protein 5 (FKBP5). FKBP5 is of particular interest as it is known to regulate sensitivity of GR and its increased levels are associated with GR resistance and prolonged stress response via impairing GR negative feedback.[61] FKBP5 also provides a rapid feedback loop for GR sensitivity, as its expression is induced upon stress exposure through the binding of the GR to GREs within the regulatory regions of the *Fkbp5* locus.[62] This regulation appears to be under neuroepigenetic control as GR has been shown to induce stable demethylation in and around GREs.[63] In line with this, chronic CORT treatment in adult mice induces an upregulation of *Fkbp5* mRNA in the HPC and hypothalamus, as well as stable demethylation at intronic GREs of the *Fkbp5* locus within these brain regions even after 4 weeks recovery.[64] Epigenetic modulation of the *Fkbp5* locus is also evident following prenatal and early life stress. For example, exposure of pregnant mouse dams to predator odor induces a sex-specific upregulation of *Fkbp5* mRNA within the amygdala of adult female offspring with stable demethylation of intronic GREs of *Fkbp5*, which may be linked to HPA axis dysfunction and anxiety-like behaviors observed in this study.[65] In humans, active demethylation of intronic GREs of the FKBP5 gene has also been suggested as a mechanism for the persistent effects of child abuse on this gene locus as observed in peripheral blood cells of child abuse victims and replicated with GR activation in a human hippocampal progenitor cell line.[66]

FKBP5 also suppresses the enzymatic activity of the maintenance DNA methyltransferase DNMT1 in human and rodent cells in vitro via impaired heterocomplex formation with CDK5 ultimately reducing global DNA methylation levels, an effect that was stimulated by the antidepressant paroxetine (a selective serotonin reuptake inhibitor) in this study.[67] FKBP5 also appears to modulate miRNA function as both FKBP51 and FKBP52 (an opposing GR-chaperone to FKBP51) interact with the miRNA-associated protein Argonaute 2 (AGO2, see Glossary) in mouse embryonic stem cells and modulate both AGO2 protein levels and mature miRNA abundance.[68] Feedback of this FKBP5-mediated regulation of miRNA function has been demonstrated in the adult rodent brain as CSDS increases RISC associated miR-15a within the amygdala with a concomitant decrease in *Fkbp5* mRNA and FKBP5 protein levels likely mediated by direct targeting of *Fkbp5* mRNA by miR-15a.[69] The authors also demonstrated that viral-mediated knock-down (KD) of miR-15a within the BLA increased FKBP5 protein levels within this region and induced an anxiogenic phenotype following CSDS in line with observations from previous studies overexpressing FKBP5 within the mouse BLA.[70] Acute dexamethasone treatment in adult humans was also demonstrated to increase miR-15a within peripheral blood cells of healthy controls and was increased within peripheral blood of childhood trauma patients indicative of its potential as a biomarker for stress-induced pathologies.[69]

Collectively, these studies emphasize that DNA methylation (both 5mC and 5hmC) and miRNAs regulate the activation and modulation of the HPA axis through a complex feedforward and feedback network within the hypothalamus and projecting corticolimbic networks via modulating genes and/or proteins of the CRF and GR/MR signaling pathways.

3. OTHER NEUROMODULATORY SYSTEMS

The central stress response and HPA axis activity are also subject to modulation by other stress-related neurotransmitter systems including dopamine, norepinephrine, serotonin and other hypothalamic neuropeptides such as vasopressin (AVP), which are subject to neuroepigenetic modulation. In early life for example, MD in mice from PND 1 to 10 increases *Avp* mRNA and decreases DNA methylation in an *Avp* gene enhancer within the PVN, which is stable from 6 weeks to 1 year of age.[71] Post-translational modification of MeCP2 immediately following MD exposure at PND 10 was shown to release this protein from the *Avp* enhancer presumably enabling altered chromatin state and active demethylation at this locus by 6 weeks of age likely mediated by the DNA oxidizing enzyme TET1.[72]

Dysfunction of serotonergic signaling has long been implicated in stress-related pathologies such as MDD, and has been suggested to increase one's vulnerability to chronic stress.[73] In adult mice, chronic stress results in stable upregulation of *htr1a* mRNA encoding the 5-HT1A receptor within the PFC and dorsal raphe (DR), which was associated with increased DNA methylation at a single CpG within an Sp1-like transcription binding site located in the promoter of the *htr1a* gene.[74] The authors suggested these effects would likely reduce serotonergic signaling within these brain regions although both serotonergic signaling and HPA axis function were not addressed in this study. Issler et al.[75] comprehensively investigated the regulation of serotonergic signaling via miR-135a, which is an abundant miRNA in serotonergic neurons within the DR of the mouse. The authors demonstrated that miR-135a directly targets both the *Slc6a4* (encoding the serotonin transporter) and *htr1a* mRNAs and OE of miR-135a within the DR increased serotonergic signaling within the adult brain and induced a pro-resiliency phenotype to CSDS in mice.[75] Interestingly, miR-135a levels were also decreased within the blood and brain of MDD patients contributing to the growing interest of peripheral miRNA levels as biomarkers for diagnostics and treatment prediction in psychiatric patients.[76,77]

Stress-related neurotrophic factors, such as brain-derived neurotrophic factor (BDNF), are also implicated in the dynamic regulation of neuronal function in response to environmental stimuli, which has led to an appreciation of the role neurotrophic factors have in stress-related psychopathologies.[78] BDNF is also implicated in the regulation of HPA axis function via the stimulation of CRF expression within the PVN[79] and prenatal and early life stress has been shown to alter the expression of *Bdnf* mRNA and decrease methylation of the *Bdnf* locus within the rat PFC.[80] Such neuroepigenetic regulation of BDNF may affect HPA axis function and/or neuronal development although this requires further investigation.

4. TRANSGENERATIONAL NEUROEPIGENETIC EFFECTS ON HPA AXIS FUNCTION

We have outlined the neuroepigenetic embedding of stress exposure throughout life that clearly alters HPA axis function and ultimately the central stress response. However, it is now considered that parental life experience may also perpetuate behavioral phenotypes and HPA axis dysfunction transgenerationally, which is thoroughly reviewed in chapter "Transgenerational epigenetics of stress" by Mansuy of this book. Briefly, initial evidence in humans suggested that exposure to trauma increases the rate of depression and behavioral deficits in the offspring of parents and grandparents exposed to traumatic life experiences.[81–83] In rodents, paternal stress[84–86] as well as maternal stress in combination with postnatal stress[87] has been demonstrated to perpetuate changes in HPA axis function as well as miRNA profiles within the brain of offspring for up to two generations, which appears to be sex and brain region-specific. Mechanistically, chronic CORT,[86] CVMS[84] and maternal stress followed by postnatal stress[87] alter miRNA profiles within the sperm of male mice, which to date by an unknown mechanism perpetuates altered miRNA profiles within the brain of offspring as well as changes in HPA axis function and behavior. Microinjection of candidate miRNAs into fertilized zygotes appears to recapitulate, to some degree, these changes following paternal stress indicative of the programming-like action of miRNAs within this context.[85,87] It is interesting to note that the differential sperm miRNAs identified in these studies share minimal overlap between all three studies suggesting that different modalities of paternal and/or maternal stress alter specific subsets of miRNAs rather than a generalized regulation of a core set of stress-sensitive miRNAs within sperm. A recent study also demonstrated that behavioral deficits and altered miRNA profiles within the HPC are present for up to three generations following maternal stress in rats, both of which are partly rescued by environmental enrichment in the offspring.[88] This study highlights that paternal and maternal transgenerational programming of stress phenotypes and HPA axis function may be dynamically regulated by environmental factors throughout the life of the offspring such that stress-induced epigenetic effects may be reversible by specific environmental stimuli, a hypothesis warranting much future investigation.

5. FUTURE DIRECTIONS

The described studies, while offering exciting insight into how stress modulates HPA axis function via neuroepigenetic mechanisms, also highlight current challenges to this field of research. The different mechanisms of stress-induced neuroepigenetic modifications highlight the fact that many such changes will likely occur during specific developmental stages at specific genomic loci (e.g., distal regulatory elements or intragenic regions), in discrete neuronal and/or non-neuronal cell types within specific neural circuits. Most studies described in this chapter focus on epigenome analysis within the promoter region of genes linked to the HPA axis, which evidently is not sufficient to investigate the majority of stress-induced epigenome changes throughout the genome requiring expanded targeting and genome-wide approaches. Moreover, all studies described employ bulk tissue dissection for molecular analysis such that the neuroepigenetic modifications discussed in this chapter are analyzed from heterogenous cell populations within a given brain region minimizing the possibility of identifying specific neuroepigenetic changes within discrete cell populations. As such, the advent and utilization of single-cell omics techniques for neuroepigenetic research opens a multitude of exciting new avenues for cell-type specific analysis of the transcriptome[89] and epigenome.[90,91] Moreover, in addition to fluorescence activated cell sorting which has been successfully used for isolating neuronal and non-neuronal cells within the brain, sophisticated transgenic mouse lines also enable the isolation of specific cell populations for cell-type specific epigenome analysis.[92] Thus, future studies will refine the focus of stress-induced epigenetic modifications within the plethora of neuronal and non-neuronal cell types within the brain to further our understanding of the molecular mechanisms regulating basal and pathological function of the HPA axis.

In this chapter we focused on neuroepigenetic modulation of the stress response but such exposure especially early in life, does not only have long-term effects on brain function but also increases risk for metabolic disorders and has long lasting effects on the immune system.[93] Indeed, in rhesus macaques, the broad impact of maternal rearing in the first year of life on DNA methylation was seen in both the brain and peripheral T cells, where GR showed epigenetic remodeling in both tissues.[94] A handful of papers have now shown changes in DNA methylation in peripheral blood cells with early trauma in humans in genes involved in HPA axis function as well as the immune response.[95–97] Of particular interest, Suderman et al. specifically reported differential methylation in promoter regions of loci encoding miRNAs in blood cells in line with preclinical studies described above showing the importance of miRNAs in modulating the stress response.[97] Although it is impossible to directly translate the preclinical findings of the neuroepigenetic effects of stress to humans, these studies highlight the fact that epigenetic regulation of HPA axis associated genes also occurs within peripheral tissue such that peripheral analysis of epigenetic modifications following stress may be informative to better understand the biological embedding of stress and impact on psychopathologies. Moreover, identification of peripheral epigenetic biomarkers for psychopathologies and/or treatment response hold immense promise for the future.[76]

A further caveat is that human studies on stress-induced epigenetic regulation to date are mostly cross sectional. Future longitudinal studies are required to establish the sequence of events leading to long lasting epigenetic changes and subsequent health outcomes. For example, the effects of positive environmental exposure are under-investigated as is the possible reversibility of stress-induced epigenetic changes. One promising example of the benefits of positive environmental exposure has shown that augmented maternal care in rats (PND 2–9) mediates dynamic epigenetic regulation of the *Crh* gene within the PVN, which appears to contribute to altered CRF signaling and HPA axis activity in adulthood and may promote a resilient phenotype in this context.[98] However, to establish true causality of these epigenetic changes we need animal models that allow specific epigenetic modification or the prevention of the establishment of such changes following stress (i.e., epigenome editing), which requires much future attention.

Epigenome editing has recently been realized both in vitro and within the rodent brain in vivo with the development of mutagenesis tools such the clustered regularly interspaced short palindromic repeats (CRISPR)-Cas9 and transcription activator-like effector nucleases (TALENs), of which modified forms can be tethered to epigenome modulating enzymes such TETs, DNMTs and transcription factors.[99–101] Such techniques provide novel opportunities to directly test causality of stress-induced neuroepigenetic changes at specific genomic loci upon HPA axis function and ultimately behavior in animal models, with some promising initial findings to date.[99,102]

6. CONCLUSIONS

Throughout this chapter, we have emphasized the association between stress-induced neuroepigenetic modifications and HPA axis function, predominantly in preclinical models. Many unanswered questions remain as to the causal links of such neuroepigenetic modifications and HPA axis function under basal and pathological conditions as well as the reversibility of such changes following exposure to stressful stimuli throughout life and across generations. Future studies will need to employ the power of novel techniques, such as those outlined in this chapter, to gain a deeper understanding of the factors that are involved in the establishment, maintenance and reversibility of such orchestrated and specific epigenetic changes, which will be an important next step in the field, and may open novel therapeutic and preventative strategies for stress-related psychiatric disorders.

GLOSSARY

Neuroepigenetic mechanisms: Focus on DNA modifications and miRNAs

DNA methylation (5mC) Covalent modification of cytosine residues where a methyl group is located predominantly at the 5th carbon of cytosine residues. 5mC is predominantly located at CG dinucleotides (or CpG sites) in mammalian cells. However, throughout development DNA methylation in the CH context (H=A, C or T) increases within neurons reaching levels of ~50% of neuronal methylcytosines in the adult brain.[103] Covalent addition of a methyl group to cytosine residues is catalyzed by the DNA methyltransferases (DNMTs) of which DNMT1 maintains the methylation pattern during cell division by copying the parent strand whereas DNMT3a and 3b are responsible for de novo DNA methylation. Although catalytically inactive, DNMT3l has been shown to increase efficiency of de novo methylation by DNMT3a and DNMT3b. DNA methylation can alter gene expression by recruiting or blocking the binding of transcriptional regulatory elements such as MeCP2.

DNA hydroxymethylation (5hmC) A distinct chemical modification of cytosine residues catalyzed by the ten-eleven translocation (TET) proteins. 5hmC occurs primarily in the CG context and is most abundant within neurons of the adult brain. TET1–3 catalyze the oxidation of 5mC to 5hmC, which can be sequentially catalyzed to 5-formylcytosine, then 5-carboxylcytosine and ultimately a non-methylated cytosine via TET proteins and DNA excision/repair-based mechanisms. Although initially proposed as an intermediate step in active demethylation, mounting evidence supports a unique functional role of stable and dynamic 5hmC on gene regulation throughout development and in adulthood in response to environmental stimuli, particularly within the brain.[46]

microRNAs (miRNAs) Predominantly, miRNA genes are transcribed by RNA polymerase II and III into primary-miRNA transcripts (pri-miRNA) from dedicated genes or within introns of other genes as individual miRNA or miRNA clusters.[104] Pri-miRNAs are cleaved by the microprocessor complex containing Drosha and DiGeorge syndrome critical region 8 (DGCR8) within the nucleus into a precursor miRNA (pre-miRNA).[105] The pre-miRNA is then transported to the cytoplasm via exportin 5 in complex with RAS-related nuclear protein (RAN)-GTP.[106] Cytoplasmic processing then involves pre-miRNA cleavage by Dicer producing a ~22 nt miRNA duplex of which one functional strand (5′ or 3′) is then loaded in complex with AGO2 into the RNA-induced silencing complex (RISC).[107] This functional strand then guides the RISC complex to the 3′UTR of the target mRNAs leading to translational repression or mRNA decay via complementary binding of the miRNA seed sequence (6–8 nucleotides at 5′ end of mature miRNA) to a seed match sequence within the 3′UTR of target genes.[108]

Common animal models of physical and psychosocial stress

Early postnatal separation Maternal separation (MS) or maternal deprivation (MD) models are conducted in rats and mice affecting critical mother–offspring contact during early life inducing aberrant behavioral phenotypes, disturbance to neuroendocrine parameters and molecular changes (e.g., epigenetics) within the brain of developing and adult rodents. Research groups employ a variety of variables for both MS and MD models often leading to discrepant behavioral and molecular findings, although many models appear to robustly affect behavioral and neuroendocrine parameters.[109] Generally, the MS and MD models employ intermittent isolation of rodent pups from their dams for varied time periods (1–8 h) often in a novel environment, which is repeated daily from PND 2 for 3–19 days. However, a single 24 h separation has also been employed at different time points between PND 2 and 14. Note that nomenclature of MS and MD models is inconsistently applied in neuroscientific fields and are often used interchangeably.

Low maternal care A model in outbred Long-Evans rats that takes advantage of significant variation in the frequency of pup licking and grooming between lactating females.

Such individual differences in licking and grooming results in a distribution of maternal care (i.e., high vs low levels of licking and grooming) in early postnatal period following which low levels of licking and grooming is associated with behavioral and HPA axis dysfunction in adult offspring.[37]

Restraint stress One of the most commonly used physical stress procedures in rodents eliciting robust HPA axis activation. Generally, rodents are placed in a cylindrical tube with ventilation holes restricting the animal's movement for varied periods of time (5–180 min). Repeated restraint stress over days to weeks has also been demonstrated to induce HPA axis dysregulation and molecular changes within the brain.

Social defeat stress A model based on ethological behaviors of aggression and social dominance in male rodents. Generally, an experimental animal (known as the intruder) is placed into the home cage of an aggressive resident for a variable period of time (>10 min) whereby the resident will attack the intruder eliciting an ethological defeat posture in the intruder animal (i.e., submissive rearing on hind legs with forepaws against body facing the aggressor). Social defeat stress can be conducted acutely or chronically. Chronic social defeat stress (CSDS) is a well characterized animal model of depression-like behavior with significant etiological, predictive, discriminative and face validity and is one of the most commonly used models to induce depression-like behavior in mice.[110]

Chronic variable mild stress Chronic variable mild stress (CVMS) is a well validated model for chronic stress and depression-like behavior in rodents with good face and predictive validity in which animals (usually adolescent or adult) are exposed to different stressor modalities throughout the exposure period.[111] Generally, stressors are presented multiple times daily for 2–8 weeks and include stressors such as wet bedding material, tilted cage, noxious noise and light/dark disturbances, which induce depression-like behavioral phenotypes and alter neuroendocrine function. Randomization of stressors throughout the exposure period is often referred to as unpredictable chronic unpredictable mild stress (UCMS).

REFERENCES

1. Dallman MF, Akana SF, Bhatnagar S, Bell ME, Strack AM. Bottomed out: metabolic significance of the circadian trough in glucocorticoid concentrations. *Int J Obes Relat Metab Disord*. 2000;24(suppl 2):S40–S46.
2. De Kloet ER, Vreugdenhil E, Oitzl MS, Joels M. Brain corticosteroid receptor balance in health and disease. *Endocr Rev*. 1998;19:269–301.
3. Heim C, Nemeroff CB. The role of childhood trauma in the neurobiology of mood and anxiety disorders: preclinical and clinical studies. *Biol Psychiatry*. 2001;49: 1023–1039.
4. Kendler KS, Karkowski LM, Prescott CA. Causal relationship between stressful life events and the onset of major depression. *Am J Psychiatry*. 1999;156:837–841.
5. Kessler RC, Davis CG, Kendler KS. Childhood adversity and adult psychiatric disorder in the US National Comorbidity Survey. *Psychol Med*. 1997;27:1101–1119.
6. Loman MM, Gunnar MR, Stress Early Experience, and Center Neurobehavioral Development. Early experience and the development of stress reactivity and regulation in children. *Neurosci Biobehav Rev*. 2010;34:867–876.
7. Lu W, Mueser KT, Rosenberg SD, Jankowski MK. Correlates of adverse childhood experiences among adults with severe mood disorders. *Psychiatr Serv*. 2008;59: 1018–1026.

8. Matheson SL, Shepherd AM, Pinchbeck RM, Laurens KR, Carr VJ. Childhood adversity in schizophrenia: a systematic meta-analysis. *Psychol Med.* 2013;43:225–238.
9. O'Connor TG, Ben-Shlomo Y, Heron J, Golding J, Adams D, Glover V. Prenatal anxiety predicts individual differences in cortisol in pre-adolescent children. *Biol Psychiatry.* 2005;58:211–217.
10. Teicher MH, Andersen SL, Polcari A, Anderson CM, Navalta CP. Developmental neurobiology of childhood stress and trauma. *Psychiatr Clin North Am.* 2002;25: 397–426:vii–viii.
11. Charmandari E, Tsigos C, Chrousos G. Endocrinology of the stress response. *Annu Rev Physiol.* 2005;67:259–284.
12. Dean J, Keshavan M. The neurobiology of depression: an integrated view. *Asian J Psychiatr.* 2017;27:101–111.
13. Mehta D, Binder EB. Gene x environment vulnerability factors for PTSD: the HPA-axis. *Neuropharmacology.* 2012;62:654–662.
14. Yehuda R. Status of glucocorticoid alterations in post-traumatic stress disorder. *Ann NY Acad Sci.* 2009;1179:56–69.
15. Guo JU, Ma DK, Mo H, et al. Neuronal activity modifies the DNA methylation landscape in the adult brain. *Nat Neurosci.* 2011;14:1345–1351.
16. Deussing JM, Jakovcevski M. Histone modifications in major depressive disorder and related rodent models. *Adv Exp Med Biol.* 2017;978:169–183.
17. Dirven BCJ, Homberg JR, Kozicz T, Henckens MJAG. Epigenetic programming of the neuroendocrine stress response by adult life stress. *J Mol Endocrinol.* 2017;59: R11–R31.
18. Weaver IC, Cervoni N, Champagne FA, et al. Epigenetic programming by maternal behavior. *Nat Neurosci.* 2004;7:847–854.
19. Provencal N, Binder EB. The effects of early life stress on the epigenome: from the womb to adulthood and even before. *Exp Neurol.* 2015;268:10–20.
20. Zhang TY, Meaney MJ. Epigenetics and the environmental regulation of the genome and its function. *Annu Rev Psychol.* 2010;61:439–466:C1–3.
21. Henckens MJ, Deussing JM, Chen A. Region-specific roles of the corticotropin-releasing factor-urocortin system in stress. *Nat Rev Neurosci.* 2016;17:636–651.
22. Ulrich-Lai YM, Herman JP. Neural regulation of endocrine and autonomic stress responses. *Nat Rev Neurosci.* 2009;10:397–409.
23. Vale W, Spiess J, Rivier C, Rivier J. Characterization of a 41-residue ovine hypothalamic peptide that stimulates secretion of corticotropin and beta-endorphin. *Science.* 1981;213:1394–1397.
24. Mueller BR, Bale TL. Sex-specific programming of offspring emotionality after stress early in pregnancy. *J Neurosci.* 2008;28:9055–9065.
25. Flandreau EI, Ressler KJ, Owens MJ, Nemeroff CB. Chronic overexpression of corticotropin-releasing factor from the central amygdala produces HPA axis hyperactivity and behavioral anxiety associated with gene-expression changes in the hippocampus and paraventricular nucleus of the hypothalamus. *Psychoneuroendocrinology.* 2012;37:27–38.
26. Chen J, Evans AN, Liu Y, Honda M, Saavedra JM, Aguilera G. Maternal deprivation in rats is associated with corticotrophin-releasing hormone (CRH) promoter hypomethylation and enhances CRH transcriptional responses to stress in adulthood. *J Neuroendocrinol.* 2012;24:1055–1064.
27. Elliott E, Ezra-Nevo G, Regev L, Neufeld-Cohen A, Chen A. Resilience to social stress coincides with functional DNA methylation of the Crf gene in adult mice. *Nat Neurosci.* 2010;13:1351–1353.
28. Elliott E, Manashirov S, Zwang R, et al. Dnmt3a in the medial prefrontal cortex regulates anxiety-like behavior in adult mice. *J Neurosci.* 2016;36:730–740.

29. Sterrenburg L, Gaszner B, Boerrigter J, et al. Chronic stress induces sex-specific alterations in methylation and expression of corticotropin-releasing factor gene in the rat. *PLoS One*. 2011;6:e28128.
30. Roeh S, Provencal N, Wiechmann T, Koedel M, Sauer S, Binder EB. HAM-TBS: high accuracy methylation measurements via targeted bisulfite sequencing. *bioRxiv*. 2017;163535.
31. Chen R, Lewis KA, Perrin MH, Vale WW. Expression cloning of a human corticotropin-releasing-factor receptor. *Proc Natl Acad Sci USA*. 1993;90:8967–8971.
32. Justice NJ, Yuan ZF, Sawchenko PE, Vale W. Type 1 corticotropin-releasing factor receptor expression reported in BAC transgenic mice: implications for reconciling ligand-receptor mismatch in the central corticotropin-releasing factor system. *J Comp Neurol*. 2008;511:479–496.
33. Lovenberg TW, Liaw CW, Grigoriadis DE, et al. Cloning and characterization of a functionally distinct corticotropin-releasing factor receptor subtype from rat brain. *Proc Natl Acad Sci USA*. 1995;92:836–840.
34. Haramati S, Navon I, Issler O, et al. MicroRNA as repressors of stress-induced anxiety: the case of amygdalar miR-34. *J Neurosci*. 2011;31:14191–14203.
35. Zhu J, Chen Z, Tian J, et al. miR-34b attenuates trauma-induced anxiety-like behavior by targeting CRHR1. *Int J Mol Med*. 2017;40:90–100.
36. Ramot A, Jiang Z, Tian JB, et al. Hypothalamic CRFR1 is essential for HPA axis regulation following chronic stress. *Nat Neurosci*. 2017;20:385–388.
37. Buschdorf JP, Meaney MJ. Epigenetics/programming in the HPA axis. *Compr Physiol*. 2015;6:87–110.
38. Weaver IC, Diorio J, Seckl JR, Szyf M, Meaney MJ. Early environmental regulation of hippocampal glucocorticoid receptor gene expression: characterization of intracellular mediators and potential genomic target sites. *Ann NY Acad Sci*. 2004;1024:182–212.
39. Turner JD, Schote AB, Macedo JA, Pelascini LP, Muller CP. Tissue specific glucocorticoid receptor expression, a role for alternative first exon usage? *Biochem Pharmacol*. 2006;72:1529–1537.
40. Hellstrom IC, Dhir SK, Diorio JC, Meaney MJ. Maternal licking regulates hippocampal glucocorticoid receptor transcription through a thyroid hormone-serotonin-NGFI-A signalling cascade. *Philos Trans R Soc Lond B Biol Sci*. 2012;367:2495–2510.
41. McGowan PO, Sasaki A, D'Alessio AC, et al. Epigenetic regulation of the glucocorticoid receptor in human brain associates with childhood abuse. *Nat Neurosci*. 2009;12:342–348.
42. Gross JA, Pacis A, Chen GG, et al. Gene-body 5-hydroxymethylation is associated with gene expression changes in the prefrontal cortex of depressed individuals. *Transl Psychiatry*. 2017;7:e1119.
43. Kriaucionis S, Heintz N. The nuclear DNA base 5-hydroxymethylcytosine is present in Purkinje neurons and the brain. *Science*. 2009;324:929–930.
44. Penn NW, Suwalski R, O'Riley C, Bojanowski K, Yura R. The presence of 5-hydroxymethylcytosine in animal deoxyribonucleic acid. *Biochem J*. 1972;126:781–790.
45. Wyatt GR, Cohen SS. The bases of the nucleic acids of some bacterial and animal viruses: the occurrence of 5-hydroxymethylcytosine. *Biochem J*. 1953;55:774–782.
46. Hack LM, Dick ALW, Provençal N. Epigenetic mechanisms involved in the effects of stress exposure: focus on 5-hydroxymethylcytosine. *Environ Epigenet*. 2016;2:dvw016.
47. Guo JU, Su Y, Zhong C, Ming GL, Song H. Hydroxylation of 5-methylcytosine by TET1 promotes active DNA demethylation in the adult brain. *Cell*. 2011;145:423–434.
48. Kaas GA, Zhong C, Eason DE, et al. TET1 controls CNS 5-methylcytosine hydroxylation, active DNA demethylation, gene transcription, and memory formation. *Neuron*. 2013;79:1086–1093.

49. Li X, Wei W, Zhao QY, et al. Neocortical Tet3-mediated accumulation of 5-hydroxymethylcytosine promotes rapid behavioral adaptation. *Proc Natl Acad Sci USA*. 2014;111:7120–7125.
50. Li S, Papale LA, Kintner DB, et al. Hippocampal increase of 5-hmC in the glucocorticoid receptor gene following acute stress. *Behav Brain Res*. 2015;286:236–240.
51. Li S, Papale LA, Zhang Q, et al. Genome-wide alterations in hippocampal 5-hydroxymethylcytosine links plasticity genes to acute stress. *Neurobiol Dis*. 2016;86: 99–108.
52. Papale LA, Li S, Madrid A, et al. Sex-specific hippocampal 5-hydroxymethylcytosine is disrupted in response to acute stress. *Neurobiol Dis*. 2016;96:54–66.
53. Papale LA, Madrid A, Li S, Alisch RS. Early-life stress links 5-hydroxymethylcytosine to anxiety-related behaviors. *Epigenetics*. 2017;12:264–276.
54. Booth MJ, Branco MR, Ficz G, et al. Quantitative sequencing of 5-methylcytosine and 5-hydroxymethylcytosine at single-base resolution. *Science*. 2012;336:934–937.
55. Yu M, Hon GC, Szulwach KE, et al. Tet-assisted bisulfite sequencing of 5-hydroxymethylcytosine. *Nat Protoc*. 2012;7:2159–2170.
56. Uchida S, Nishida A, Hara K, et al. Characterization of the vulnerability to repeated stress in Fischer 344 rats: possible involvement of microRNA-mediated down-regulation of the glucocorticoid receptor. *Eur J Neurosci*. 2008;27:2250–2261.
57. Deng M, Tufan T, Raza MU, Jones TC, Zhu MY. MicroRNAs 29b and 181a down-regulate the expression of the norepinephrine transporter and glucocorticoid receptors in PC12 cells. *J Neurochem*. 2016;139:197–207.
58. Mannironi C, Camon J, De Vito F, et al. Acute stress alters amygdala microRNA miR-135a and miR-124 expression: inferences for corticosteroid dependent stress response. *PLoS One*. 2013;8:e73385.
59. Sober S, Laan M, Annilo T. MicroRNAs miR-124 and miR-135a are potential regulators of the mineralocorticoid receptor gene (NR3C2) expression. *Biochem Biophys Res Commun*. 2010;391:727–732.
60. Jin J, Kim SN, Liu X, et al. miR-17-92 cluster regulates adult hippocampal neurogenesis, anxiety, and depression. *Cell Rep*. 2016;16:1653–1663.
61. Rein T. FK506 binding protein 51 integrates pathways of adaptation: FKBP51 shapes the reactivity to environmental change. *Bioessays*. 2016;38:894–902.
62. Makkonen H, Kauhanen M, Paakinaho V, Jaaskelainen T, Palvimo JJ. Long-range activation of FKBP51 transcription by the androgen receptor via distal intronic enhancers. *Nucleic Acids Res*. 2009;37:4135–4148.
63. Thomassin H, Flavin M, Espinas ML, Grange T. Glucocorticoid-induced DNA demethylation and gene memory during development. *EMBO J*. 2001;20:1974–1983.
64. Lee RS, Tamashiro KL, Yang X, et al. Chronic corticosterone exposure increases expression and decreases deoxyribonucleic acid methylation of Fkbp5 in mice. *Endocrinology*. 2010;151:4332–4343.
65. St-Cyr S, Abuaish S, Sivanathan S, McGowan PO. Maternal programming of sex-specific responses to predator odor stress in adult rats. *Horm Behav*. 2017;94:1–12.
66. Klengel T, Mehta D, Anacker C, et al. Allele-specific FKBP5 DNA demethylation mediates gene-childhood trauma interactions. *Nat Neurosci*. 2013;16:33–41.
67. Gassen NC, Fries GR, Zannas AS, et al. Chaperoning epigenetics: FKBP51 decreases the activity of DNMT1 and mediates epigenetic effects of the antidepressant paroxetine. *Sci Signal*. 2015;8:ra119.
68. Martinez NJ, Chang HM, Borrajo Jde R, Gregory RI. The co-chaperones Fkbp4/5 control Argonaute2 expression and facilitate RISC assembly. *RNA*. 2013;19: 1583–1593.
69. Volk N, Paul ED, Haramati S, et al. MicroRNA-19b associates with Ago2 in the amygdala following chronic stress and regulates the adrenergic receptor beta 1. *J Neurosci*. 2014;34:15070–15082.

70. Hartmann J, Wagner KV, Gaali S, et al. Pharmacological inhibition of the psychiatric risk factor FKBP51 has anxiolytic properties. *J Neurosci*. 2015;35:9007–9016.
71. Murgatroyd C, Patchev AV, Wu Y, et al. Dynamic DNA methylation programs persistent adverse effects of early-life stress. *Nat Neurosci*. 2009;12:1559–1566.
72. Murgatroyd C, Spengler D. Polycomb binding precedes early-life stress responsive DNA methylation at the Avp enhancer. *PLoS One*. 2014;9:e90277.
73. Albert PR, Francois BL. Modifying 5-HT1A receptor gene expression as a new target for antidepressant therapy. *Front Neurosci*. 2010;4:35.
74. Le Francois B, Soo J, Millar AM, et al. Chronic mild stress and antidepressant treatment alter 5-HT1A receptor expression by modifying DNA methylation of a conserved Sp4 site. *Neurobiol Dis*. 2015;82:332–341.
75. Issler O, Haramati S, Paul ED, et al. MicroRNA 135 is essential for chronic stress resiliency, antidepressant efficacy, and intact serotonergic activity. *Neuron*. 2014;83: 344–360.
76. Issler O, Chen A. Determining the role of microRNAs in psychiatric disorders. *Nat Rev Neurosci*. 2015;16:201–212.
77. Lopez JP, Lim R, Cruceanu C, et al. miR-1202 is a primate-specific and brain-enriched microRNA involved in major depression and antidepressant treatment. *Nat Med*. 2014;20:764–768.
78. Krishnan V, Nestler EJ. The molecular neurobiology of depression. *Nature*. 2008;455: 894–902.
79. Jeanneteau FD, Lambert WM, Ismaili N, et al. BDNF and glucocorticoids regulate corticotrophin-releasing hormone (CRH) homeostasis in the hypothalamus. *Proc Natl Acad Sci USA*. 2012;109:1305–1310.
80. Roth TL, Lubin FD, Funk AJ, Sweatt JD. Lasting epigenetic influence of early-life adversity on the BDNF gene. *Biol Psychiatry*. 2009;65:760–769.
81. Dubowitz H, Black MM, Kerr MA, et al. Type and timing of mothers' victimization: effects on mothers and children. *Pediatrics*. 2001;107:728–735.
82. Miranda JK, de la Osa N, Granero R, Ezpeleta L. Maternal experiences of childhood abuse and intimate partner violence: psychopathology and functional impairment in clinical children and adolescents. *Child Abuse Negl*. 2011;35:700–711.
83. Ullmann E, Barthel A, Licinio J, Petrowski K, Bornstein SR, Strauss B. Increased rate of depression and psychosomatic symptoms in Jewish migrants from the post-Soviet-Union to Germany in the 3rd generation after the Shoa. *Transl Psychiatry*. 2013;3:e241.
84. Rodgers AB, Morgan CP, Bronson SL, Revello S, Bale TL. Paternal stress exposure alters sperm microRNA content and reprograms offspring HPA stress axis regulation. *J Neurosci*. 2013;33:9003–9012.
85. Rodgers AB, Morgan CP, Leu NA, Bale TL. Transgenerational epigenetic programming via sperm microRNA recapitulates effects of paternal stress. *Proc Natl Acad Sci USA*. 2015;112:13699–13704.
86. Short AK, Fennell KA, Perreau VM, et al. Elevated paternal glucocorticoid exposure alters the small noncoding RNA profile in sperm and modifies anxiety and depressive phenotypes in the offspring. *Transl Psychiatry*. 2016;6:e837.
87. Gapp K, Jawaid A, Sarkies P, et al. Implication of sperm RNAs in transgenerational inheritance of the effects of early trauma in mice. *Nat Neurosci*. 2014;17:667–669.
88. McCreary JK, Erickson ZT, Hao Y, Ilnytskyy Y, Kovalchuk I, Metz GA. Environmental intervention as a therapy for adverse programming by ancestral stress. *Sci Rep*. 2016;6:37814.
89. Tasic B, Menon V, Nguyen TN, et al. Adult mouse cortical cell taxonomy revealed by single cell transcriptomics. *Nat Neurosci*. 2016;19:335–346.
90. Buenrostro JD, Wu B, Litzenburger UM, et al. Single-cell chromatin accessibility reveals principles of regulatory variation. *Nature*. 2015;523:486–490.

91. Luo C, Keown CL, Kurihara L, et al. Single-cell methylomes identify neuronal subtypes and regulatory elements in mammalian cortex. *Science*. 2017;357:600–604.
92. Mo A, Mukamel EA, Davis FP, et al. Epigenomic signatures of neuronal diversity in the mammalian brain. *Neuron*. 2015;86:1369–1384.
93. Danese A, Pariante CM, Caspi A, Taylor A, Poulton R. Childhood maltreatment predicts adult inflammation in a life-course study. *Proc Natl Acad Sci USA*. 2007;104: 1319–1324.
94. Provencal N, Suderman MJ, Guillemin C, et al. The signature of maternal rearing in the methylome in rhesus macaque prefrontal cortex and T cells. *J Neurosci*. 2012;32: 15626–15642.
95. Mehta D, Klengel T, Conneely KN, et al. Childhood maltreatment is associated with distinct genomic and epigenetic profiles in posttraumatic stress disorder. *Proc Natl Acad Sci USA*. 2013;110:8302–8307.
96. Suderman M, Borghol N, Pappas JJ, et al. Childhood abuse is associated with methylation of multiple loci in adult DNA. *BMC Med Genomics*. 2014;7:13.
97. Weder N, Zhang H, Jensen K, et al. Child abuse, depression, and methylation in genes involved with stress, neural plasticity, and brain circuitry. *J Am Acad Child Adolesc Psychiatry*. 2014;53:417–24:e5.
98. Singh-Taylor A, Molet J, Jiang S, et al. NRSF-dependent epigenetic mechanisms contribute to programming of stress-sensitive neurons by neonatal experience, promoting resilience. *Mol Psychiatry*. 2018;23:648–657.
99. Heller EA, Hamilton PJ, Burek DD, et al. Targeted epigenetic remodeling of the Cdk5 gene in nucleus accumbens regulates cocaine- and stress-evoked behavior. *J Neurosci*. 2016;36:4690–4697.
100. Liu XS, Wu H, Ji X, et al. Editing DNA methylation in the mammalian genome. *Cell*. 2016;167:233–47:e17.
101. Morita S, Noguchi H, Horii T, et al. Targeted DNA demethylation in vivo using dCas9-peptide repeat and scFv-TET1 catalytic domain fusions. *Nat Biotechnol*. 2016;34:1060–1065.
102. Heller EA, Cates HM, Pena CJ, et al. Locus-specific epigenetic remodeling controls addiction- and depression-related behaviors. *Nat Neurosci*. 2014;17:1720–1727.
103. Lister R, Mukamel EA, Nery JR, et al. Global epigenomic reconfiguration during mammalian brain development. *Science*. 2013;341:1237905.
104. Borchert GM, Lanier W, Davidson BL. RNA polymerase III transcribes human microRNAs. *Nat Struct Mol Biol*. 2006;13:1097–1101.
105. Lee Y, Ahn C, Han J, et al. The nuclear RNase III Drosha initiates microRNA processing. *Nature*. 2003;425:415–419.
106. Yi R, Qin Y, Macara IG, Cullen BR. Exportin-5 mediates the nuclear export of pre-microRNAs and short hairpin RNAs. *Genes Dev*. 2003;17:3011–3016.
107. Chendrimada TP, Gregory RI, Kumaraswamy E, et al. TRBP recruits the Dicer complex to Ago2 for microRNA processing and gene silencing. *Nature*. 2005;436: 740–744.
108. Bartel DP. MicroRNAs: target recognition and regulatory functions. *Cell*. 2009;136: 215–233.
109. Molet J, Maras PM, Avishai-Eliner S, Baram TZ. Naturalistic rodent models of chronic early-life stress. *Dev Psychobiol*. 2014;56:1675–1688.
110. Golden SA, Covington III HE, Berton O, Russo SJ. A standardized protocol for repeated social defeat stress in mice. *Nat Protoc*. 2011;6:1183–1191.
111. Khemissi W, Farooq RK, Le Guisquet AM, Sakly M, Belzung C. Dysregulation of the hypothalamus-pituitary-adrenal axis predicts some aspects of the behavioral response to chronic fluoxetine: association with hippocampal cell proliferation. *Front Behav Neurosci*. 2014;8:340.

CHAPTER SEVEN

Wnt Signaling in the Hippocampus in Relation to Neurogenesis, Neuroplasticity, Stress and Epigenetics

Ramona A.J. Zwamborn*, Clara Snijders†, Ning An†, Alix Thomson†, Bart P.F. Rutten‡, Laurence de Nijs†,1

*Department of Neurology and Neurosurgery, Brain Center Rudolf Magnus, University Medical Center Utrecht, Utrecht, The Netherlands
†School for Mental Health and Neuroscience (MHeNS), Department of Psychiatry and Neuropsychology, Faculty of Health, Medicine and Life Sciences, Maastricht University, Maastricht, The Netherlands
‡Department of Psychiatry and Neuropsychology, School for Mental Health and Neuroscience (MHeNs), European Graduate School of Neuroscience (EURON), Faculty of Health, Medicine and Life Sciences, Maastricht University Medical Centre +, Maastricht, The Netherlands
[1]Corresponding author: e-mail address: laurence.denijs@maastrichtuniversity.nl

Contents

Abstract

The Wnt signaling pathway has been recognized as an important pathway, extending its function throughout the lifespan. Evidence suggests that dysfunctional Wnt signaling in the adult brain leads to aberrant neurogenesis, synaptogenesis, modulation of mature synapses and neurotransmitter release in the hippocampus. Due to the involvement of Wnt proteins in hippocampal functioning, altered Wnt signaling has been

Progress in Molecular Biology and Translational Science, Volume 158
ISSN 1877-1173
https://doi.org/10.1016/bs.pmbts.2018.04.005

suggested to be an important factor in the pathophysiology of mood disorders. Interestingly, the effects of mood-stabilizing drugs are believed to work through interactions with Wnt molecules, and epigenetic mechanisms have been shown to interact with components of the Wnt pathway and impact mechanisms such as synaptic plasticity. This can affect learning and memory formation, in addition to various behavioral outcomes in individuals, when they are faced with stressful or conflict situations. This review will discuss the integrated role of Wnt signaling in the context of appropriate stress response, which is believed to be mediated by adult hippocampal neurogenesis and plasticity. Current knowledge regarding the role of Wnt signaling in mood disorders and antidepressant medication effect will be covered. Finally, the interplay between Wnt signaling and epigenetic mechanisms will be discussed along with their combined potential to impact neuroplasticity.

1. INTRODUCTION

Signaling pathways are important facets of biological life. Among others, signaling pathways provide critical cell-to-cell communication, regulate proper cell activity and are involved in tissue homeostasis and regeneration.[1] Among the large variety of signaling pathways, Wnt signaling pathways have been recognized as important pathways extending their function throughout the lifespan. They are involved in various cellular and physiological processes, such as cell fate specification, cell proliferation and cell migration. Over the past decade, the Wnt signaling pathways have received a lot of attention, especially for their implication in embryonic development and in a variety of human pathologies such as developmental defects,[2,3] several types of cancer,[4,5] neurodegenerative disorders[6–8] and stress-related mood disorders.[9–11] This review aims to provide a comprehensive overview of the Wnt signaling pathways, their involvement in adult hippocampal neurogenesis and plasticity, their contribution to stress-related mood disorders in adulthood and pharmacological intervention, and finally their interaction with epigenetic factors.

2. THE Wnt SIGNALING PATHWAYS: AN OVERVIEW

Wnt proteins (Wnts) are highly conserved lipid-modified glycoproteins that can act as short or long range signaling molecules.[12] They are characterized by a hydrophobic signal peptide, highly conserved cysteine residues, the absence of additional transmembrane domains, and the presence of N-linked glycosylation sites.[12] To activate intracellular signal transduction, Wnts bind with high affinity to the Frizzled receptor (FZD) by covalent attachment of palmitic acid on the first cysteine residue, a process which is referred to as palmitoylation.[13] The FZD receptor is a seven-transmembrane receptor with three basic regions: (1) the N-terminal that participates in the binding of agonist (e.g., Wnts) or antagonist (e.g., secreted Frizzled-related protein family [sFRP], and members of the Dickkopf [Dkk] family[14]), (2) a transmembrane central core with hydrophobic alpha-helices that span the lipid bilayers, and (3) three intracellular loops with a C-terminal "tail" for downstream signaling in the cytoplasm[12] and three extracellular loops that participate in palmitoylation and glycosylation.[15] Besides the binding of FZD, co-receptors such as Ryk, Ror1,2 and lipoprotein receptor-related protein (LRP)-5/6 are necessary for the activation of the pathway.[12,16] These co-receptors have multiple non-overlapping binding sites for different classes of Wnts.[17] In total, 19 Wnts and 12 FZD receptors have been identified in vertebrates.[12] At least 12 of the Wnts have been evolutionarily and functionally conserved. For example, 11 of the Wnts are present in both Cnidaria (sea anemone) and vertebrates, indicating that these Wnts were already present before the divergence of bilaterians and cnidarians.[18] Interestingly, each of the conserved Wnts found in Cnidaria displayed a distinct expression pattern which matched the expression profile of the vertebrates.[18] This conservation along evolution elegantly demonstrates the diverse and important role of Wnt proteins throughout species.

The diversity of the Wnt pathways originates from the fact that the various Wnt ligands and FZDs can be subdivided into distinct subclasses based on their biological readout and downstream pathways. Namely, Wnt ligands regulate three main pathways; the canonical Wnt/β-catenin pathway and the non-canonical or β-catenin-independent pathway which can be divided into the Wnt/planar cell polarity (PCP) pathway and the Wnt/calcium (Ca^{2+}) pathway (Fig. 1). Although the non-canonical pathway will briefly be discussed, this review will mainly focus on the canonical Wnt/β-catenin pathway because of its critical function in the formation of neuronal circuits.[14] However, it is worth noting that both pathways have been suggested to antagonize each other by competing for FZD binding sites.[19–22] For example, Wnts which are known to primarily stimulate non-canonical PCP signaling compete for and inhibit the binding of Wnts activating the canonical β-catenin-dependent pathway.[23]

2.1 β-Catenin-Independent Pathway

As mentioned, the non-canonical β-catenin-independent pathway can be subdivided into the Wnt/PCP and the Wnt/Ca^{2+} pathway (Fig. 1). The PCP pathway has been proposed to affect cell polarization, morphogenetic

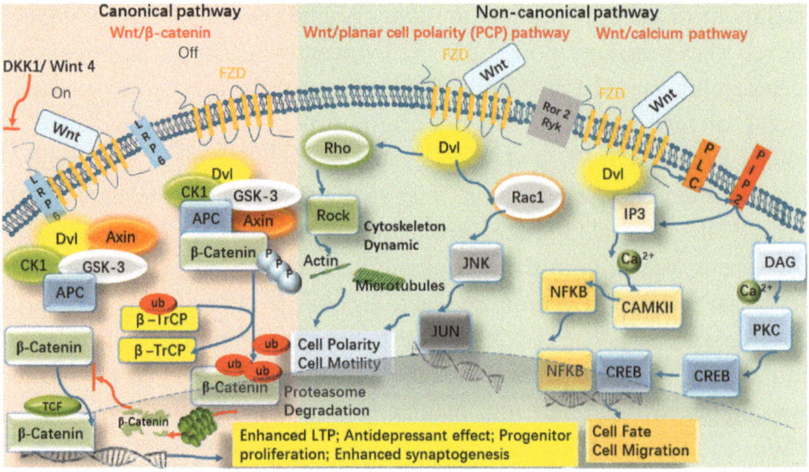

Fig. 1 Overview of canonical and non-canonical Wnt signaling pathways. The canonical Wnt/β-catenin pathway is activated by the binding of Wnt ligand to FZD receptor, which induces the recruitment of Dvl leading to the phosphorylation of the co-receptor LRP5/6. This results in the inhibition of the "β-catenin destruction complex" formed by Axin, GSK-3β, APC and CK1 which no longer phosphorylates β-catenin, resulting in the translocation of β-catenin to the nucleus where it activates transcription mediated by TCF. In the absence of Wnt, β-catenin is phosphorylated by the destruction complex and degraded by ubiquitin-mediated proteolysis in the cytosol. The non-canonical Wnt/planar cell polarity pathway activation results in the induction of small GTPases Rho and Rac1, which in turn activate Rock and JNK respectively, leading either to direct actin polarization or transcriptional activation via JUN. The non-canonical Wnt/calcium activation leads to the production of IP3 and DAG derived from membrane-bound PIP2 via the activation of PLC. Activation of IP3 and DAG result in the release of Ca^{2+} from ER activating CAMKII and PKC which in turn activate CREB and NF-κB. Both CREB and NF-κB will translocate to the nucleus activating transcription. For abbreviations, see the main text.

movements and motility during gastrulation.[24] However, the specific signaling mechanisms in vertebrates are poorly understood. Although they have yet to be characterized, it is known that signaling centers send out polarizing signals which are then transduced through the FZD receptor with co-receptors such as PTK7 or Glypican. This leads to the local activation ofDisheveled (Dvl) and the small GTPases RHOA and RAC1 which in turn activate RHO family GTPases such as ROCK and JUN-N-terminal kinase (JNK).[12,25,26] The activation of RHO GTPases and JNK further leads to actin polymerization and microtubule stabilization which regulates cell polarity, cell motility and morphogenetic movements during embryonic development[12] and will thus not be covered in this review focusing on adult Wnt signaling. For a more elaborated review on this pathway, readers are referred to the review of Sokol.[25]

The Wnt/Ca^{2+} pathway is activated by Wnt ligands binding to FZD receptors which results in the activation of phospholipase C (PLC) and triggers an increase in the concentration of the intracellular signaling molecules inositol 1,4,5-triphosphate (IP3) and 1,2 diacylglycerol (DAG) derived from membrane-bound phospholipid phosphatidyl inositol 4,5-bisphosphate (PIP2) (Fig. 1).[27] The activation of both IP3 and DAG results in an increase in intracellular Ca^{2+} from the endoplasmic reticulum (ER), leading to the activation of protein kinase C (PKC) and CaMKII.[24,28] Both CaMKII and PKC activate a number of downstream effectors such as the nuclear transcription factors NF-κB and CREB. Activation of NF-κB or CREB is known to be involved in cell fate and cell migration during embryonic development, neurodegeneration and arousing inflammatory responses.[27] For a more detailed description of this pathway, readers are referred to more comprehensive reviews such as De.[27]

2.2 β-Catenin-Dependent Signaling

Signaling of the canonical Wnt/β-catenin pathway functions by modifying the stabilization of β-catenin (Fig. 1). In the absence of Wnt, β-catenin is targeted for degradation by a multi-protein destruction complex including the tumor suppressor protein adenomatous polyposis coli (APC), casein kinase 1 (CK1), glycogen synthase kinase 3 (GSK-3α and GSK-3β), and Axin.[29,30] Axin is proposed as the scaffold protein that directly interacts with other core components of the destruction complex.[31] Within that complex, β-catenin is phosphorylated by GSK-3 on serine 33 and 37 and threonine 41 which requires phosphorylation of β-catenin by CK1.[31] The phosphorylation of β-catenin triggers the recruitment of ubiquitin E3 β-TrCP, causing the ubiquitination and proteasomal degradation of β-catenin.[32]

Upon activation of the pathway by the binding of Wnt, CK1 and GSK-3 do not phosphorylate Axin-bound β-catenin.[12] The un-phosphorylated β-catenin is therefore not recognized by the protein β-TrCP, is not ubiquitinated, and not targeted for destruction by the proteasome.[32] This results in the accumulation of β-catenin and its translocation to the nucleus where it interacts with TCF/LEF1 transcription factors to regulate the expression of Wnt target genes.[33] The following models have attempted to explain the mechanisms by which Wnt signaling inhibits β-catenin degradation; (1) the disruption of Axin/GSK-3 interactions diminishes β-catenin phosphorylation and results in the stabilization of β-catenin.[34] (2) The inhibition of GSK-3 phosphorylation of β-catenin by Wnts via phosphorylated PPPSPXS motifs of the co-receptor LRP5/6, leads to stabilization of β-catenin.[35,36] (3) The abolishment of β-TrCP recruitment to β-catenin by Wnts blocks β-catenin ubiquitination, and results in the stabilization of β-catenin.[37] Finally, (4) the membrane sequestration of Axin1/ GSK-3 complex induced by Wnts binding the co-receptor LRP5/6.[33,38–40]

Recently, a new model has been proposed combining the latter two models. This framework postulates that Wnt binding causes the relocation of the destruction complex to the membrane, which inhibits β-catenin ubiquitination and leads to free and transcriptionally active β-catenin accumulation.[41] More elaborately, upon activation of the FZD, the cytoplasmic part of FZD facilitates the interaction between the LRP6 tail and Axin by phosphorylation of the LRP6 tail[33] which is regulated by two separate kinases, namely GSK-3 and CK1γ. GSK-3 phosphorylates serine in the PPPSP motif found in Axin, APC and LRP6.[39,42] CK1γ kinase phosphorylates the same proteins on residues adjacent to the PPPSP motif.[43] Moreover, it is known that Dvl further facilitates the interaction between the LRP tail and Axin, because the DIX domain in Dvl is similar to a DIX region in Axin, resulting in direct binding of the two domains.[44–46]

The Wnt signaling can be activated by several anta- and agonists such as Dickopf-related proteins (DKK), SFRP, Sclerostin and Wnt inhibitory factor (WIF) or R-spondin (RSPOs) family and Norrin.[12] The importance of extracellular regulation of Wnts in their functional roles should not be overlooked, as disruption of these signals can result in altered cellular function. For example, the loss of Dickkopf-1 (Dkk-1) increases the number of neural progenitors in the hippocampus.[47] Moreover, a lentiviral knockdown of sFRP3 in the dentate gyrus increased canonical Wnt signaling, neural progenitor proliferation and neuronal development.[48] These studies indicate the extracellular regulation of the Wnt could affect activation of the stem cell population, which could further affect the central nervous system (CNS) over a longer period of time. For a more extended review on this topic, see Cruciat and Niehrs.[49]

In addition to the anta- and agonistic regulation described above, environmental factors can have a major impact on Wnt signaling. For example, it is known that exposure to an enriched environment results in increased hippocampal cell survival,[50] hippocampal synapse number, and mossy fiber terminals, accompanied by an increase in Wnt expression.[51] Empirically, the addition of Wnt7 to the CA3 region of the hippocampus mimicked the effects of exposure to an enriched environment on synapse number and large mossy fiber terminals,[51,52] providing support for the link between environmental factors and Wnt signaling in the hippocampus. In contrast to exposure to an enriched environment, exposure to chronic or acute stress can decrease hippocampal neurogenesis and cell survival[53–55] which is hypothesized to be regulated by Wnt signaling.[56] It is thought that stress decreases canonical Wnt proteins and their receptors in a neurogenic niche in the hippocampus. This decrease in Wnts decreases progenitor proliferation and differentiation[56] and dendritic spine morphogenesis,[57] processes which are essential for normal hippocampal functioning. The next paragraphs will further consolidate on the functional effects of Wnt signaling on neurogenesis and plasticity in relation to stress and epigenetics in the adult CNS.

3. Wnt SIGNALING IN RELATION TO HIPPOCAMPAL NEUROGENESIS, NEUROPLASTICITY, STRESS AND EPIGENETICS

In the nervous system, Wnt signaling has an important function in neural development and in the formation of neuronal circuits. It has been shown that Wnts are involved in the regulation of self-renewal, maintenance and differentiation of neural progenitor cells throughout life,[8,58,59] and that they are important in regulating synaptic development and function through the formation and maturation of pre- and post-synaptic sites.[60]

Recently, an increasing number of studies has found an association between exposure to a chronic or acute stressful event and aberrant Wnt signaling, accompanied by a decrease in adult neurogenesis.[61,62] Similarly, electrophysiological recordings in hippocampal slices showed that blocking Wnt signaling impairs long-term potentiation (LTP), indicating its involvement in neuronal plasticity.[63] Interestingly, epigenetic mechanisms have also been found to regulate the progression of hippocampal neurogenesis.[33] The epigenetic regulation of Wnt has mainly been described in the context of leukemia and the hematopoietic system,[64–66] therefore the question now arises as to whether and how Wnt signaling and epigenetic factors coordinate in establishing neurogenesis and neuronal plasticity along with a proper stress response in the context of brain functioning.

Deregulation of Wnts has been implicated in mood disorders such as major depressive disorder and bipolar disorder.[9–11] This was reported by several studies, which have shown that administrating various antidepressants to rats resulted in an increased expression of hippocampal Wnts. This increased expression of Wnts was accompanied by a decrease in depressed-like responses such as latency to feed in the novelty suppressed feeding test.[67,68] These and similar results suggest that several components of the Wnt signaling pathways are involved in the efficacy of antidepressant treatments.

The following sections will further explore this crosstalk between Wnt signaling, neurogenesis, neuroplasticity, stress response and mood disorders, as well as the epigenetic mechanisms involved.

3.1 Wnt Signaling and Hippocampal Neurogenesis

Years ago, the idea of neuronal cells being regularly added to the adult brain to form functional connections was largely rejected by the scientific community,[69] despite accumulating evidence proving support for the existence of adult neurogenesis.[70,71] Nowadays, the existence of adult neurogenesis in rodents and humans is more widely accepted. In adult rodents, it was found that thousands of new neurons are generated daily, forming functional connections that have the ability to respond to environmental inputs and contribute to tissue homeostasis and general brain function.[72] In humans, the existence of adult neurogenesis was demonstrated by the incorporation of the thymidine analog bromodeoxyuridine (BrdU) in the hippocampus of adult cancer patients to assess tumor growth.[73] Adult neurogenesis is primarily restricted to two regions, namely the subgranular zone (SGZ) in the dentate gyrus of the hippocampus and the subventricular zone of the lateral ventricles.[9] This review will mainly focus on hippocampal neurogenesis, as this region has the highest rate of neurogenesis and disruption in this region has been mostly associated with mood and anxiety disorders.

Adult hippocampal neurogenesis is believed to play a critical role in object recognition memory,[74] pattern separation,[75] learning and memory formation,[72] fear conditioning and cognitive flexibility.[76,77] Hippocampal neurogenesis is coordinated by intrinsic and extrinsic factors, that manage the progression of neurogenesis and preserve the stem cell pool.[78] Molecular factors which are known to regulate hippocampal neurogenesis include Notch, sonic hedgehog (Shh), bone morphogenetic proteins, growth and neurotrophic factors, neurotransmitters and Wnts.[78,79]

The first study to provide voluble evidence of the involvement of Wnts in adult hippocampal neurogenesis was the study of Maretto and colleagues.[80] This study identified the expression of β-catenin by using a BAT-gal reporter mouse expressing a β-catenin-activated transgene with nuclear β-galactosidase under the control of TCF/Lef.[80] Using this transgenic mouse model, it was determined that the canonical Wnt pathway is active in the SGZ and the dentate granule cell layer in the adult hippocampus.[56] In addition, this study found that canonical Wnt signaling is involved in regulating hippocampal neurogenesis. They found that, after injecting a single dose of BrdU into the adult BAT-gal mice and observing the Wnt/β-catenin pathway activity in BrdU-positive cells, these labeled cells showed morphological features of proliferating neuroblasts. Furthermore, cell culture analyses revealed that β-catenin activation promoted neural stem cell proliferation and increased cell count, suggesting that Wnt signaling promotes proliferation and self-renewal.[81] The importance of Wnts in vivo has also been shown by peripheral administration of the GSK-3β inhibitor lithium, which was shown to stimulate proliferation and neuronal fate specification in a mouse model of Alzheimer's disease.[82] Furthermore, inhibition of the autonomous Wnt signaling by either Axin or truncated cadherin promoted neurogenesis and depleted the population of multipotent precursors.[83]

Given these findings, it is assumed that Wnt signaling has a critical function in regulating the delicate balance between stem cell maintenance and differentiation during hippocampal adult neurogenesis.[83,84]

3.2 Wnt Pathways and Hippocampal Neuroplasticity

In the adult brain, in addition to neurogenesis, preexisting neural circuits have the ability to undergo modifications in dendritic complexity, efficacy of excitatory and inhibitory synaptic transmission and/or synapse number.[10] The formation of entirely new neural connections through neurogenesis is termed neuronal plasticity. Hippocampal plasticity can be modulated by various factors including serotonin, BDNF and mTOR.[10] However, over the years it has become clear that Wnt ligands hold the potential to modulate synaptic hippocampal plasticity.[57,85,86]

There are several ways in which Wnts are believed to regulate hippocampal plasticity. First, Wnts are believed to regulate pre-synaptic terminals formation and function. For example, transfection of constitutively active β-catenin and Axin in hippocampal cell cultures increased the number of pre-synaptic terminals per area of dendrites.[87] Second, Wnt signaling has been suggested to play a role in clustering and recycling of synaptic vesicles.[88] It has been shown that Wnt7a regulates the normal recycling route of synaptic vesicles trough Dvl1, and deficiency of both proteins reduces the miniature excitatory post-synaptic current frequency, which could indicate a defect in neurotransmitter release.[89] Third, both the canonical and the non-canonical Wnt pathways are believed to be involved in synaptic transmission by the influx of Ca^{2+},[90] which is generally accompanied by the activation of CAMKII, triggering actin cytoskeleton reorganization to regulate dendritic growth.[91] Finally, Wnt signaling is further believed to regulate dendritic spine morphology. Wnt5a has been shown to stimulate PSD95 clustering, which is known to increase the number of dendritic spines[92] and functional synapses.[93]

Together, these results suggest that Wnts play a role in establishing and maintaining synaptic plasticity through various mechanisms.

3.3 Wnts and Stress-Induced Aberrant Hippocampal Neurogenesis and Neuroplasticity

Over the past few years, many studies have reported that stress exposure could alter hippocampal structure and synaptic plasticity[94] by decreasing dendritic branching, dendritic spine density,[95,96] and cell proliferation or adult neurogenesis,[97,98] leading to a reduction in hippocampal volume.[99] Chronic stress has been found to delay the maturation and survival of new neurons,[100] to induce the remodeling of dendritic neurites and synapses[101] as well as producing aberrant vesicle reorganization and clustering in the dentate gyrus of the hippocampus.[102] Given the documented role of Wnts in neurogenesis and neuroplasticity, Wnt signaling has been proposed to be one of the underlying mechanisms by which stress alters hippocampal functioning.

Under stressful conditions, the hypothalamic–pituitary–adrenal (HPA) axis is activated. Corticotropin-releasing hormone (CRH) is secreted by the paraventricular nucleus of the hypothalamus, which activates the release of adrenocorticotropic hormone (ACTH) by the pituitary gland. ACTH stimulates the release of glucocorticoids (GC) by the adrenal glands, which is considered to alter hippocampal neurogenesis via the activation of Wnt pathways (Fig. 2). This has been supported by several studies observing an association between GC expression and Wnt signaling.[61,103] For example, it has been found that the administration of dexamethasone, an agonist of the GC receptor, resulted in GSK-3β activation, which ultimately decreased hippocampal precursor proliferation.[103] Moreover, it has been shown that activation of the GC receptor was correlated with an altered expression of Wnt4,[104] leading to a decrease in neurogenesis. Furthermore, Matrisciano and colleagues found that mice receiving a GC injection showed increased DKK1 levels and, consequently, reduced Wnt signaling

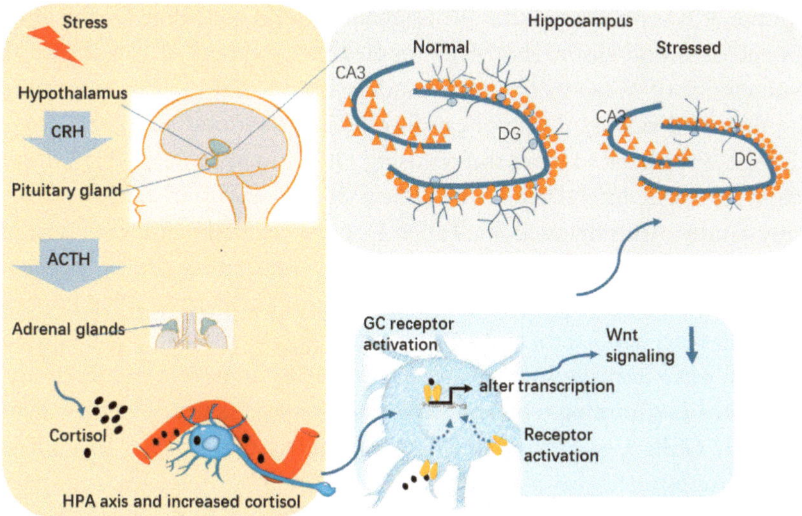

Fig. 2 Stress-induced aberrant hippocampal neurogenesis mediated through Wnt signaling. The activation of the stress response via the HPA axis induces the release of glucocorticoids (GC) in blood, which is considered to reduce hippocampal neurogenesis via the decrease of Wnt pathways activation.

in the hippocampus.[61] They also demonstrated that restraint stress in mice increased DKK1 hippocampal level which decreased hippocampal volume.[61] Finally, the same study found that transgenic mice deficient for DKK1 showed increased neurogenesis in response to stress compared to control mice.[61]

This notion is supported by the study of Gogolla et al. showing that mice exposed to positive behavioral experiences, such as an enriched environment, displayed an increase in the number of hippocampal synapses and dendritic spines, which was mediated through Wnt signaling.[51] Interestingly, accumulating evidence shows that antidepressants promote hippocampal proliferation of progenitors and increase the number of integrated new neurons through the action of Wnts.[53,82,105] These results indicate that Wnt signaling could be at least one of the underlying mechanisms mediating the crosstalk between stress and its associated functional changes in the hippocampus.

3.4 Aberrant Wnt Signaling Mediates Depressive- and Anxiety-Like Behavior

Exposure to stress is a strong driver of mood disorders such as major depressive disorder, post-traumatic stress disorder (PTSD) and bipolar disorder.[106] Given the involvement of Wnts in hippocampal functioning and in stress response, it was postulated that Wnt signaling could contribute to the development of mood disorders. Many studies have started to investigate the involvement of Wnts in behavior modulation related to stress processing. The first evidence of the implication of Wnts in mood and stress-related behavior was obtained from studies using Dvl1 mutant mice, which display defects in social interactions and deficits in the acoustic startle reflex.[107] Many studies (summarized in Table 1) have validated the involvement of Wnt signaling on depressive- and anxiety-like behavior. To highlight a few important findings, aberrant expression of GSK-3β in mice has been associated with depression and anxiety-like behaviors,[112,115] while GSK-3α knock-in mice displayed a decrease in exploratory activity, immobility time in the forced swimming test (FST) and aggressive behavior.[116] This suggests that both GSK-3 isoforms induce similar behavioral effects. In contrast, it was demonstrated that overexpression of GSK-3β in mice leads to hyperactivity in the open field (OF) test and to an increase in the acoustic startle response.[110] Lastly, viral-mediated knockdown of hippocampal Fz6 in rodents increased depressive- and anxiety-like behaviors as measured by the sucrose preference test and elevated plus maze.[117] This supports the hypothesis that Wnt signaling is likely involved in regulating mood-related behavior.

In addition, several studies have pointed to the involvement of adult neurogenesis in mood-related disorders. Snyder and colleges demonstrated that neurogenesis-deficient mice show depressive-like behavior such as increased food avoidance on novelty suppressed feeding (NSF), increased total immobility on the FST and a reduced sucrose preference on the sucrose test.[118] Another study found that a reduction of synaptogenesis and axonal growth cone collapse mediated by β-catenin resulted in decreased immobile time in the FST and hyperactivity in the OF.[111] Together, these studies suggest that the formation and regulation of synapses and neurons, which is at least partly mediated by Wnts, is crucial for the processing of stress responses. Furthermore, disruption of the Wnt signaling pathway results in behavioral responses that correspond with the behavioral expression of several mood-related phenotypes and possibly mood disorders in humans.

3.5 Wnts in the Pharmacological Intervention of Mood Disorders

The involvement of the Wnt pathway in the regulation of mood disorders suggests that components of the Wnt pathway could be an important target for pharmacological intervention in psychiatric disorders. Microarray analysis demonstrated an increase in FZD, β-catenin and TCF in the hippocampus

Table 1 Summary of Key Studies Demonstrating the Relationship Between Wnt Signaling and Depressive/Anxiety-Like Behavior

	Animal Model	Genetic/Systemic Regulation	Functional Effect	References
Fear memory	C57Bl/6J	100 ng/side amygdala infusion of Dkk and Wnt1	Decreased long-term fear memory consolidation	108
	C57Bl/6J	Inhibition of β-catenin expression	Decreased long-term fear memory consolidation	109
Depressive- and anxiety-like behavior	Heterozygous GSK mice on FVB/NTac background	Transgenic overexpressing of GSK-3β	Hyperactivity in the open field test and increased acoustic startle response	110
	C57Bl/6J with GSK-3β heterozygous knock-out	Decreased expression of GSK-3β	Reduced immobility time on FST and frequency in hole pokes	111
	C57BL/129SvJ chimeric knock-out mice	Pharmacological or genetic inhibition of GSK-3β	Reduced amphetamine induced hyperactivity in the FST	112
	C57Bl/6J	Viral inhibition of GSK-3β in Nac	Decreased sucrose consumption	113
	C57Bl/6J	Viral overexpression of Dvl in Nac	Deceased immobility in FST and increased social avoidance	113
	Male Sprague-Dawley rats	Viral overexpression of hippocampal Wnt2	Decreased number of escape failure in the learned helplessness and an increase in sucrose consumption	67
	C57Bl/6J with hippocampal floxed β-catenin mice	Decreased expression of β-catenin	Increased immobility time in TST	114
	C57Bl/6J	Overexpression of β-catenin in hippocampus	Reduced amphetamine induced hyperactivity in the FST	115

FST, forced swimming test; TST, tail suspension test; NSF, novelty suppressed feeding.

of rats following chronic treatment of antidepressants.[67,119] Moreover, the deletion of Wnt antagonist sFRP3 has been shown to induce antidepressant-like behavioral responses corresponding to those induced by antidepressants.[48] Along with similar studies, this suggests that mood-stabilizing drugs establish at least part of their therapeutic action through altering canonical Wnt signaling, and consequently adult hippocampal neurogenesis (Refs. 105,120; see Table 2 for a complete overview of these studies). An example of such an antidepressant is fluoxetine, which is known to upregulate β-catenin. Its administration has been shown to result in advanced maturation of immature neurons and enhanced neurogenesis-dependent LTP in the hippocampus.[121] Furthermore, its behavioral effects were abolished upon X-ray irradiation of the hippocampus, indicating the necessity of proper adult neurogenesis to exert its antidepressant effects.[134,135] The importance of neurogenesis in antidepressant action was further confirmed by lithium treatment, which has been shown to promote adult hippocampal progenitor development through GSK-3β inhibition and β-catenin upregulation.[83] Interestingly, this increased hippocampal proliferation was abolished by RNAi-mediated inhibition of β-catenin.[126] In contrast to this, David and colleagues found that abolishing neurogenesis in mice blocked the effects of fluoxetine and imipramine when subjected to the NSF test, while fluoxetine treatment still affected their behavior in the OF and the FST tests.[123] This suggests the existence of both neurogenesis-dependent and -independent mechanisms of antidepressant action.

Several mood regulators, such as lithium or vortioxetine, have been implicated in the regulation of dendritic arborization or the maturation and functional integration of new neurons[121] through the interact with elements of the stress response and Wnts signaling[136–138] Furthermore, some antidepressants known to influence the serotonergic and noradrenergic system are believed to modulate the Wnt pathway. For example, an upregulation of Wnt receptor gene *Fz9* was found after administration of atomoxetine and venlafaxine but not citalopram,[67] which suggests that the serotonergic and noradrenergic systems differentially regulate the expression components of the Wnt signaling pathway.

All in all, Wnt signaling is at the crossroad of multiple transduction pathways, resulting in the regulation of neurotransmitter responses, neuroplasticity and adult neurogenesis. The involvement of Wnt signaling in these processes is thought to be implicated in context appropriate stress responses. The interaction of Wnts with other cellular pathways could add to the diversity of functions and the involvement of Wnt signaling in several mood and anxiety disorders.

Table 2 Summary of Key Studies Demonstrating an Effect of Mood-Stabilizing Drugs on Wnt Signaling

Antidepressants	Treatment	Cellular Effect	Behavioral Effects	Animal Model	Study
SSRIs, TCAs	Chronic fluoxetine or imipramine administration	Reduces sFRP3 expression	Reduced immobility time in FST and TST	C57Bl/6J mice	48
	Chronic fluoxetine administration	Reduced proliferation	Reduces latency in NSF test. Improved coat score and decreased grooming latency	129/Sv mice	121
	Chronic fluoxetine administration	X-irradiated neurogenesis	Inhibited neurogenesis suppresses behavioral response to chronic fluoxetine treatment in NSF test	129/Sv mice	120
	Chronic fluoxetine administration	Increased Wnt3a and cell proliferation	N.A.	Male Sprague-Dawley rats	122
	Chronic fluoxetine treatment suppressed the effects of chronic corticosterone administration	Decreased neurogenesis	Reversed depressive behavior by fluoxetine treatment	C57BL/6Ntac, β-arrestin2 +/− mice	123
	2 weeks of CRS and fluoxetine administration	revers CRS-induced decreased expression of Wnt2 and Wnt3 in the ventral hippocampus	Reversed depressive behavior	Male C57BL/6 mice	124
Lithium	Chronic lithium administration	Increased β-catenin and proliferation	Improved contextual fear conditioning and novel object recognition tests	Ts65Dn mice	125
	Chronic lithium administration	Dose-dependent upregulation of proliferation and β-catenin	N.A.	Hippocampal female Fisher 344 rats	126
	Chronic lithium administration	Decreased dendritic length and branching	No difference in stress behavior in the OF or the plus maze	Adult male Sprague-Dawley rats	127

Continued

Table 2 Summary of Key Studies Demonstrating an Effect of Mood-Stabilizing Drugs on Wnt Signaling—cont'd

Antidepressants	Treatment	Cellular Effect	Behavioral Effects	Animal Model	Study
	Chronic valproic acid administration	Increase in Wnt3α and β-catenin and NSCs differentiation	N.A.	NSCs of Sprague-Dawley rats	128
	Chronic valproic acid administration	Reduced GSK-3β and increased β-catenin hippocampus	Reduced spatial learning, memory in MWM and increased repetitive behavior.	Male Wistar rats	62
Venlafaxine	Chronic and high dose venlafaxine	Increased β-catenin and proliferative	N.A.	Adult male albino Wistar rats	129
Electroconvulsive therapy	Chronic electroconvulsion	Dose-dependent increase in neurogenesis	N.A.	Male Wistar rats	130
	Chronic electroconvulsion	Upregulation of β-catenin and Wnt2	N.A.	Male Sprague-Dawley rats	131
	Chronic electroconvulsion	Dose-dependent increase in proliferation	Locomotor activity in OF and spontaneous alteration behavior in a Y-maze were not influenced	Sprague-Dawley rats	132
Vortioxetine	Chronic vortioxetine	Increased proliferation and LTP	Enhanced short-term episodic memory and novel object recognition	Male Sprague-Dawley rats	133

SRIs, selective serotonin reuptake inhibitors; TCAs, tricyclic and tetracyclic antidepressants; N.A., not available; FST, forced swimming test; TST, tail suspension test; NSF, novelty suppressed feeding; MWM, Morris water maze; CRS, chronic restraint stress.

3.6 Wnt Signaling and Epigenetics

The previous paragraphs suggest a functional role for Wnts in stress-related mood disorders. However, recent evidence sheds light on another mechanism that is involved in the onset and development of stress-induced mood disorder, namely epigenetic regulation.[139] Epigenetic regulation is defined as the changes in gene expression that are not coded in the DNA sequence itself. These mechanisms include histone modifications, DNA methylation and post-transcriptional regulation by non-coding RNAs such as microRNAs (miRNAs). It is believed that epigenetic regulation can alter chromatin structure, which, in turn, can regulate gene expression. It is known that epigenetic changes form a molecular link between environmental factors and genetic expression.

This link between genes and the environment makes epigenetic mechanisms interesting candidates for mediating the deleterious effect of stress exposure in the development of mood disorders.[140] The importance of epigenetic regulation in the etiology of mood disorders is reinforced by the observation of epigenetic differences in monozygotic twins who are discordant for depression[141] and bipolar disease.[142] Differences in DNA methylation, a form of epigenetic regulation whereby a methyl group is added to the fifth carbon atom of the nucleotide cytosine, have been widely associated with multiple mood disorders.[139] For example, our group recently profiled genome-wide blood DNA methylation in relation to the development of PTSD in two longitudinal prospective military cohorts. We demonstrated a decrease in DNA methylation levels in 12 genomic regions that were associated with the development of PTSD symptoms. Pathway analysis identified Wnts as being a top enriched signaling pathway showing differential DNA methylation profiles associated with PTSD symptoms scores.[143] Similarly, significant enrichments of differences in blood DNA methylation were found within genes involved in Wnt signaling in mood disorder patients.[144] In schizophrenia, an upregulation of HDAC1 and its epigenetic corepressor, DAXX, was found. Interestingly, this study also found decreased expression of glutamate decarboxylase 67 (GAD67), a gene that is hypothesized to be involved in the Wnt signaling pathway.[145] It is thought that the decreased expression of GAD67 observed in schizophrenia might be associated with epigenetic regulation. Finally, miRNAs, which are short, non-coding RNA molecules known to play a role in post-transcriptional regulation of gene expression, have been shown to be involved in antidepressant response by targeting the Wnt signaling pathway.[146–149]

Mood disorders are partly thought to arise because of an impairment in the ability to make distinctions between new information and similar traumatic memories,[150] a process which is thought to be regulated by adult neurogenesis and synaptic plasticity. Epigenetic mechanisms have been found to be involved in memory formation and proper hippocampal functioning,[151] and along with the explicit role of Wnt signaling in memory formation and plasticity point to an interaction between epigenetic mechanisms and components of the Wnt pathway. Furthermore, epigenetic changes could help explain the environmental effects on Wnt signaling, for example, exposure to an enriched or a deprived environment leads to differences in hippocampal cell survival,[50,55] which is accompanied by altered Wnt expression.[56]

Interestingly, a binding protein of a downstream element in the Wnt signaling pathway, CREB binding protein (CBP), which has endogenous histone acetylase activity, has been found involved in memory formation. It has been demonstrated that mice lacking one allele of CBP showed impaired contextual and cued fear memory along with disrupted novel object recognition.[152] Further experiments showed a direct interaction between miR-431 and *Kremen1*, which is a high-affinity receptor for DKK1. More specifically, miR-431 was found to promote regenerative axon growth through silencing of *Kremen1*.[153,154] Moreover, several miRNAs (miR-296, -124, and -101) were found to alter GSK-3β brain levels and inhibit the Wnt signaling pathway.[155] Similarly, miR-142-3p was found to inhibit Wnt/β-catenin signaling.[156] It has also been shown that Wnt5a exposure holds the potential to lower miR-101b levels in hippocampal neurons, which further affects expression levels of cyclooxygenase-2 (COX2), an enzyme involved in the inflammatory response that has recently been found to play a role in neuronal plasticity.[157,158] Finally, miRNAs have recently been suggested to regulate several components of the Wnt signaling pathways, which in turn have been found to influence miRNA activity.[159,160] These results highlight the importance to gain a deeper understanding of miRNA regulation of these and other signaling pathways in order to assess their potential to be used as diagnostic markers or therapeutic agents for various human diseases. The impact of miRNAs on Wnt pathway signaling in health and disease conditions was recently reviewed elsewhere.[161,162]

Given these interactions between epigenetic mechanisms and Wnt signaling, it could be hypothesized that environmental effects, such as stress, trigger both epigenetic and Wnt signaling molecules to interact and affect hippocampal neuroplasticity. This can ultimately have lasting effects on behavior, which may play a role in psychiatric illnesses such as mood

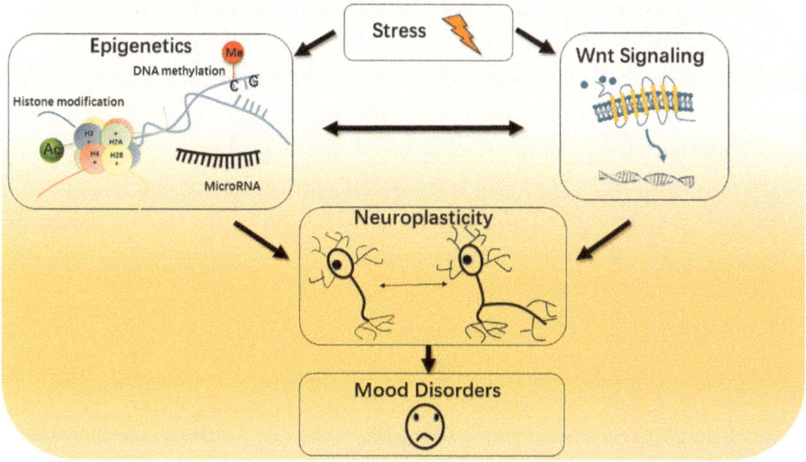

Fig. 3 Crosstalk between stress, Wnt signaling, epigenetic, neuroplasticity and mood disorders. Both epigenetic factors and proteins of the Wnt signaling pathway have been shown to impact on hippocampal neuroplasticity upon exposure to stress. It was demonstrated that epigenetic mechanisms could regulate certain components of the Wnt pathway and vice versa. This can ultimately play a role in psychiatric illnesses including mood disorders.

disorders (Fig. 3). These interactions between the environment and Wnt signaling could potentially provide renewed pathways for the treatment of mood disorders.

4. CONCLUSION AND PROSPECTS

Wnts are considered to be critical components of hippocampal functioning, by regulating neurogenesis and assisting with the integration of these new neurons into existing networks. Exposure to acute or chronic stress is known to inhibit these hippocampal functions by the suppression of hippocampal plasticity; including synaptogenesis, modulation of mature synapses and neurotransmitter release. The latter changes are known to result in interference between ambiguous or uncertain threats, such as overgeneralization, consequently leading to distorted context appropriate stress responses. Furthermore, targeting the canonical Wnt pathway in animal models has shown therapeutic efficacy in mood disorders, through fine-tuning of the Wnt pathway. These animal models indicate the importance of the Wnt pathway in the regulation of a context appropriate stress response.

It is noteworthy that many components of the Wnt signaling pathway act as signaling hubs, regulating multiple signaling pathways.[84,163,164]

For example, GSK-3 is known to regulate signaling pathways such as Sonic Hedgehog, Notch, and growth factor signaling, as well as phosphorylating over 100 substrates which all influence cell survival in the hippocampus.[165,166] Therefore, further research in this field will yield a more comprehensive picture of the regulation of Wnts on aberrant stress responses. Moreover, research concerning selective inhibition or activation of the Wnt pathway should be carried out more extensively. It is generally accepted that the hippocampus is not the only brain structure involved in mood disorders. Additional cerebral structures such as the amygdala, the frontal cortex and the thalamus have also been shown to be involved, revealing further structural abnormalities and misbalanced neurotransmitters.[167,168]

In addition, it would be interesting to further investigate Wnt-mediated plasticity, as this could lead to more rapid action antidepressant effect; such as ketamine through the blockade of NMDA receptors via GSK-3β.[169,170] This could significantly decrease the time that is required for most of the current antidepressants acting on neurogenesis.

Finally, long-lasting elevated cortisol levels are proposed to alter epigenetic patterns in neural stem and progenitor cells, which could lead to quiescence and a reduction of neurogenesis.[171] It would therefore be interesting to investigate the epigenetic pattern of hippocampal Wnt genes. A good starting point would be to investigate GSK-3, as these kinases have been extensively implicated in mood disorders and the functioning of several mood regulating drugs.[68,113,165,172] To be more precise, based on recent research it is recommended to start with GSK-3β, as an elegant study found that GSK-3β but not GSK-3α knock-in mice showed impaired hippocampal progenitor proliferation, mood regulation, and novel object recognition.[173] However, we must keep in mind that the above-mentioned strategies aimed at the stimulation of hippocampal neurogenesis and plasticity may only improve specific hippocampal-related functional deficits, e.g., overgeneralization.

To conclude, more than a quarter century of research on Wnt signaling has significantly enhanced our understanding of the Wnt signaling pathway, resulting in the identification of the canonical Wnt pathway as a key regulatory pathway in development and mood disorders. However, many questions concerning this pathway remain unanswered. We believe that further research aimed at uncovering the molecular mechanisms by which Wnts regulate diverse aspects of structural and functional neural circuit formation will provide a new basis for the development of novel therapeutic targets in stress-related mood disorders.

REFERENCES

1. Gordon MD, Nusse R. Wnt signaling: multiple pathways, multiple receptors, and multiple transcription factors. *J Biol Chem*. 2006;281(32):22429–22433.
2. Berndt JD, Moon RT. Cell biology. Making a point with Wnt signals. *Science*. 2013;339(6126):1388–1389.
3. Kumar M, Camlin NJ, Holt JE, Teixeira JM, McLaughlin EA, Tanwar PS. Germ cell specific overactivation of WNT/βcatenin signalling has no effect on folliculogenesis but causes fertility defects due to abnormal foetal development. *Sci Rep*. 2016;6.
4. Anastas JN, Moon RT. WNT signalling pathways as therapeutic targets in cancer. *Nat Rev Cancer*. 2013;13(1):11–26.
5. Yang K, Wang X, Zhang H, et al. The evolving roles of canonical WNT signaling in stem cells and tumorigenesis: implications in targeted cancer therapies. *Lab Invest*. 2016;96(2):116–136.
6. Harvey K, Marchetti B. Regulating Wnt signaling: a strategy to prevent neurodegeneration and induce regeneration. *J Mol Cell Biol*. 2014;6(1):1–2.
7. Arenas E. Wnt signaling in midbrain dopaminergic neuron development and regenerative medicine for Parkinson's disease. *J Mol Cell Biol*. 2014;6(1):42–53.
8. Chow H-M, Tse K-H, Herrup K. Senescent neurons in the Alzheimer's brain kill nearby healty neurons by blocking their WNT lifeline: the continuing sage of the zombie apocalypse. *Alzheimer's Dement J Alzheimer's Assoc*. 2016;12(7):P658.
9. Schoenfeld TJ, Gould E. Stress, stress hormones, and adult neurogenesis. *Exp Neurol*. 2012;233(1):12–21.
10. Pilar-Cuéllar F, Vidal R, Díaz A, et al. Neural plasticity and proliferation in the generation of antidepressant effects: hippocampal implication. *Neural Plast*. 2013;2013(5): 537265.
11. Duman RS, Voleti B. Signaling pathways underlying the pathophysiology and treatment of depression: novel mechanisms for rapid-acting agents. *Trends Neurosci*. 2012;35(1):47–56.
12. Niehrs C. The complex world of WNT receptor signalling. *Nat Rev Mol Cell Biol*. 2012;13(12):767–779.
13. Janda CY, Waghray D, Levin AM, Thomas C, Garcia KC. Structural basis of Wnt recognition by Frizzled. *Science*. 2012;337(6090):59–64.
14. Inestrosa NC, Varela-Nallar L. Wnt signalling in neuronal differentiation and development. *Cell Tissue Res*. 2015;359(1):215–223.
15. Wang H, Liu T, Malbon CC. Structure-function analysis of Frizzleds. *Cell Signal*. 2006;18(7):934–941.
16. He X, Semenov M, Tamai K, Zeng X. LDL receptor-related proteins 5 and 6 in Wnt/beta-catenin signaling: arrows point the way. *Development*. 2004;131(8): 1663–1677.
17. Gong Y, Bourhis E, Chiu C, et al. Wnt isoform-specific interactions with coreceptor specify inhibition or potentiation of signaling by LRP6 antibodies. *PLoS One*. 2010;5(9):e12682.
18. Kusserow A, Pang K, Sturm C, et al. Unexpected complexity of the Wnt gene family in a sea anemone. *Nature*. 2005;433(7022):156–160.
19. Smit L, Baas A, Kuipers J, Korswagen H, van de Wetering M, Clevers H. Wnt activates the Tak1/Nemo-like kinase pathway. *J Biol Chem*. 2004;279(17):17232–17240.
20. Reya T, Clevers H. Wnt signalling in stem cells and cancer. *Nature*. 2005;434(7035): 843–850.
21. Grumolato L, Liu G, Mong P, et al. Canonical and noncanonical Wnts use a common mechanism to activate completely unrelated coreceptors. *Genes Dev*. 2010;24(22): 2517–2530.

22. Kühl M, Geis K, Sheldahl LC, Pukrop T, Moon RT, Wedlich D. Antagonistic regulation of convergent extension movements in Xenopus by Wnt/beta-catenin and Wnt/Ca2+ signaling. *Mech Dev*. 2001;106(1–2):61–76.
23. Sato A, Yamamoto H, Sakane H, Koyama H, Kikuchi A. Wnt5a regulates distinct signalling pathways by binding to Frizzled2. *EMBO J*. 2009;29(1):41–54.
24. Wang Y. Wnt/planar cell polarity signaling: a new paradigm for cancer therapy. *Mol Cancer Ther*. 2009;8(8):2103–2109.
25. Sokol SY. Spatial and temporal aspects of Wnt signaling and planar cell polarity during vertebrate embryonic development. *Semin Cell Dev Biol*. 2015;42:78–85.
26. Berger H, Breuer M, Peradziryi H, Podleschny M, Jacob R, Borchers A. PTK7 localization and protein stability is affected by canonical Wnt ligands. *J Cell Sci*. 2017;30:1890–1903. jcs-198580.
27. De A. Wnt/Ca2+ signaling pathway: a brief overview. *Acta Biochim Biophys Sin (Shanghai)*. 2011;43(10):745–756.
28. Kuhl M. Ca2+/calmodulin-dependent protein kinase II is stimulated by Wnt and frizzled homologs and promotes ventral cell fates in Xenopus. *J Biol Chem*. 2000;275(17):12701–12711.
29. Hart MJ, de los Santos R, Albert IN, Rubinfeld B, Polakis P. Downregulation of β-catenin by human Axin and its association with the APC tumor suppressor, β-catenin and GSK3β. *Curr Biol*. 1998;8(10):573–581.
30. Behrens J. Functional interaction of an axin homolog, conductin, with -catenin, APC, and GSK3. *Science*. 1998;280(5363):596–599.
31. Liu C, Li Y, Semenov M, et al. Control of beta-catenin phosphorylation/degradation by a dual-kinase mechanism. *Cell*. 2002;108(6):837–847.
32. Aberle H, Bauer A, Stappert J, Kispert A, Kemler R. Beta-catenin is a target for the ubiquitin-proteasome pathway. *EMBO J*. 1997;16(13):3797–3804.
33. Clevers H, Nusse R. Wnt/β-catenin signaling and disease. *Cell*. 2012;149(6):1192–1205.
34. Liu X, Rubin JS, Kimmel AR. Rapid, Wnt-induced changes in GSK3beta associations that regulate beta-catenin stabilization are mediated by Galpha proteins. *Curr Biol*. 2005;15(22):1989–1997.
35. Wu G, Huang H, Garcia Abreu J, He X. Inhibition of GSK3 phosphorylation of beta-catenin via phosphorylated PPPSPXS motifs of Wnt coreceptor LRP6. *PLoS One*. 2009;4(3):e4926.
36. Cselenyi CS, Jernigan KK, Tahinci E, Thorne CA, Lee LA, Lee E. LRP6 transduces a canonical Wnt signal independently of Axin degradation by inhibiting GSK3's phosphorylation of beta-catenin. *Proc Natl Acad Sci USA*. 2008;105(23):8032–8037.
37. Li VSW, Ng SS, Boersema PJ, et al. Wnt signaling through inhibition of β-catenin degradation in an intact Axin1 complex. *Cell*. 2012;149(6):1245–1256.
38. Kim S-E, Huang H, Zhao M, et al. Wnt stabilization of β-catenin reveals principles for morphogen receptor-scaffold assemblies. *Science*. 2013;340(6134):867–870.
39. Zeng X, Tamai K, Doble B, et al. A dual-kinase mechanism for Wnt co-receptor phosphorylation and activation. *Nature*. 2005;438(7069):873–877.
40. Mao J, Wang J, Liu B, et al. Low-density lipoprotein receptor-related protein-5 binds to Axin and regulates the canonical Wnt signaling pathway. *Mol Cell*. 2001;7(4):801–809.
41. Bengoa-Vergniory N, Kypta RM. Canonical and noncanonical Wnt signaling in neural stem/progenitor cells. *Cell Mol Life Sci*. 2015;72(21):4157–4172.
42. MacDonald BT, Yokota C, Tamai K, Zeng X, He X. Wnt signal amplification via activity, cooperativity, and regulation of multiple intracellular PPPSP motifs in the Wnt co-receptor LRP6. *J Biol Chem*. 2008;283(23):16115–16123.

43. Davidson G, Wu W, Shen J, et al. Casein kinase 1 gamma couples Wnt receptor activation to cytoplasmic signal transduction. *Nature*. 2005;438(7069):867–872.
44. Schwarz-Romond T, Merrifield C, Nichols BJ, Bienz M. The Wnt signalling effector Dishevelled forms dynamic protein assemblies rather than stable associations with cytoplasmic vesicles. *J Cell Sci*. 2005;118(Pt 22):5269–5277.
45. Chen W, ten Berge D, Brown J, et al. Dishevelled 2 recruits beta-arrestin 2 to mediate Wnt5A-stimulated endocytosis of Frizzled 4. *Science*. 2003;301(5638):1391–1394.
46. Fiedler M, Mendoza-Topaz C, Rutherford TJ, Mieszczanek J, Bienz M. Dishevelled interacts with the DIX domain polymerization interface of Axin to interfere with its function in down-regulating β-catenin. *Proc Natl Acad Sci USA*. 2011;108(5): 1937–1942.
47. Seib DRM, Corsini NS, Ellwanger K, et al. Loss of Dickkopf-1 restores neurogenesis in old age and counteracts cognitive decline. *Cell Stem Cell*. 2013;12(2):204–214.
48. Jang M-H, Bonaguidi MA, Kitabatake Y, et al. Secreted frizzled-related protein 3 regulates activity-dependent adult hippocampal neurogenesis. *Cell Stem Cell*. 2013;12(2): 215–223.
49. Cruciat C-M, Niehrs C. Secreted and transmembrane wnt inhibitors and activators. *Cold Spring Harb Perspect Biol*. 2013;5(3):a015081.
50. Kempermann G, Kuhn HG, Gage FH. More hippocampal neurons in adult mice living in an enriched environment. *Nature*. 1997;386(6624):493–495.
51. Gogolla N, Galimberti I, Deguchi Y, Caroni P. Wnt signaling mediates experience-related regulation of synapse numbers and mossy fiber connectivities in the adult hippocampus. *Neuron*. 2009;62(4):510–525.
52. Salinas PC, Zou Y. Wnt signaling in neural circuit assembly. *Annu Rev Neurosci*. 2008;31:339–358.
53. Mineur YS, Belzung C, Crusio WE. Functional implications of decreases in neurogenesis following chronic mild stress in mice. *Neuroscience*. 2007;150(2):251–259.
54. Thomas RM, Hotsenpiller G, Peterson DA. Acute psychosocial stress reduces cell survival in adult hippocampal neurogenesis without altering proliferation. *J Neurosci*. 2007;27(11):2734–2743.
55. Lucassen PJ, Oomen CA. Stress, hippocampal neurogenesis and cognition: functional correlations. *Front Biol*. 2016;11(3):182–192.
56. Lie D-C, Colamarino SA, Song H-J, et al. Wnt signalling regulates adult hippocampal neurogenesis. *Nature*. 2005;437(7063):1370–1375.
57. Park M, Shen K. WNTs in synapse formation and neuronal circuitry. *EMBO J*. 2012;31(12):2697–2704.
58. Munji RN, Choe Y, Li G, Siegenthaler JA, Pleasure SJ. Wnt signaling regulates neuronal differentiation of cortical intermediate progenitors. *J Neurosci*. 2011;31(5): 1676–1687.
59. Hirabayashi Y, Itoh Y, Tabata H, et al. The Wnt/beta-catenin pathway directs neuronal differentiation of cortical neural precursor cells. *Development*. 2004;131(12): 2791–2801.
60. Oliva CA, Vargas JY, Inestrosa NC. Wnt signaling: role in LTP, neural networks and memory. *Ageing Res Rev*. 2013;12(3):786–800.
61. Matrisciano F, Busceti CL, Bucci D, et al. Induction of the Wnt antagonist Dickkopf-1 is involved in stress-induced hippocampal damage. *PLoS One*. 2011;6(1):e16447.
62. Zhang Y, Sun Y, Wang F, Wang Z, Peng Y, Li R. Downregulating the canonical Wnt/β-catenin signaling pathway attenuates the susceptibility to autism-like phenotypes by decreasing oxidative stress. *Neurochem Res*. 2012;37(7):1409–1419.
63. Chen J, Park CS, Tang S-J. Activity-dependent synaptic Wnt release regulates hippocampal long term potentiation. *J Biol Chem*. 2006;281(17):11910–11916.

64. Bennett LB, Taylor KH, Arthur GL, Rahmatpanah FB, Hooshmand SI, Caldwell CW. Epigenetic regulation of WNT signaling in chronic lymphocytic leukemia. *Epigenomics*. 2010;2:53–70.
65. Chim CS, Fung TK, Wong KF, Lau JS, Liang R. Infrequent Wnt inhibitory factor-1 (Wif-1) methylation in chronic lymphocytic leukemia. *Leuk Res*. 2006;30(9): 1135–1139.
66. Chim CS, Pang R, Fung TK, Choi CL, Liang R. Epigenetic dysregulation of Wnt signaling pathway in multiple myeloma. *Leukemia*. 2007;21(12):2527.
67. Okamoto H, Voleti B, Banasr M, et al. Wnt2 expression and signaling is increased by different classes of antidepressant treatments. *Biol Psychiatry*. 2010;68(6): 521–527.
68. Kaidanovich-Beilin O, Milman A, Weizman A, Pick CG, Eldar-Finkelman H. Rapid antidepressive-like activity of specific glycogen synthase kinase-3 inhibitor and its effect on beta-catenin in mouse hippocampus. *Biol Psychiatry*. 2004;55(8): 781–784.
69. Gross CG. Neurogenesis in the adult brain: death of a dogma. *Nat Rev Neurosci*. 2000;1(1):67–73.
70. Kaplan MS, Hinds JW. Neurogenesis in the adult rat: electron microscopic analysis of light radioautographs. *Science*. 1977;197(4308):1092–1094.
71. Altman J, Das GD. Autoradiographic and histological evidence of postnatal hippocampal neurogenesis in rats. *J Comp Neurol*. 1965;124(3):319–335.
72. Deng W, Aimone JB, Gage FH. New neurons and new memories: how does adult hippocampal neurogenesis affect learning and memory? *Nat Rev Neurosci*. 2010;11(5):339–350.
73. Eriksson PS, Perfilieva E, Björk-Eriksson T, et al. Neurogenesis in the adult human hippocampus. *Nat Med*. 1998;4(11):1313–1317.
74. Jessberger S, Clark RE, Broadbent NJ, et al. Dentate gyrus-specific knockdown of adult neurogenesis impairs spatial and object recognition memory in adult rats. *Learn Mem*. 2009;16(2):147–154.
75. Nakashiba T, Cushman JD, Pelkey KA, et al. Young dentate granule cells mediate pattern separation, whereas old granule cells facilitate pattern completion. *Cell*. 2012;149(1):188–201.
76. Saxe MD, Battaglia F, Wang J-W, et al. Ablation of hippocampal neurogenesis impairs contextual fear conditioning and synaptic plasticity in the dentate gyrus. *Proc Natl Acad Sci USA*. 2006;103(46):17501–17506.
77. Anacker C, Hen R. Adult hippocampal neurogenesis and cognitive flexibility—linking memory and mood. *Nat Rev Neurosci*. 2017.
78. Wu MV, Hen R. The young and the restless: regulation of adult neurogenesis by Wnt signaling. *Cell Stem Cell*. 2013;12(2):139–140.
79. Faigle R, Song H. Signaling mechanisms regulating adult neural stem cells and neurogenesis. *Biochim Biophys Acta*. 2013;1830(2):2435–2448.
80. Maretto S, Cordenonsi M, Dupont S, et al. Mapping Wnt/beta-catenin signaling during mouse development and in colorectal tumors. *Proc Natl Acad Sci USA*. 2003;100(6):3299–3304.
81. Qu Q, Sun G, Li W, et al. Orphan nuclear receptor TLX activates Wnt/beta-catenin signalling to stimulate neural stem cell proliferation and self-renewal. *Nat Cell Biol*. 2010;12(1):31–40-9.
82. Fiorentini A, Rosi MC, Grossi C, Luccarini I, Casamenti F. Lithium improves hippocampal neurogenesis, neuropathology and cognitive functions in APP mutant mice. *PLoS One*. 2010;5(12):e14382.
83. Wexler EM, Paucer A, Kornblum HI, Palmer TD, Plamer TD, Geschwind DH. Endogenous Wnt signaling maintains neural progenitor cell potency. *Stem Cells*. 2009;27(5):1130–1141.

84. Hussaini SMQ, Choi C-I, Cho CH, Kim HJ, Jun H, Jang M-H. Wnt signaling in neuropsychiatric disorders: ties with adult hippocampal neurogenesis and behavior. *Neurosci Biobehav Rev*. 2014;47:369–383.
85. Budnik V, Salinas PC. Wnt signaling during synaptic development and plasticity. *Curr Opin Neurobiol*. 2011;21(1):151–159.
86. Purro SA, Ciani L, Hoyos-Flight M, Stamatakou E, Siomou E, Salinas PC. Wnt regulates axon behavior through changes in microtubule growth directionality: a new role for adenomatous polyposis coli. *J Neurosci*. 2008;28(34):8644–8654.
87. Davis EK, Zou Y, Ghosh A. Wnts acting through canonical and noncanonical signaling pathways exert opposite effects on hippocampal synapse formation. *Neural Dev*. 2008;3:32.
88. Wilbrecht L, Holtmaat A, Wright N, Fox K, Svoboda K. Structural plasticity underlies experience-dependent functional plasticity of cortical circuits. *J Neurosci*. 2010;30(14): 4927–4932.
89. Hall AC, Lucas FR, Salinas PC. Axonal remodeling and synaptic differentiation in the cerebellum is regulated by WNT-7a signaling. *Cell*. 2000;100(5):525–535.
90. Cerpa W, Gambrill A, Inestrosa NC, Barria A. Regulation of NMDA-receptor synaptic transmission by Wnt signaling. *J Neurosci*. 2011;31(26):9466–9471.
91. Cohen S, Greenberg ME. Communication between the synapse and the nucleus in neuronal development, plasticity, and disease. *Annu Rev Cell Dev Biol*. 2008;24: 183–209.
92. Chen C-M, Orefice LL, Chiu S-L, et al. Wnt5a is essential for hippocampal dendritic maintenance and spatial learning and memory in adult mice. *Proc Natl Acad Sci USA*. 2017;114:E619–E628.
93. Farías GG, Alfaro IE, Cerpa W, et al. Wnt-5a/JNK signaling promotes the clustering of PSD-95 in hippocampal neurons. *J Biol Chem*. 2009;284:15857–15866.
94. Kessler RC. The effects of stressful life events on depression. *Annu Rev Psychol*. 1997;48:191–214.
95. Watanabe Y, Gould E, McEwen BS. Stress induces atrophy of apical dendrites of hippocampal CA3 pyramidal neurons. *Brain Res*. 1992;588(2):341–345.
96. Christian KM, Miracle AD, Wellman CL, Nakazawa K. Chronic stress-induced hippocampal dendritic retraction requires CA3 NMDA receptors. *Neuroscience*. 2011;174:26–36.
97. Ferragud A, Haro A, Sylvain A, Velázquez-Sánchez C, Hernández-Rabaza V, Canales JJ. Enhanced habit-based learning and decreased neurogenesis in the adult hippocampus in a murine model of chronic social stress. *Behav Brain Res*. 2010;210(1): 134–139.
98. Gould E, Tanapat P. Lesion-induced proliferation of neuronal progenitors in the dentate gyrus of the adult rat. *Neuroscience*. 1997;80(2):427–436.
99. Golub Y, Kaltwasser SF, Mauch CP, et al. Reduced hippocampus volume in the mouse model of posttraumatic stress disorder. *J Psychiatr Res*. 2011;45(5):650–659.
100. Fuchs E, Flügge G. Adult neuroplasticity: more than 40 years of research. *Neural Plast*. 2014;2014(5):541870.
101. Sousa N, Lukoyanov NV, Madeira MD, Almeida OF, Paula-Barbosa MM. Reorganization of the morphology of hippocampal neurites and synapses after stress-induced damage correlates with behavioral improvement. *Neuroscience*. 2000;97(2): 253–266.
102. Magarinos AM, Verdugo JMG, McEwen BS. Chronic stress alters synaptic terminal structure in hippocampus. *Proc Natl Acad Sci USA*. 1997;94(25):14002–14008.
103. Boku S, Nakagawa S, Masuda T, et al. Glucocorticoids and lithium reciprocally regulate the proliferation of adult dentate gyrus-derived neural precursor cells through GSK-3beta and beta-catenin/TCF pathway. *Neuropsychopharmacology*. 2009;34(3): 805–815.

104. Datson NA, van der Perk J, de Kloet ER, Vreugdenhil E. Identification of corticosteroid-responsive genes in rat hippocampus using serial analysis of gene expression. *Eur J Neurosci*. 2001;14(4):675–689.
105. Malberg JE, Eisch AJ, Nestler EJ, Duman RS. Chronic antidepressant treatment increases neurogenesis in adult rat hippocampus. *J Neurosci*. 2000;20(24): 9104–9110.
106. Mehta D, Klengel T, Conneely KN, et al. Childhood maltreatment is associated with distinct genomic and epigenetic profiles in posttraumatic stress disorder. *Proc Natl Acad Sci USA*. 2013;110(20):8302–8307.
107. Lijam N, Paylor R, McDonald MP, et al. Social interaction and sensorimotor gating abnormalities in mice lacking Dvl1. *Cell*. 1997;90(5):895–905.
108. Maguschak KA, Ressler KJ. Wnt signaling in amygdala-dependent learning and memory. *J Neurosci*. 2011;31(37):13057–13067.
109. Maguschak KA, Ressler KJ. Beta-catenin is required for memory consolidation. *Nat Neurosci*. 2008;11(11):1319–1326.
110. Prickaerts J, Moechars D, Cryns K, et al. Transgenic mice overexpressing glycogen synthase kinase 3beta: a putative model of hyperactivity and mania. *J Neurosci*. 2006;26(35):9022–9029.
111. O'Brien WT, Harper AD, Jové F, et al. Glycogen synthase kinase-3beta haploinsufficiency mimics the behavioral and molecular effects of lithium. *J Neurosci*. 2004;24(30):6791–6798.
112. Beaulieu J-M, Sotnikova TD, Yao W-D, et al. Lithium antagonizes dopamine-dependent behaviors mediated by an AKT/glycogen synthase kinase 3 signaling cascade. *Proc Natl Acad Sci USA*. 2004;101(14):5099–5104.
113. Wilkinson MB, Dias C, Magida J, et al. A novel role of the WNT-dishevelled-GSK3β signaling cascade in the mouse nucleus accumbens in a social defeat model of depression. *J Neurosci*. 2011;31(25):9084–9092.
114. Gould TD, O'Donnell KC, Picchini AM, Dow ER, Chen G, Manji HK. Generation and behavioral characterization of beta-catenin forebrain-specific conditional knock-out mice. *Behav Brain Res*. 2008;189(1):117–125.
115. Gould TD, Dow ER, O'Donnell KC, Chen G, Manji HK. Targeting signal transduction pathways in the treatment of mood disorders: recent insights into the relevance of the Wnt pathway. *CNS Neurol Disord Drug Targets*. 2007;6(3):193–204.
116. Kaidanovich-Beilin O, Lipina TV, Takao K, et al. Abnormalities in brain structure and behavior in GSK-3alpha mutant mice. *Mol Brain*. 2009;2(1):35.
117. Voleti B, Tanis KQ, Newton SS, Duman RS. Analysis of target genes regulated by chronic electroconvulsive therapy reveals role for Fzd6 in depression. *Biol Psychiatry*. 2012;71(1):51–58.
118. Snyder JS, Soumier A, Brewer M, Pickel J, Cameron HA. Adult hippocampal neurogenesis buffers stress responses and depressive behaviour. *Nature*. 2011;476(7361): 458–461.
119. Pilar-Cúellar F, Vidal R, Díaz A, et al. Signaling pathways involved in antidepressant-induced cell proliferation and synaptic plasticity. *Curr Pharm Des*. 2014;20(23): 3776–3794.
120. Santarelli L. Requirement of hippocampal neurogenesis for the behavioral effects of antidepressants. *Science*. 2003;301(5634):805–809.
121. Wang J-W, David DJ, Monckton JE, Battaglia F, Hen R. Chronic fluoxetine stimulates maturation and synaptic plasticity of adult-born hippocampal granule cells. *J Neurosci*. 2008;28(6):1374–1384.
122. Pinnock SB, Blake AM, Platt NJ, Herbert J. The roles of BDNF, pCREB and Wnt3a in the latent period preceding activation of progenitor cell mitosis in the adult dentate gyrus by fluoxetine. *PLoS One*. 2010;5(10):e13652.

123. David DJ, Samuels BA, Rainer Q, et al. Neurogenesis-dependent and -independent effects of fluoxetine in an animal model of anxiety/depression. *Neuron.* 2009;62(4): 479–493.
124. Zhou WJ, Xu N, Kong L, et al. The antidepressant roles of Wnt2 and Wnt3 in stress-induced depression-like behaviors. *Transl Psychiatry.* 2016;6(9):e892.
125. Contestabile A, Greco B, Ghezzi D, Tucci V, Benfenati F, Gasparini L. Lithium rescues synaptic plasticity and memory in Down syndrome mice. *J Clin Invest.* 2013;123(1): 348–361.
126. Wexler EM, Geschwind DH, Palmer TD. Lithium regulates adult hippocampal progenitor development through canonical Wnt pathway activation. *Mol Psychiatry.* 2008;13(3):285–292.
127. Wood GE, Young LT, Reagan LP, Chen B, McEwen BS. Stress-induced structural remodeling in hippocampus: prevention by lithium treatment. *Proc Natl Acad Sci USA.* 2004;101(11):3973–3978.
128. Wang L, Liu Y, Li S, Long Z-Y, Wu Y-M. Wnt signaling pathway participates in valproic acid-induced neuronal differentiation of neural stem cells. *Int J Clin Exp Pathol.* 2015;8(1):578–585.
129. Mostany R, Valdizan E, Pazos A. A role for nuclear β-catenin in SNRI antidepressant-induced hippocampal cell proliferation. *Neuropharmacology.* 2008;55:18–26.
130. Madsen TM, Treschow A, Bengzon J, Bolwig TG, Lindvall O, Tingström A. Increased neurogenesis in a model of electroconvulsive therapy. *Biol Psychiatry.* 2000;47(12): 1043–1049.
131. Madsen T, Newton S, Eaton M. Chronic electroconvulsive seizure up-regulates β-catenin expression in rat hippocampus: role in adult neurogenesis. *Biol Psychiatry.* 2003;54:1006–1014.
132. Nakamura K, Ito M, Liu Y, Seki T, Suzuki T, Arai H. Effects of single and repeated electroconvulsive stimulation on hippocampal cell proliferation and spontaneous behaviors in the rat. *Brain Res.* 2013;1491:88–97.
133. Bétry C, Etiévant A, Pehrson A, Sánchez C, Haddjeri N. Effect of the multimodal acting antidepressant vortioxetine on rat hippocampal plasticity and recognition memory. *Prog Neuropsychopharmacol Biol Psychiatry.* 2015;58:38–46.
134. Perera TD, Dwork AJ, Keegan KA, et al. Necessity of hippocampal neurogenesis for the therapeutic action of antidepressants in adult nonhuman primates. *PLoS One.* 2011;6(4):e17600.
135. Rola R, Fishman K, Baure J, et al. Hippocampal neurogenesis and neuroinflammation after cranial irradiation with (56)Fe particles. *Radiat Res.* 2008;169(6):626–632.
136. Nishimura W, Yao I, Iida J. Interaction of synaptic scaffolding molecule and β-catenin. *J Neurosci.* 2002;22(3):757–765.
137. Bamji S, Shimazu K, Kimes N. Role of β-catenin in synaptic vesicle localization and presynaptic assembly. *Neuron.* 2003;40:719–731.
138. Speese SD, Budnik V. Wnts: up-and-coming at the synapse. *Trends Neurosci.* 2007;30(6):268–275.
139. McGowan PO, Kato T. Epigenetics in mood disorders. *Environ Health Prev Med.* 2008;13(1):16–24.
140. Schmitt A, Malchow B, Hasan A, Fallkai P. The impact of environmental factors in severe psychiatric disorders. *Front Neurosci.* 2014;8:19.
141. Cordova-Palomera A, Fatjo-Vilas M, Gasto C, Navarro V, Krebs MO, Fananas L. Genome-wide methylation study on depression: differential methylation and variable methylation in monozygotic twins. *Transl Psychiatry.* 2015;5(4):e557.
142. Dempster EL, Pidsley R, Schalkwyk LC, et al. Disease-associated epigenetic changes in monozygotic twins discordant for schizophrenia and bipolar disorder. *Hum Mol Genet.* 2011;20:4786–4796, ddr416.

143. Rutten BPF, Vermetten E, Vinkers CH, et al. Longitudinal analyses of the DNA methylome in deployed military servicemen identify susceptibility loci for post-traumatic stress disorder. *Mol Psychiatry*. 2017.
144. Walker RM, Sussmann JE, Whalley HC, et al. Preliminary assessment of pre-morbid DNA methylation in individuals at high genetic risk of mood disorders. *Bipolar Disord*. 2016;18(5):410–422.
145. Benes FM, Lim B, Matzilevich D, Walsh JP, Subburaju S, Minns M. Regulation of the GABA cell phenotype in hippocampus of schizophrenics and bipolars. *Proc Natl Acad Sci USA*. 2007;104(24):10164–10169.
146. Vasu MM, Anitha A, Takahashi T, et al. Fluoxetine increases the expression of miR-572 and miR-663a in human neuroblastoma cell lines. *PLoS One*. 2016;11(10): e0164425.
147. Zhou R, Yuan P, Wang Y, et al. Evidence for selective microRNAs and their effectors as common long-term targets for the actions of mood stabilizers. *Neuropsychopharmacol Off Publ Am Coll Neuropsychopharmacol*. 2009;34(6):1395.
148. Enatescu VR, Papava I, Enatescu I, et al. Circulating plasma micro RNAs in patients with major depressive disorder treated with antidepressants: a pilot study. *Psychiatry Investig*. 2016;13(5):549–557.
149. Lopez JP, Fiori LM, Cruceanu C, et al. MicroRNAs 146a/b-5 and 425-3p and 24-3p are markers of antidepressant response and regulate MAPK/Wnt-system genes. *Nat Commun*. 2017;8.
150. Maren S, Phan KL, Liberzon I. The contextual brain: implications for fear conditioning, extinction and psychopathology. *Nat Rev Neurosci*. 2013;14(6):417–428.
151. Hunter RG, McCarthy KJ, Milne TA, Pfaff DW, McEwen BS. Regulation of hippocampal H3 histone methylation by acute and chronic stress. *Proc Natl Acad Sci USA*. 2009;106(49):20912–20917.
152. Alarcón JM, Malleret G, Touzani K, et al. Chromatin acetylation, memory, and LTP are impaired in CBP+/− mice: a model for the cognitive deficit in Rubinstein-Taybi syndrome and its amelioration. *Neuron*. 2004;42(6):947–959.
153. Mao B, Wu W, Davidson G, Marhold J. Kremen proteins are Dickkopf receptors that regulate Wnt/beta-catenin signalling. *Nature*. 2002;417(6889):664.
154. Wu D, Murashov AK. MicroRNA-431 regulates axon regeneration in mature sensory neurons by targeting the Wnt antagonist Kremen1. *Front Mol Neurosci*. 2013;6.
155. Liu Y, Huang T, Zhao X, Cheng L. MicroRNAs modulate the Wnt signaling pathway through targeting its inhibitors. *Biochem Biophys Res Commun*. 2011;408(2):259–264.
156. Hu T, Phiwpan K, Guo J, et al. MicroRNA-142-3p negatively regulates canonical Wnt signaling pathway. *PLoS One*. 2016;11(6):e0158432.
157. Codocedo JF, Inestrosa NC. Wnt-5a-regulated miR-101b controls COX2 expression in hippocampal neurons. *Biol Res*. 2016;49(1):9.
158. Cowley TR, Fahey B, O'Mara SM. COX-2, but not COX-1, activity is necessary for the induction of perforant path long-term potentiation and spatial learning in vivo. *Eur J Neurosci*. 2008;27(11):2999–3008.
159. Inui M, Martello G, Piccolo S. MicroRNA control of signal transduction. *Nat Rev Mol Cell Biol*. 2010;11(4):252.
160. Schepeler T. Emerging roles of microRNAs in the Wnt signaling network. *Crit Rev Oncog*. 2013;18(4).
161. Onyido EK, Sweeney E, Nateri AS. Wnt-signalling pathways and microRNAs network in carcinogenesis: experimental and bioinformatics approaches. *Mol Cancer*. 2016;15(1):56.
162. Song JL, Nigam P, Tektas SS, Selva E. microRNA regulation of Wnt signaling pathways in development and disease. *Cell Signal*. 2015;27(7):1380–1391.

163. Hiester BG, Galati DF, Salinas PC, Jones KR. Neurotrophin and Wnt signaling cooperatively regulate dendritic spine formation. *Mol Cell Neurosci.* 2013;56:115–127.
164. Fortress AM, Frick KM. Hippocampal Wnt signaling: memory regulation and hormone interactions. *Neuroscientist.* 2015;22(3):278–294.
165. Kim W-Y, Snider WD. Functions of GSK-3 signaling in development of the nervous system. *Front Mol Neurosci.* 2011;4:44.
166. Kaidanovich-Beilin O, Woodgett JR. GSK-3: functional insights from cell biology and animal models. *Front Mol Neurosci.* 2011;4:40.
167. Savitz JB, Rauch SL, Drevets WC. Clinical application of brain imaging for the diagnosis of mood disorders: the current state of play. *Mol Psychiatry.* 2013;18(5):528–539.
168. Russo SJ, Nestler EJ. The brain reward circuitry in mood disorders. *Nat Rev Neurosci.* 2013;14(9):609–625.
169. Li N, Lee B, Liu R, Banasr M, Dwyer J. mTOR-dependent synapse formation underlies the rapid antidepressant effects of NMDA antagonists. *Science.* 2010;329(5994):959–964.
170. Beurel E, Song L, Jope RS. Inhibition of glycogen synthase kinase-3 is necessary for the rapid antidepressant effect of ketamine in mice. *Mol Psychiatry.* 2011;16(11):1068–1070.
171. Yang X, Ewald ER, Huo Y, et al. Glucocorticoid-induced loss of DNA methylation in non-neuronal cells and potential involvement of DNMT1 in epigenetic regulation of Fkbp5. *Biochem Biophys Res Commun.* 2012;420(3):570–575.
172. Freyberg Z, Ferrando SJ, Javitch JA. Roles of the Akt/GSK-3 and Wnt signaling pathways in schizophrenia and antipsychotic drug action. *Am J Psychiatry.* 2010;167(4):388–396.
173. Pardo M, Abrial E, Jope RS, Beurel E. GSK3β isoform-selective regulation of depression, memory and hippocampal cell proliferation. *Genes Brain Behav.* 2016;15:348–355.

CHAPTER EIGHT

Neuroepigenetics of Neurotrophin Signaling: Neurobiology of Anxiety and Affective Disorders

Mathilde Règue-Guyon*, Laurence Lanfumey*, Raymond Mongeau[†,1]
*Inserm UMR S894, Centre de Psychiatrie et Neuroscience, Université Paris Descartes, Paris, France
[†]EA 4475, Pharmacologie de la Circulation Cérébrale, Faculté de Pharmacie, Université Paris Descartes, Paris, France
[1]Corresponding author: e-mail address: raymond.mongeau@parisdescartes.fr

Contents

Progress in Molecular Biology and Translational Science, Volume 158
ISSN 1877-1173
https://doi.org/10.1016/bs.pmbts.2018.03.002

Abstract

This review examines the epigenetic of neurotrophin signaling in anxiety, affective disorders and related symptoms associated with drug addiction, in particular alcoholism. It is first important to understand the epigenetics of aversion memories, as they are so central to anxiety and affective disorders symptomology. The crucial role of neurotrophins in memory formation, in particular the brain-derived nerve growth factor (BDNF), is explored at the physiological and behavioral levels. Numerous studies describe how various epigenetic phenomena, mainly histone acetylation, histone methylation, DNA methylation, but also other less known epigenetic phenomena such as histone poly[ADP]-ribosylation and 5-HT2C receptor pre-mRNA editing, exert significant regulatory roles in aversion memory and fear extinction memory formation. Other models of anxiety and affective disorders, that use stress or transgenic constructs directed at elements of the stress axis or the serotonergic system, are then explored in relation with the epigenetic of neurotrophin signaling. Epigenetic marks differentially change according to brain areas. In the hippocampus for example, anxious or chronically stressed animals tend to show epigenetic changes that are at the opposite of those observed after memory consolidation following a brief aversive stimulus. Behaviorally relevant epigenetic changes have been found to be reversible by drug treatments. Surprisingly, moderate alcohol consumption may trigger, on the long term, changes of BDNF expression and of its epigenetic elements that are somehow similar to those produced by antidepressant drugs. However, alcohol withdrawal associated with anxiety symptoms has not yet been very well explored. Overall, it appears that a multidisciplinary view on the epigenetics of neurotrophin secretion might bring innovative treatments to psychiatric diseases involving stress and fear memories.

1. INTRODUCTION

The etiology of anxiety and affective disorders has traditionally been focused on genetic factors. More recently, evidence has been accumulating indicating that epigenetic factors, independent of DNA sequence variations, determine the course of affective disorders, panic disorder (PD), post-traumatic stress disorder (PTSD), social anxiety disorder (SAD) as well as withdrawal anxiety following drug addiction. Drugs, chronic stress as well as traumatic memories encode persistent epigenetic reprogramming that change brain interaction with the environment and lead to dysfunctional behaviors. Exposure to substance or environmental conditions generating anxiety and affective symptoms may trigger epigenetic dysregulation of synaptic plasticity through neurotrophic factors. Neurotrophic factors are small peptides or proteins that are essential for growth, differentiation of neurons and development of the central nervous system (CNS). In the adult CNS, they are also involved in neuron survival, in brain repair following brain injuries and in the formation of long-term memory. Many neurotrophic factors, such as the brain-derived nerve growth factor (BDNF), exert their "trophic effects" on neurons and other cells by tyrosine kinase receptors signaling.

Neurotrophic factor epigenetic regulations should be viewed alongside various other epigenetic alterations that may account for distinct aspects of anxiety and affective disorders. For example, SAD was shown to be associated with DNA hypomethylation at the oxytocin receptor gene involved in prosocial effects.[1] In turn, the corticotropin releasing hormone (CRH) is crucially involved in the hypothalamic–pituitary–adrenal stress (HPA) axis and PD symptoms are associated with hypomethylation of the CRH1 receptor promoter.[2] Catecholaminergic hyperactivity has also often been involved in anxiety. Polymorphism (Met/Met) at the catechol-*O*-methyltransferase enzyme, critical for the catabolic regulation of synaptic dopamine, was shown to be involved in anxiety predisposition and the Met/Met genotype is associated with DNA methylation at four CpG sites, two of which were associated with impaired fear inhibition.[3] There exist also a predisposition to affective disorders and alcohol dependence involving the short promoter version of the serotonergic reuptake carrier (S-5-HTTLPR) gene, which exhibits a tendency toward increased methylation of CpG islands.[4]

Beyond chromatin changes and DNA methylation, the term "epigenetic" refers to a variety of molecular modifications that have long-term effects on genes programs without changing DNA sequences. The 5-HT2C receptor is the only 5-HT receptor being the target of a peculiar epigenetic regulation, pre-mRNA editing. Numerous studies show that suicide may be triggered by an increased 5-HT2C receptor pre-mRNA editing, regulating 5-HT2C receptor alternative splicing and 5-HT2C receptor density, which themselves are found dysregulated in the brain of individuals who committed suicide.[5–8] Anxiety and affective disorders are thus associated with a wide range of epigenetic changes not only at the neurotrophin level, but also at the HPA axis and the monoaminergic system levels.

The risk of developing anxiety and affective disorders is increased through epistatic interaction between S-5-HTTLPR and the Val66Met polymorphism of BDNF, resulting in altered BDNF secretion.

S-5-HTTLPR and BDNF-Met predispose to epigenetic aberrations, which additionally contribute to make individuals more susceptible to environmental adversities.[9] Numerous studies on patients revealed aberrant DNA methylation of the BDNF gene: the promoter of *Bdnf* exon IV was found hypermethylated in post-mortem brains from individuals who committed suicide.[10] Other studies found hypermethylation of the *Bdnf* promoter in blood of individuals that had depression or suicidal ideation, suggesting that methylation could be a biomarker of suicide risk.[11,12] Hypermethylation of the *Bdnf* promoter has been found in association with multiple related conditions including late-life depression,[13] post-stroke depression,[14] depressive disorder generated by the acute coronary syndrome,[15] dysphoric symptoms triggered by alcohol withdrawal[16] and PTSD.[17] The brain atrophies and reduced white matter integrity, often found in these pathological conditions, may well be associated with the *Bdnf* promoter hypermethylation.[18,19]

Common epigenetic modifications at neurotrophic genes therefore do exist in various pathologies related to depression and anxiety. So far, the literature on trophic factors in human is almost completely focused on the *Bdnf* gene. The latter exhibits several functional exons, which make its epigenetic analysis more complex. Furthermore, the relationship between specific symptoms and gene repression operated by DNA methylation on neurotrophins' secretion can be understood only through exploration of the differential epigenetic marks into specific brain areas. Such information is rather scarce in the clinical literature. One also wonders how much these epigenetic changes relate to specific functions, symptoms or co-morbidities. For example, *Bdnf* epigenetic changes may be triggered by alcohol addiction frequently associated with anxiety and relapses in PTSD patients.[17]

The relationship between alcohol addiction and anxiety is also interesting from the point of view of persistent memory changes occurring in both human addiction and PTSD, claimed to have common epigenetic and neuroanatomical basis at the level of extinction circuits.[20] Animal studies also show that epigenetic alterations, such as increased 5-HT2CR pre-mRNA editing in the nucleus accumbens (NAc) and the decrease in DNA methylation of the *Bdnf* gene in the hippocampus may result from chronic alcohol consumption,[21,22] while cocaine addiction prior to chronic stress potentiate depressive-like behaviors in mice through decreased levels of histone H3 dimethylation regulating accumbal BDNF signaling.[23]

The human studies listed above relate only to methylation and 5-HT2CR editing changes. Other crucial and more transient epigenetic alterations, such as chromatin remodeling through histone acetylation in relation with trophic factors for example, have only been investigated so far in relevant animal models. Pathological conditions can be modeled using stresses or relevant transgenic constructs at the level of the serotonergic or the HPA axis levels. We will thus examine all these topics, and in the first section, how epigenetic and trophic factors regulate the formation as well as the extinction of aversion memory, since it is crucial to understand anxiety and cognitive symptoms.[24,25] Second, we review the evidence on epigenetic regulation of neurotrophins in relation to other animal models of anxiety and affective disorders. Finally, we explore addiction in relation with the epigenetics of neurotrophins, in association with consequent anxiety and depressive symptoms. To conclude, we provide a comprehensive framework, and describe the current challenges and perspectives.

2. EPIGENETIC CHANGES AT NEUROTROPHINS GENES REGULATING AVERSION MEMORIES

2.1 Involvement of Neurotrophins in Synaptic Plasticity and Memory

Among all neurotransmitters and trophic factors, glutamate and BDNF have been among the most involved in memory. BDNF directly depolarize neurons, enhances glutamatergic transmission and induces phosphorylation of NMDA receptor subunits via its TrkB receptor.[26] Both glutamate and BDNF exert positive regulations on long-term potentiation (LTP), the electrophysiological model of reference to study memory formation at the cellular level. Consistently, there is a marked impairment of LTP in mutant mice lacking BDNF, and this is rescued by recombinant BDNF application. The endogenous BDNF required for LTP formation origins from presynaptic neurons.[27] The synaptic plasticity function of neurotrophins occurs through a modulation of neuron growth and dendrites arborization, and as far as BDNF is concerned, this is neural activity dependent.[28] BDNF-dependent LTP formation involves protein synthesis[29] and translation, the final step of DNA to protein synthesis, is increased by BDNF via multiple intracellular signaling pathways including PI3-K and Akt, kinases involved in cell growth, differentiation, survival and intracellular trafficking. Other neurotrophins, such as the insulin-like growth factor, IGF-II, also modulate LTP and memory formation through *de novo* protein synthesis, a process dependent on the glycogen synthase kinase 3 (GSK3), an enzyme active in a number of central intracellular signaling pathways.[30]

2.2 Neurotrophins Involvement in Conditioned Aversive Memories

Neurotrophins are well known to be required for the consolidation of aversion memories. In the passive avoidance test (in which animals learn to abstain from a particular response to avoid an aversive stimulus), treatments either with an anti-BDNF antibody or with *Bdnf* antisense mRNA produce memory impairments in conjunction with a reduction of LTP and ERK signaling.[31,32] However, the precise role of TrkB receptors in avoidance memory remains to be established.[33] The most used animal model to assess the relation between TrkB receptors, neurotrophins, epigenetic and aversion memory is the fear-conditioning paradigm, in which an aversive foot shock is associated with the conditioned stimulus (CS), that is, either a context or a cue (such as a tone). The animal is then placed back into the same context (the cage where it has first received foot shocks) to assess contextual fear or into a new context to assess conditioned fear expression to a cue. Neurotrophins expressions were explored after either contextual or cue fear consolidation or after acquisition of fear extinction, during which the CS is repeatedly presented without aversive consequences, so as to extinguish conditioned fear.

BDNF expression is required as much for acquisition, consolidation as extinction of fear, although its exact role may change depending on brain areas. BDNF TrkB receptors are required for acquisition and consolidation of conditioned fear.[34,35] A higher level of BDNF in the amygdala during the consolidation period correlates with a greater fear response to a CS presentation 24 h later.[36] Fear memories are stored in the amygdala, while processing of contextual information, necessary for contextual fear acquisition and its extinction, require the hippocampus. Interestingly, hippocampus-specific deletion of BDNF impairs fear extinction,[37] while hippocampal BDNF favors acquisition of extinction memory.[38] In interconnection with the hippocampus, the most crucial area for fear extinction processes is the infralimbic area of the prefrontal cortex.[39,40] BDNF is one of the key players in establishing fear extinction memory, since its infusion into the infralimbic cortex induces by itself fear extinction.[38,41]

Several other neurotrophins, beside BDNF, have been implicated in fear memory. A study shows that hippocampal IGF-2 is up-regulated after contextual fear extinction, while microinjection of IGF-2 into the hippocampus facilitates extinction.[42] In turn, mutant mice lacking the GFRa2 receptor of glial cell line-derived neurotrophic factor (GDNF), have impaired contextual fear conditioning.[43] In contrast, rats naturally selected to exhibit low levels of conditioned fear have greater hippocampal expression of fibroblast growth factor-2.[44] Therefore, there are neurotrophins interplays in various brain areas for encoding aversion and extinction memories.

2.3 Histone Modifications Occurring During Aversive Memory Regulation

Neurotrophins secretion are crucial for protein synthesis and dendrites remodeling necessary for memory trace formation, but epigenetic factors intervene in converting short-term fear memory into persistent and robust long-term memory. Epigenetic mechanisms are thus upstream of neurotrophic alterations, and determine which and where gene products are available for expression of learning. In pathological conditions, such as anxiety, epigenetic mechanisms may trigger the formation of excessively strong and persistent fear memories.

Increased histone acetylation is generally associated with memory formation, and consistently, HDAC inhibitors enhance fear memory consolidation. In keeping with this view, microinjection of an HDAC inhibitor into the hippocampus enhances contextual fear, but does not alter cue auditory fear memory,[45] while infusing an HDAC inhibitor into the amygdala does enhance cue auditory fear.[46] Consistently, histone acetylation increases in both the amygdala and the hippocampus after fear conditioning (Fig. 1).[46,57] There is a complex pattern of histone acetylation among various brain areas emerging soon after fear conditioning. Hippocampal histone H3 (not H4) was found more acetylated 1 h, but not 24 h, after contextual fear conditioning.[57] Other histone modifications, beside acetylation, are important for fear memory consolidation. Histone H3 is phosphorylated by ERK/MAPK following contextual fear conditioning, corresponding to specific gene activation in the hippocampus.[58] Epigenetic marks at various histone residues, such as methylation at lysine residues, can either activate or repress transcription during fear conditioning. Indeed, trimethylation of histone H3 lysine 4, permissive to transcription, and dimethylation of histone H3 lysine 9, repressing transcription are both increased in the hippocampus following contextual fear.[59]

2.4 Histone Modifications Lead to Neurotrophins Regulation of Fear Memories

An important question is whether any of these histone modifications resulting in increase gene expression account for neurotrophins secretion during aversive memory formation. Up until now, the literature appears

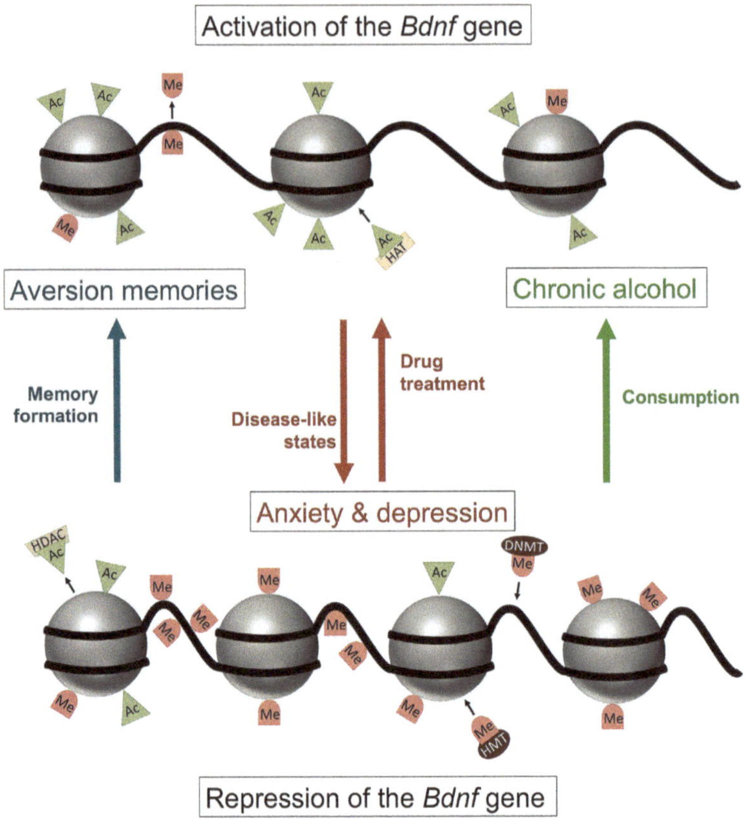

Fig. 1 Epigenetic changes at hippocampal *Bdnf* gene, in relation with the regulation of aversion memories, in models of anxiety and depression and with chronic alcohol intake. Epigenetic changes may be different depending on brain areas. In the hippocampus, aversion memory formation is associated with an increase in acetylation, notably at histone H3,[47,48] and decreased DNA methylation[49] at the level of the *Bdnf* gene. This leads to DNA decompaction and activation of the Bdnf gene. Second, models using stress procedures (ELS, SDS, chronic stress, predator exposure) are consistently associated with increased DNA and histone methylation.[50–53] Transgenic models (GR-i mice or the SERT KO mice) also display increase methylation of the *Bdnf gene*, associated with dysregulations in the gene expression of several DNMT.[54,55] Finally, chronic ethanol consumption is linked to both types of epigenetic modifications. Chronic ethanol intake is associated with increased marks of activation at histone H3[56] at several *Bdnf* promoters and with decreased DNA methylation of the *Bdnf* gene.[21] An alteration in the mRNA expression of Dnmt3b in the hippocampus is also detected after chronic ethanol.[21] Interestingly, *Bdnf* promoter 4 methylation seems to be increased in the serum of alcohol-dependent patients, but the latter is decreased during alcohol withdrawal.[16] It would be interesting to know whether anxiety symptoms associated with alcohol withdrawal in rodents is associated with similar epigenetic changes at the hippocampal level.

exclusively focused on *Bdnf*. Together with an increase of *Bdnf* exon IV and exon IX (reflecting total BDNF secretion) in the hippocampus appearing 2 h after contextual fear conditioning, there is an increase acetylation of H3 at the *Bdnf* promoter 4 (Fig. 1).[47] Furthermore, histone H3 phosphoacetylation at *Bdnf* promoter 4 was observed and this, together with histone acetylation, was argued to positively contribute to regulate total BDNF during conditioning.[47]

In a different paradigm, a single prolonged stress (SPS) reduces, rather than increase, hippocampal *Bdnf* exon IV in parallel with a decrease H3 acetylation at *Bdnf* promoter 4, a change that occurs 2 h after SPS, but that does not persist 24 h after SPS. Interestingly, SPS enhances contextual fear conditioning, potentiates *Bdnf* exon IV transcription and increases H3 acetylation at *Bdnf* promoter 4 in the hippocampus (Fig. 1).[48] Based on this evidence, Fuchikami and colleagues argued that the enhanced levels of BDNF occurring through histone acetylation, triggered by a traumatic-like stress, and observed during fear memory consolidation is somehow relevant with the persistent fear memory observed in anxious patients.[48]

The other key brain area for fear extinction, the frontal cortex, was investigated for changes in acetylation patterns.[60] As in the hippocampus with contextual fear, frontal cortex histone H3 (but not H4) is more acetylated at *Bdnf* promoter P1 and P4 after cue fear consolidation. In turn, extinction training reduces acetylation of H3 at P1, and has no effect on the increased acetylation that occurs with conditioning at P4. *Bdnf* exon IV is increased after fear extinction in relation with these changes, compared to either naïve mice or mice that had only fear conditioning. With extinction appears also an increase in H4 acetylation that accounts for the fact that *Bdnf* exon IV level is higher after fear conditioning followed by extinction (compared to fear conditioning alone).[60]

2.5 DNA Methylation Is Differentially Altered in Brain Areas With Fear Consolidation

DNA methylation generally inhibits gene expression by preventing transcription factors from binding to promoter regions. Blocking this process results in fear learning impairment. Brain expression of DNA methyltransferase (DNMT) was measured following fear conditioning. There is an increase of DNMT in the lateral amygdala during auditory fear conditioning.[46] This might be surprising, as one would expect less gene repression during memory formation. Furthermore, DNMT increases in the hippocampus with contextual fear memory.[61] Contextual fear persists for many months, and this most likely involves epigenetic changes in prefrontal cortical areas (such as the anterior cingulate and prelimbic cortices), rather than in the hippocampus. Indeed, the cortex encodes "remote" compared to "recent" contextual fear memories that are rather hippocampus dependent. DNA methylation in the prefrontal cortex can persist at specific promoter regions for at least 1 month.[62] Microinjection of a DNMT inhibitor into the anterior cingulate cortex prevents DNA methylation and disrupts in conjunction the stability of remote contextual fear memory.[62]

2.6 DNA Methylation Changes Lead to Neurotrophic Regulation of Fear Memories

Although observations of global DNMT up-regulations are revealing, a crucial question in relation with synaptic plasticity changes, is whether increases or decreases of methylation at specific neurotrophins genes exist during aversive memory formation. In one study[47] contextual fear conditioning was found to be associated with hypomethylation at several *Bdnf* promoter sites. This contrasts with the increase methylation expected from global gene studies described above. Furthermore, contextual fear conditioning is associated with an increase in *Bdnf* exon IV expression, in line with the repressive action exerted by methylation on gene expression. Context exposure alone (in absence of fear conditioning) does not change *Bdnf* exon IV expression nor does it change *Bdnf* promoter 4 methylation. However, some active demethylation at several CpG sites within exon IV were found after context exposure alone, despite the fact that demethylation of exon IV is generally greater at CpG site with contextual fear conditioning vs. context exposure alone.[47] Another study[49] has shown similar transcriptional changes in the *Bdnf* gene that persist at least 24 h after conditioning, suggesting that hippocampal BDNF is required for the establishment of long-lasting memory storage. Indeed, hypomethylation of a particular CpG island in a *Bdnf* exon is detected at 0.5 h and maintained for up to 24 h after contextual fear conditioning, and these changes are associated with *Bdnf* exons up-regulation (Fig. 1). However, no further increases in *Bdnf* exons or DNA demethylation are observed following fear memory re-consolidation.[49]

In parallel with these hypomethylation changes in hippocampus-dependent memory, an increased DNA methylation of *Bdnf* exon IV and a concomitant decrease in mRNA expression within the medial prefrontal cortex were associated with resistance to learn fear extinction in female mice.[63] In addition, recently Ju and colleagues[64] reported hypermethylation of *Bdnf* gene and decreased BDNF mRNA expression in the prefrontal cortex and the hippocampus, following inescapable foot shocks.[64]

2.7 Drug-Induced Changes in the Epigenetic Regulation of Memory and Neurotrophins

2.7.1 NMDA Receptors

To answer the question whether glutamatergic drugs can reverse any of these specific epigenetic changes at neurotrophins genes, Ju et al.[64] studied the effect of long-term treatment with ketamine. This drug triggers BDNF release via its glutamatergic NMDA receptor antagonist action. Ketamine combined with extinction training prevented the spontaneous re-emergence of fear. These therapeutic-like changes occurred in conjunction with a normalization of DNA methylation at the *Bdnf* gene.[64] Glutamatergic NMDA receptors have long been known to be crucial for memory acquisition and consolidation.[65] In their study on consolidation of contextual fear, Lubin et al. show that NMDA receptor blockade prevents the hippocampal increase of total *Bdnf* gene expression (exon IX) as well as that of exon IV, while blocking the memory-associated *Bdnf* DNA hypomethylation observed at exon IV.[47] Therefore, DNA methylation clearly is a nuclear target of NMDA receptor activation during fear memory formation.

2.7.2 DNMT and HDAC Inhibitors

Directly reducing methylation patterns with DNMT inhibitors also impairs memory. More specifically in relation with the neural circuit of fear, intra-amygdala infusion of a DNMT inhibitor following retrieval of an auditory conditioned fear impairs long-term, but not short-term memory, as well as the long-term retention of auditory-evoked field potentials in the amygdala.[66] Intra-amygdala infusion of a DNMT inhibitor similarly impairs fear conditioning-related up-regulation of H3 acetylation, while pre-treatment with a HDAC inhibitor rescues the DNMT inhibitor-induced memory consolidation impairment.[46] Similarly, contextual fear memory is modulated in the hippocampus with micro-infusions of a DNMT inhibitor. It induces a memory consolidation deficit, along with a deficit in LTP, and interestingly, both deficits are prevented by raising histone acetylation levels prior to DNMT inhibition.[67] Pharmacological studies thus reveal that DNA methylations act in concert with histone modifications to regulate memory formation.

Blocking histone acetylation is detrimental to memory formation. By itself, infusions of an HDAC inhibitor into the amygdala facilitates cue fear memory,[46] while infusion into the hippocampus facilitates contextual fear memory, without affecting cue-induced fear.[45] Furthermore, blocking HAT activity either systemically, or through infusion into the amygdala impairs as much consolidation as re-consolidation of auditory fear.[68,69]

As stated earlier, the infralimbic cortex is involved in fear extinction and targeting histone acetylation in that structure alters extinction. Consistently, HDAC2 expression is decreased in the infralimbic cortex after fear extinction learning.[70] The prefrontal cortex is not the only target to modulate fear extinction. Blockers of HDAC activity enhance fear extinction memory when microinjected as much in the hippocampus as in the infralimbic cortex.[71,72]

The question that arises at this point is how drugs acting on histone acetylation and methylation change conditioned behaviors through neurotrophins gene expression. Hippocampal infusions of a DNMT inhibitor alters *Bdnf* DNA methylation and triggers changes in *Bdnf* exon IV levels indicating that a reduction of DNA methylation is sufficient to regulate *Bdnf* transcripts in the hippocampus, thereby reducing the consolidation of contextual fear.[47] This might seem counterintuitive (see Fig. 1), but in this case the DNMT inhibitor also decreased histone acetylation through cross-talk between DNA methylation and histone acetylation.[47] Furthermore, medial frontal cortex infusion of a DNMT inhibitor interferes with LTP-associated demethylation, the associated *Bdnf* expression and the induction of LTP itself.[73] Beside fear extinction, the prefrontal cortex is also involved in trace conditioning, a form of cue fear conditioning in which presentation of the CS and the unconditioned stimulus (US) is separated by a relatively long inter-stimulus interval (20 s). In the latter study, Sui et al. observed in conjunction with the reduction in *Bdnf* gene methylation, a reduction of trace conditioning following fronto-cortical infusion of the DMNT inhibitor.[73] In contrast, acetylation of H3 and H4 at the *Bdnf* gene occurring during LTP induction with high frequency stimulation can be increased with an HDAC inhibitor, and this similarly enhances frontal cortex dependent trace fear conditioning.[73] Frontal cortex dependent extinction of conditioned fear, leading to H4 acetylation at the *Bdnf* P4 gene and associated increase in *Bdnf* gene expression in that brain area, were shown to be modulated with the HDAC inhibitor valproate. This drug enhances the extinction memory, potentiates the effect of a weak fear extinction training, the associated *Bdnf* exon IV expression and H4 acetylation around the *Bdnf* gene.[60] HDAC inhibitors can induce other histone modifications beside acetylation. Inhibition of HDACs with sodium butyrate results in increased H3K4 trimethylation and decreased H3K9 dimethylation associated with contextual fear consolidation.[74]

2.7.3 Innovative Epigenetic Drug Targets

While histone acetylation and methylation are well recognized pharmacological targets regulating gene expression during learning, other epigenetic factors such as histone poly[ADP]-ribosylation, sumoylation, and ubiquitination, involved in memory consolidation in *Aplysia* and *Drosophila*[75,76] might also provide innovative pharmacological targets in mammals. There are, in particular, important cross-talks between poly [ADP]-ribosylation of histones, via the poly[ADP]-ribose polymerase (PARP-1), and other epigenetic marks including DNMT regulation of DNA methylation patterns and acetylation of histone H3 and H4 at the level of specific promoters.[77–79] It has long been known that PARP-1 is activated by DNA damage, as it occurs during traumatic brain injury for example, which by itself has important consequences on trophic factors secretion and fear memory.[80] Recently, it was also shown that PARP-1 activity is triggered under physiological conditions, in absence of DNA damage, to regulate epigenetically driven events such as memory.[75,81] PARP-1 controls neurotrophins activity, probably by loosening the highly condensed chromatin structure via poly[ADP]-ribosylation of histone, which renders DNA accessible to transcription and repair.[82] PARP-1 activation has been shown, more recently, to be required for long-term fear memory formation in mice.[83,84] Mice tested in a passive avoidance task display poly[ADP]-ribosylation of histone H1 in the cerebral cortex and the hippocampus immediately after fear conditioning. Moreover, intracerebroventricular injection of a PARP inhibitor during learning impairs long-term (but not short-term) fear memory formation.[83] Microinjection of PARP inhibitors into the dorsal hippocampus blocks both the consolidation and the re-consolidation of contextual fear. Similar PARP inhibitors into the prefrontal cortex were without effect on the formation of contextual fear, but effectively blocked the long-term consolidation of fear extinction.[84] In turn, a PARP inhibitor into the hippocampus was without effect on extinction of contextual fear.[84] Therefore, PARP inhibition potentially targets epigenetic cross-talks between neurotrophic factors and aversion memories.

Beside conditioned fear, the non-associative aspects of anxiety-related behaviors are generated by neurobiological mechanisms that display similarly relevant epigenetic targets. Other relevant animal models are thus necessary to explore these aspects.

3. EPIGENETIC CHANGES AT NEUROTROPHINS GENES IN OTHER ANIMAL MODELS OF ANXIETY AND AFFECTIVE DISORDERS

Animal models meeting validity criteria for anxiety and affective disorders can be broadly divided in two categories: models using repeated stress procedures and transgenic animals. Despite their different origins, these models have common behavioral and pathophysiological characteristics. Animals may present various anxiety- and depressive-like behaviors, as assessed through well-established behavioral paradigms, dysregulation of the immune system and of the corticosteroid reactivity, in parallel with learning and memory impairments. Neurotrophins and especially BDNF have often been found altered in these animal models.

Regarding models involving a stress procedure, most studies used one of the following paradigms: the social defeat stress (SDS) model, also referred to as the resident-intruder test, which uses aggressive social conflicts between rodents of the same species to generate subordination psychological stress. The chronic mild or ultra-mild stress (CMS/CUMS) models are well-validated models of depression that use repeated mild stress to generate pathology-relevant conditions, such as anhedonia. Predator exposure and early-life stress (ELS) models have also been used to mimic traumatic situations that later generate anxiety and depression-like disorders. Concerning genetic models, they generally are based on a single mutation to study the implication of a particular gene of interest in anxiety and affective disorders. In the context of epigenetic studies, we will focus here on three rodent mutant models: (i) rodents with a genetic knock-out of the serotonin transporter (SERT), since the serotonergic system has long been involved in anxiety and affective disorders, (ii) mice with an impaired glucocorticoid receptor expression (GR-i), as depressed patients often have reduced GR-mediated negative feedback inhibition of HPA axis activity, and (iii) mice expressing only the fully edited VGV isoform of the 5-HT2C receptor, since as stated earlier, increased 5-HT2C editing was found in the brain of anxious or depressed individuals who committed suicide.

3.1 Epigenetic Modifications Observed in Animal Models of Anxiety and Affective Disorders

Stressful life events have been shown to cause long-lasting modifications in the expression of specific genes.[85,86] Epigenetic mechanisms are implicated in these modifications by leading to stable but reversible changes in genes expression. To begin with, early-life stress caused by poor maternal care causes several epigenetic changes (see chapter "Neuroepigenetics of prenatal psychological stress" by Claes) and anxiety and affective behaviors at the adult age. In particular, low maternal care was linked to increased DNA methylation at the GR promoter gene, leading to decreased GR expression and therefore dysregulated negative feedback of the HPA axis, a biological marker of depression.[86,87] Chronic stress also profoundly impacts DNA methylation patterns. In SDS, subordination decreases DNA methylation at the promoter of the *Crf* gene, causing an up-regulation of CRF (corticotropin releasing factor).[88] Interestingly, this epigenetic modification was not detected in mice resilient to chronic social defeat.[88] DNA methylation alterations were also detected at the *Fkbp5* gene promoter, a factor impacting the HPA axis by regulating GR sensitivity, in both blood and brain of stressed animals (see chapter "Central neuroepigenetic regulation of the hypothalamic–pituitary–adrenal axis" by Dick and Provencal).[89] Regarding the serotonergic system, chronic stress (in the CUMS paradigm) induces increased DNA methylation at a specific promoter, the −681 CpG promoter site.[90] This particular DNA methylation was shown to disturb the action of a transcriptional repressor (Sp4) correlated with enhanced 5-HT1A receptors mRNA expression. These changes occurring in midbrain and prefrontal cortical areas, are expected to decrease serotonergic transmission.[90] Consequently, several lines of evidence indicate that depressogenic stresses induce changes in DNA methylation that affect serotonergic transmission and HPA axis regulation.

Chronic stress (in the SDS paradigm) has also been extensively demonstrated to induce histone H3 and H4 hypoacetylation in the hippocampus (Fig. 1),[91–94] but in contrast, an increase in global H3 acetylation in the amygdala and the prefrontal cortex.[91,95] The SDS-induced change in the hippocampus is correlated to an increase HDAC5 expression.[93] Dysregulations in the levels of HDACs expression were observed in other chronic stress models, including a modification in HDAC2 level after CUMS[96] and of several HDACs in adult animals submitted to ELS.[97]

Reversing these epigenetic alterations generated by chronic stress may provide treatments with therapeutic efficacy for anxiety and affective disorders. In particular, HDAC inhibitors, such as trichostatin or valproic acid, have been shown to have anxiolytic-like effects in various models.[98,99]

3.2 Epigenetic Modifications Affecting Neurotrophins in Models of Anxiety and Affective Disorders

Several regulatory CpG islands of the *Bdnf* gene are preferential epigenetic targets for DNA methylation in relation with anxiety and affective behaviors. ELS generates a significant increase in DNA methylation at regulatory regions of the *Bdnf* gene in the hippocampus (Fig. 1) and the prefrontal cortex.[50] Chronic predator exposure produces a robust hypermethylation within the dorsal CA1 (but not CA3) region of the hippocampus at *Bdnf* promoter 4, and this correlates with decreased *Bdnf* mRNA expression (Fig. 1).[51] Another neurotrophic factor, GDNF, was also found modified by stress exposure. DNA methylation and subsequent MeCP2 binding were increased at *Gdnf* promoter, along with an increased expression of DNMT1 and DNMT3a upon chronic stress exposure.[100]

The protein complexes associated with MeCP2 at the *Gdnf* promoter were reported to be distinct between strains of mice with different susceptibility to anxiety- and depression-like behaviors. In stress-susceptible BALB/c mice, MeCP2 seems to interact with HDAC2, inducing a decreased H3 acetylation that represses *Gdnf* transcription. In C57BL/6 mice, a stress-resilient strain, MeCP2 appears to be associated with CREB, which transcriptionally activates the *Gdnf* gene, increasing GDNF expression.[100] Overall, stress exposure modulates DNA methylation and neurotrophins expression with common mechanisms although it creates strain-dependent outcomes.

Epigenetic modifications were similarly observed in mutant animal models of depression that have impaired HPA axis regulation. It has often been reported that GR-i mice have, beside the expected decrease in GR, have lower levels of *Bdnf* mRNA in the hippocampus,[54,101] particularly of *Bdnf* exon IV.[54,102] This dysregulation was correlated to alterations in the epigenetic machinery. Modified expression of *Dnmt1* and *Dnmt3a* indicated altered DNA methylation while changes in the histone methyltransferases Mll3 and Mll4, pointed toward dysregulation in histone post-transcriptional modifications.[54]

Decreased *Bdnf* mRNA expression was also observed in the hippocampus and the prefrontal cortex of the anxious-like SERT KO mice. The *Bdnf* gene promoter was more methylated (Fig. 1), along with increased mRNA levels of *Dnmt1* and decreased mRNA levels of the demethylating enzyme Gadd45b.[55] Interestingly, deletion of Gadd45b was shown to specifically abolish DNA demethylation of *Bdnf*, perturbating its production and functional activities.[103] To add up to the link between stress-induced epigenetic regulation of *Bdnf* via Gadd45b, it was demonstrated recently that CUMS lead to reduced expression of the Gadd45 family enzymes.[104] Histone modification dysregulations were also reported in several animal models of anxiety and depression: long-lasting repressive H3K27 hypermethylation is associated with lower levels of *Bdnf* gene expression after SDS, an alteration that proved difficult to reverse pharmacologically.[52]

Although it does not reverse the repressive hypermethylated state at H3K27, chronic treatment with the tricyclic antidepressant drug imipramine can restore *Bdnf* levels in animals submitted to SDS, an effect achieved in association with a down-regulation of HDAC5 and an up-regulation of acetylation at histone H3.[52] In the exact same way, ELS exposure was associated with a long-lasting down-regulation in *Bdnf* mRNA levels mediated by H3K27 hypermethylation and H3 hypoacetylation.[53] Here again, *Bdnf* mRNA levels were rescued by chronic treatment with the selective serotonin reuptake inhibitor fluoxetine, which up-regulated H3 acetylation without changing methylation of H3K27.[53] Finally, in the learned helplessness model of depression, decreased *Bdnf* expression was accompanied by an increase HDAC5 binding at *Bdnf* promoter 4.[105] Treatments with either imipramine or a HDAC inhibitor decreased HDAC5 occupancy at *Bdnf* promoter, normalizing *Bdnf* levels and improving depressive-like symptoms.[105]

3.3 Epigenetic Regulation Through Non-coding RNA and RNA Editing

Epigenetics allow regulation of genetically encoded information by modifying genes transcription without affecting DNA sequence. As such, environmental inputs can finely tune neurotrophins transcription through non-coding RNA and RNA editing.

Silencing gene expression by using microinjections of antisense RNA probes has long been used as a technique to investigate genes function at several biological levels. For instance, infusion of BDNF antisense sequences in the amygdala decreased BDNF levels and generated anxiety-like behaviors.[106] But since then, the presence of endogenous antisense transcripts has been detected in vivo. These transcripts are classified as long non-coding RNA. Regulatory long non-coding RNA for BDNF (BDNF-AS) was demonstrated to regulate *Bdnf* gene transcription as its depletion induces up-regulation of *Bdnf* mRNA levels. Indeed, BDNF-AS depletion induces a decreased binding of a histone lysine methyltransferase, EZH2, leading to H3K27 hypomethylation. Consequently, long non-coding BDNF-AS inhibits *Bdnf* mRNA transcription by recruiting at least EZH2 at the *Bdnf* gene promoter, inducing a repressive chromatin state.[107]

Joint actions between two different epigenetic mechanisms are not rare. An important amount of mRNA editing actually occurs in non-coding RNA.[108] Moreover, some non-coding RNAs, such as microRNA miR-29b, were shown to interact with DNA methylation via a negative regulation of DNMTs.[109] It is worth noting that non-coding RNA epigenetic mechanisms also affect the serotonergic system in relation with BDNF: in humans, microRNA miR-22 was shown to be associated with anxiety, in particular PD, and to regulate both BDNF and 5-HT2C receptor expression.[110] In animals, the microRNA miR-16 was shown to modify SERT expression[111] and the small nucleolar RNA HbII-52 is involved in the regulation of 5-HT2C receptors.[112,113]

5-HT2C receptors have been extensively described as a key factor underlying anxiety and affective disorders, in particular through pre-mRNA editing.[114] The pre-mRNA undergoes adenosine-to-inosine editing, at five different sites (A, B, C, D and E) of exon V in the receptor sequence and is mediated by adenosine deaminases (ADAR). Stress increases 5-HT2CR pre-mRNA editing in mice[115] and ELS enhances 5-HT2CR pre-mRNA editing in adult animals.[116] Globally, these data indicate links between epigenetic modifications through non-coding RNA and 5-HT2CR editing, as well as between anxio-depressive behaviors and 5-HT2CR editing. Furthermore, deletion of 5-HT2C receptors increases BDNF expression.[117] Kawahara and colleagues created a genetically modified mice line which express only the fully edited (VGV) isoform of 5-HT2CR.[118] We have shown that, as a result of altered splicing processes, full edition of the receptor induced a massive expression of 5-HT2CR in various limbic areas.[7] These VGV mice were shown to have both anxiety- and depression-like behavioral profiles.[7,119] More specifically, they display anxiety, aggressive behaviors and strong fear reactions in response to an innately aversive stimulus.[7] VGV mice also have severe fear memory impairments and exaggerated fear reactions.[120] They display fear extinction deficits, together with contextual fear generalization, features all relevant with PTSD. They also display dysregulated *Bdnf* mRNA expression levels in brain structures implicated in the fear circuitry (hippocampus, amygdala), consistent with BDNF involvement in memory consolidation (see Section 2.2). Paroxetine, the first line treatment of PTSD, ameliorates both fear behaviors and *Bdnf* mRNA levels in VGV mice.[120]

Interestingly, 5-HT2C pre-mRNA editing may also underlie alcohol addiction in mice.[22] Exploring animal models of anxiety and affective disorders together with those of addiction might thus reveal common epigenetic targets regulating neurotrophins genes expression.

4. EPIGENETIC CHANGES AT NEUROTROPHINS GENES IN ADDICTION

4.1 Common Features Between Addiction and Anxio-Depressive Disorders

Anxiety and substance use disorders are among the most frequent psychiatric problems, and their co-occurrence reaches a high prevalence (for review see Refs. 121,122). Addiction is one of the most common comorbid psychiatric disorders among individuals with generalized anxiety disorder.[123] Several studies found that adults with PTSD were two to four times more likely to have a comorbid substance use disorders compared to those without PTSD,[124] while a history of PTSD predisposes individuals to elevated rates of drug abuse and dependence.[125] Interestingly, among all drugs of abuse, alcohol is especially associated with anxiety disorders.[126] Like stresses, drugs may imprint epigenetic changes within the neural circuits of fear.

Although the common pathophysiological mechanisms underlying addiction and anxiety disorders are still poorly understood, they implicate some of the same brain areas discussed earlier in relation with memory and stress, but also the ventral tegmental area (VTA) and the NAc.[127] In these structures, neurotrophins are also known to play a major role in addiction.[128] Changes in BDNF mRNA and protein have been reported in the VTA and NAc following administration of addictive drugs such as ethanol (see Ref. 129) or cocaine, a potent dopaminergic reuptake inhibitor.[130] Several lines of evidence demonstrates that dopamine, best known for its role in reward and addiction, is also regulating fear learning, anxiety and affective disorders.[131]

As described in the preceding parts of this chapter, neurotrophin levels and related epigenetic markers are modified by fear memories and depressogenic stresses. Whether similar regulations take place during addiction, and in particular during ethanol addiction, will be examined in the following section.

4.2 The Role of Neurotrophins in Addiction

Substance use disorders have been characterized as diseases of maladaptive plasticity,[132] and diverse neurotrophins, which have been strongly associated to various addictive drug consumption, have been proposed as putative molecular mediators of their long-lasting effects.[133,134] In particular, changes in BDNF mRNA and protein levels have been observed in many brain regions following administration or withdrawal of various classes of addictive compounds in rodents.[128] While acute cocaine administration could induce a transient increase in BDNF protein levels and activate TrkB-mediated signaling in the NAc of rats, BDNF infusions into the NAc can increase in turn cocaine self-administration.[135] Acute opiate administration in rodents also increase BDNF mRNA levels in many regions such as the NAc, the VTA, the orbitofrontal and medial prefrontal cortex.[128,136,137] Withdrawal after chronic exposure to opiates lead to a rapid and prolonged increase in BDNF mRNA and TrkB mRNA in the locus coeruleus.[138] In the VTA, BDNF protein and mRNA levels are increased in opiate-dependent rats.[139] In addition, neurotrophin-4 mRNA and proteins levels were also shown to be increased in various rat brain areas such as the brainstem and the cerebellum after chronic treatment with ecstasy.[134]

Altogether, these preclinical data support the notion that neurotrophins signaling, and especially BDNF-dependent activation of TrkB receptors, plays a key role in molecular adaptations leading to the development and persistence of addictive behavior in rodents. In humans, genetic analysis suggest that the BDNF Val66Met polymorphism may contribute to alcohol dependence vulnerability in terms of executive performance and reward scores[140,141] and to the low serum BDNF levels found in alcoholic patients.[142]

In contrast, several studies showed that moderate consumption of alcohol could increase BDNF levels in several parts of the brain such as the striatum[129] and the hippocampus.[21,56] Systemic acute alcohol administration, as well as 4 weeks of voluntary 10% alcohol consumption in an unlimited two-bottle choice access paradigm significantly increases *Bdnf* mRNA levels in mice dorsal striatum.[143,144] A similar increase of *Bdnf* mRNA expression was observed in rats consuming ethanol in an operant self-administration paradigm.[145] Ethanol not only modified BDNF levels in the reward system, but also other neurotrophins such as GDNF. Indeed, systemic administration of ethanol increased GDNF expression and protein levels in the VTA. Additionally, GDNF levels are elevated after an ethanol-drinking session in rats that consumed ethanol in an intermittent-access two-bottle choice procedure for a short, but not a long period suggesting that during initial ethanol-drinking experiences, GDNF in the VTA increases and protects against the development of excessive ethanol intake.[133]

Although the dopaminergic reward system is one of the major targets for drug addiction, drugs of abuse may impact other brain areas discussed earlier. In the hippocampus of mice that had freely consumed ethanol for 3 weeks, it was also found that BDNF expression was significantly up-regulated at both mRNA and protein levels.[56] Interestingly, chronic ethanol consumption enhanced P-ERK/ERK and P-AKT/AKT ratios leading to an increase CREB activation. This effect was associated to the BDNF TrkB activation since ethanol had no effect under ANA-12, a selective TrkB receptor inhibitor.[21] Similarly, an increase in hippocampal BDNF expression was reported in rats after 6 and 12 weeks (but not after 18 and 24 weeks) of repeated episodic ethanol consumption.[146] In the latter study, not only BDNF, but also the nerve growth factor and neurotrophin-3 were shown to increase following episodic ethanol exposure. Note, however, that the booster effect of ethanol on neurotrophins levels is not consistently reported in the literature.[147–150]

Ethanol consumption also induced marked changes in the expression of other genes implicated in neural development, leading to TrkB-mediated neurogenesis in the dentate gyrus.[21,56,151,152] Although these new neurons most likely become incorporated normally in the hippocampal network,[152] it seems that the ethanol-increase in neurotrophins-signaling and neurogenesis hardly prevent the ethanol-induced impairments in learning, behavioral flexibility and memory capacities.[21] Nevertheless, wherever the brain areas where it may occur, the mesocorticolimbic system or the hippocampus, the increases in neurotrophins observed after moderate alcohol intake may constitute a protection mechanism against the development of excessive ethanol intake,[129,133] and an adaptive process to counteract the cognitive deficits induced by ethanol.[21]

4.3 Addiction Modifies Various Epigenetic Marks

Addiction, as most complex diseases, involves variations at many different genes contributing to various levels of risk or protection. Evidence from family, adoption and twin studies converge about the relevance of genetic factors in the development of drug addictions, including alcohol abuse. Nevertheless, there are only few genetic predictors available that account for a small portion of the genetic variance in liability (for a review see Ref. 153).

In addition to genetic, environmental factors can modulate the risk of substance abuse across lifespan. Drugs induce complex patterns of epigenetic modifications involving DNA methylation and histone acetylation or methylation.[154] In the case of cocaine addiction, histone covalent modifications occur at gene promoters transcriptionally regulated by this compound. These genes are directly implicated in cocaine-induced neuroplasticity underlying addiction-like behaviors.[155] In addition, genome-wide approaches reveal that cocaine affects gene expression via histone acetylation, suggesting that the enzymes involved (HATs and HDACs) are pivotal for cocaine-dependent effects.[156] The role of DNA methylation, and downstream targets of DNA demethylation, in incubation of cocaine craving has also been demonstrated. Indeed withdrawal periods, as well as cue-induced cocaine seeking, have been shown to be associated in particular, with an enhancement of DNA methylation alterations in NAc.[157]

Acetylation of histone H3 and H4 appears also to play a central role in transcriptional responses involved in addiction to amphetamine and its derivative methamphetamine (METH). Specifically, a single METH injection was reported to induce both increased acetylation of histone 4 (H4K5 and H4K8) and decreased acetylation of histone 3 (H3K9, H3K18) and histone 4 (H4K16) in the NAc. In turn, chronic METH may specifically alter the level of histone 3 trimethylation (H3K4me3) on the promoter of genes responsible for long-lasting adaptations.[154] Microarray analysis also demonstrated that METH could affect the expression of genes that act in complexes with HDACs as either activators or repressors.[154]

Several epigenetic alterations have also been observed under alcohol dependence and in animal models of alcoholism associated with emotional disorders such as anxiety. Using a global chromatin immunoprecipitation assay, Zhou et al. reported significant changes in H3K4me3 abundance in a large number of genes in the hippocampus of alcoholics.[158] H3K4me3 and H3K27me2 levels were also found significantly higher in the hippocampus of mice under chronic ethanol exposure.[56]

In rats, a hyperacetylation of histone H3 in the amygdala and cortical areas was associated with alcohol dependence.[159] Furthermore, acute administration of ethanol has been reported to increase histone H3 and H4 acetylation and to decrease HDAC activity in the rat amygdala, while withdrawal from chronic ethanol exposure produces the opposite effects.[160,161] In contrast, ethanol-preferring rats have been shown to display higher nuclear HDAC activity and HDAC2 protein levels and lower acetylation of H3K9 in the amygdala compared to non-preferring rats.[162] The same study also indicated how acute ethanol exposure decreases amygdaloid HDAC activity and HDAC2 protein levels, increases global and gene (*Bdnf* and *Arc*)-specific histone acetylation, and attenuates anxiety-like behaviors in ethanol-preferring rats, but have no effects in non-preferring rat.[162] Chronic ethanol consumption in a free choice paradigm in mice was also found to induce marked changes in the expression of genes implicated in histone post-translational modifications in the mouse hippocampus, including decreases of *Hdac1*, *Hdac2*, and *Hdac5* mRNA expressions. The decreased expression of *Hdac2* was also evidenced by HDAC2 immunostaining specifically in the dentate gyrus of the hippocampus.[56] All these data converged to a key role of HDAC-mediated epigenetic mechanisms in alcoholism.

DNA methylation is also regulated by ethanol consumption. Several studies indicate that ethanol exposure during early gestation can cause changes in DNA methylation and gene expression in the CNS of offspring.[163] Fetal ethanol exposure augments DNA methyltransferase *Dnmt3a*, *Dnmt3b*, and *Dnmt3l* transcripts and DNMT3A protein levels.[164] In adult mice, we also reported that chronic ethanol consumption lead to brain reduction in *Dnmt3b* mRNA expression and a rise of *Mecp2*.[21]

4.4 Epigenetic Modifications Affecting Neurotrophins in Addiction

There are still few studies that have focused on epigenetic regulation of neurotrophins genes, and in particular on Bdnf in relation with substance abuse (see Ref. 165 for a review), and most of them are related to cocaine addiction. Under a one-week withdrawal period from chronic cocaine administration, an increase in H3 acetylation (but not H4) at the *Bdnf* promoter 2 in the rat striatum has been observed.[166] Chronic administration of cocaine in rats caused robust induction of H3 acetylation (but not H4) at *Bdnf* promoters 2 and 3 in the NAc, and this was reverted by HDAC4 overexpression in this same region.[167]

Among the other few studies with different addiction drugs, prenatal exposure to maternal smoking was shown to be associated with higher rates of DNA methylation at the *Bdnf* promoter 6 in blood cells of adolescents.[168] In the case of alcohol abuse, alterations in the methylation of *Bdnf* exon IV during alcohol withdrawal in patients have recently been reported.[16] This study showed that the mean methylation of the *Bdnf* promoter 4 was significantly increased in alcohol-dependent patients compared to healthy controls. However, methylation at *Bdnf* promoter 4 was decreased during alcohol withdrawal, and the data suggested that changes in BDNF methylation in alcohol-dependent patients during alcohol withdrawal could be associated with withdrawal symptoms, such as depression and anxiety.

At the preclinical level, as already reported above, chronic ethanol consumption was found to induce marked changes in the hippocampal expression of genes implicated in neural development and histone post-translational modifications. Transcripts encoding *Bdnf* exons II, III and VI were up-regulated, whereas those from *Bdnf* VIII exon and *Hdacs* were down-regulated by ethanol compared to water consumption. These ethanol-induced changes were associated with enrichment in both acetylated H3 at *Bdnf* promoter P6 and trimethylated H3 at P2 and P3. Conversely, acetylated H3 at P3 and P8 and trimethylated H3 at P8 were decreased in ethanol-exposed mice.[56] Analysis of the methylation of various CpG islands within the *Bdnf* gene showed a repressing effect of ethanol consumption on DNA methylation at *Bdnf* gene level in the hippocampus (Fig. 1).[21] Although to our knowledge, no other analyses have been conducted on specific neurotrophin gene methylation under alcohol consumption, some clinical and preclinical studies have reported global methylation modulation by ethanol (see Ref. 169). Ethanol also impairs the methyl donor group S-adenosylmethionine (SAM) synthesis by inhibiting the enzymes that regulates its production[170] and consequently impairs DNA methylation, directly under SAM's control.

4.5 Treating Addiction Associated Anxiety via Epigenetically Targeted Drugs

Small-molecules, inhibitors of epigenetic enzymes such as HDACs and DNMTs, may reduce voluntary ethanol consumption. A reducing effect of decitabine (a DNMT inhibitor) on ethanol consumption has been reported in two types of ethanol consumption in mice (the binge-like drinking in the dark and chronic intermittent every other day drinking).[171] SAM prevented the increased drinking caused by chronic intermittent ethanol, while the HDAC inhibitor, trichostatin A, facilitated ethanol drinking induced by chronic intermittent ethanol exposure.[172] In addition, HDAC inhibitors MS-275 and sodium butyrate were reported to decrease ethanol self-administration and consumption in rodent models.[159,173] The HDAC inhibitor valproic acid has also been reported to decrease ethanol preference and consumption in a two-bottle choice paradigm in rats.[174] Valproic acid, commercialized as an anticonvulsant drug, has been proposed to treat alcohol withdrawal and to prevent alcohol relapse in alcoholic patients.[175] Consequently, drugs that specifically target either HDAC or DNA methylation could potentially help tackle alcoholism associated with anxiety.

5. CONCLUSION

Although the literature is rather complex and much work remains to unveil the role of epigenetic in the regulation of neurotrophins in anxiety and affective disorders, the recent research has made some actors stand out, such as the acetylation of histone H3 which modulate neurotrophins levels and neural plasticity. The consolidation of negative memories in anxious patients may well be mediated by the actions of neurotrophins in areas such as the hippocampus, the frontal cortex and the amygdala, in conjunction with the acetylation of histones H3 at various *Bdnf* promoters, such as *Bdnf* promoter 4. For example, alcohol consumption may facilitate acetylation in these brain areas and make aversive memories more persistent. Decreased acetylation of histones at neurotrophin promoters may thus well be a strategy for reducing certain symptoms in anxiety disorders, as much as decreasing acetylation of histone at *Bdnf* promoter P1 in the frontal cortex would set in place the conditions favoring fear extinction.[60] However, the acetylation patterns are not so easy to interpret in models of anxiety and affective disorders, because acetylations tend to globally increase after repeated stress in areas such as the amygdala, but to decrease in the hippocampus (Fig. 1), most likely through increase of HDACs. This occurs also specifically at hippocampal histone H3 *Bdnf* promoter 4, although stress does rather potentiate fear memories and BDNF expression by increasing H3 acetylation.[48] Chronic antidepressant treatments may prevent the reduction of BDNF levels and the deacetylation at H3 in the hippocampus generated by a history of stress. Nevertheless, in conditions of moderate consumption, even alcohol may produce such therapeutic-like effects (Fig. 1).[21]

Similarly to antidepressant drugs, we have seen that HDAC inhibitors, including valproate, generally improve anxiety or depression-like behaviors together with BDNF levels. More specifically, valproate enhances fear extinction through expression of *Bdnf* exon IV and H4 acetylation at the *Bdnf* gene.[60] Valproate and other HDAC inhibitors are also candidate drugs to treat alcohol dependence and associated anxiety. Nevertheless, as with many other drugs, the therapeutic window is limited, since these inhibitors also reinforce the consolidation of negative events. We will have to await human studies on acetylation of histones in depression and anxiety, to understand whether neurotrophins can be therapeutically manipulated through epigenetic tools such as HDAC inhibitors.

In turn, many clinical studies revealed hypermethylation of DNA at specific *Bdnf* gene sites in the brain and in the periphery of anxious, depressed or suicidal individuals. Although, again, moderate alcohol consumption may reduce DNA methylation, alcoholics suffering withdrawal depression and anxiety also display hypermethylation at *Bdnf* promoter 4 generating low levels of BDNF.[16] The literature reviewed on animal models of depression and PTSD is also consistent with hypermethylation at the *Bdnf* gene leading to decreased brain levels of BDNF, as a result of chronic or traumatic stresses. Other neurotrophins may be altered in the same manner by stress, but the information on those is rather scarce at this point. Formation of aversion memories, which requires increased BDNF secretion for consolidation, appears to rather engage a demethylation of *Bdnf* gene. In turn, individuals that are resistant to fear extinction might be so as a result of hypermethylation of *Bdnf* genes in brain areas such as the frontal cortex. DNMT inhibitors might thus emerge as innovative treatments for anxiety and affective disorders.

Finally, in interaction with DNA methylation, there are many other epigenetic targets, some reviewed here, that will become interesting to consider for treatment of these psychiatric diseases, such as microRNAs, 5-HT2C receptor pre-mRNA editing and histone poly[ADP]-ribosylation. Despite the bulk of information accumulated so far, investigations regarding the epigenetics of neurotrophins remain in their infancy. The hope is to be able to reset at the epigenetic level neurotrophins endogenous secretion differentially among brain areas and in a time wise manner, so as to provide cures for various disorders that have enduring impacts on cognition and emotions.

REFERENCES

1. Gottschalk MG, Domschke K. Novel developments in genetic and epigenetic mechanisms of anxiety. *Curr Opin Psychiatry*. 2016;29:32–38.
2. Schartner C, Ziegler C, Schiele MA, et al. CRHR1 promoter hypomethylation: an epigenetic readout of panic disorder? *Eur Neuropsychopharmacol*. 2017;27:360–371.
3. Norrholm SD, Jovanovic T, Smith AK, et al. Differential genetic and epigenetic regulation of catechol-O-methyltransferase is associated with impaired fear inhibition in posttraumatic stress disorder. *Front Behav Neurosci*. 2013;7:30.
4. Philibert RA, Sandhu H, Hollenbeck N, Gunter T, Adams W, Madan A. The relationship of 5HTT (SLC6A4) methylation and genotype on mRNA expression and liability to major depression and alcohol dependence in subjects from the Iowa Adoption Studies. *Am J Med Genet B Neuropsychiatr Genet*. 2008;147B:543–549.
5. Dracheva S, Patel N, Woo DA, Marcus SM, Siever LJ, Haroutunian V. Increased serotonin 2C receptor mRNA editing: a possible risk factor for suicide. *Mol Psychiatry*. 2008;13:1001–1010.
6. Dracheva S, Chin B, Haroutunian V. Altered serotonin 2C receptor RNA splicing in suicide: association with editing. *Neuroreport*. 2008;19:379–382.
7. Martin CBP, Ramond F, Farrington DT, et al. RNA splicing and editing modulation of 5-HT(2C) receptor function: relevance to anxiety and aggression in VGV mice. *Mol Psychiatry*. 2013;18:656–665.

8. Pandey GN, Dwivedi Y, Ren X, et al. Regional distribution and relative abundance of serotonin(2c) receptors in human brain: effect of suicide. *Neurochem Res*. 2006;31: 167–176.
9. Ignácio ZM, Réus GZ, Abelaira HM, Quevedo J. Epigenetic and epistatic interactions between serotonin transporter and brain-derived neurotrophic factor genetic polymorphism: insights in depression. *Neuroscience*. 2014;275:455–468.
10. Keller S, Sarchiapone M, Zarrilli F, et al. Increased BDNF promoter methylation in the Wernicke area of suicide subjects. *Arch Gen Psychiatry*. 2010;67:258–267.
11. Kang H-J, Kim J-M, Lee J-Y, et al. BDNF promoter methylation and suicidal behavior in depressive patients. *J Affect Disord*. 2013;151:679–685.
12. Song Y, Miyaki K, Suzuki T, et al. Altered DNA methylation status of human brain derived neurotrophis factor gene could be useful as biomarker of depression. *Am J Med Genet B Neuropsychiatr Genet*. 2014;165B:357–364.
13. Januar V, Ancelin M-L, Ritchie K, Saffery R, Ryan J. BDNF promoter methylation and genetic variation in late-life depression. *Transl Psychiatry*. 2015;5:e619.
14. Kim J-M, Stewart R, Kang H-J, et al. A longitudinal study of BDNF promoter methylation and genotype with poststroke depression. *J Affect Disord*. 2013;149:93–99.
15. Kim J-M, Stewart R, Kang H-J, et al. BDNF methylation and depressive disorder in acute coronary syndrome: the K-DEPACS and EsDEPACS studies. *Psychoneuroendocrinology*. 2015;62:159–165.
16. Heberlein A, Büscher P, Schuster R, et al. Do changes in the BDNF promoter methylation indicate the risk of alcohol relapse? *Eur Neuropsychopharmacol*. 2015;25: 1892–1897.
17. Kim TY, Kim SJ, Chung HG, Choi JH, Kim SH, Kang JI. Epigenetic alterations of the BDNF gene in combat-related post-traumatic stress disorder. *Acta Psychiatr Scand*. 2017;135:170–179.
18. Choi S, Han K-M, Won E, Yoon B-J, Lee M-S, Ham B-J. Association of brain-derived neurotrophic factor DNA methylation and reduced white matter integrity in the anterior corona radiata in major depression. *J Affect Disord*. 2015;172:74–80.
19. Na K-S, Won E, Kang J, et al. Brain-derived neurotrophic factor promoter methylation and cortical thickness in recurrent major depressive disorder. *Sci Rep*. 2016;6:21089.
20. Pizzimenti CL, Lattal KM. Epigenetics and memory: causes, consequences and treatments for post-traumatic stress disorder and addiction. *Genes Brain Behav*. 2015;14:73–84.
21. Stragier E, Martin V, Davenas E, et al. Brain plasticity and cognitive functions after ethanol consumption in C57BL/6J mice. *Transl Psychiatry*. 2015;5:e696.
22. Watanabe Y, Yoshimoto K, Tatebe H, et al. Enhancement of alcohol drinking in mice depends on alterations in RNA editing of serotonin 2C receptors. *Int J Neuropsychopharmacol*. 2014;17:739–751.
23. Covington HE, Maze I, Sun H, et al. A role for repressive histone methylation in cocaine-induced vulnerability to stress. *Neuron*. 2011;71:656–670.
24. Andero R, Choi DC, Ressler KJ. BDNF-TrkB receptor regulation of distributed adult neural plasticity, memory formation, and psychiatric disorders. *Prog Mol Biol Transl Sci*. 2014;122:169–192.
25. Kuhn M, Popovic A, Pezawas L. Neuroplasticity and memory formation in major depressive disorder: an imaging genetics perspective on serotonin and BDNF. *Restor Neurol Neurosci*. 2014;32:25–49.
26. Levine ES, Dreyfus CF, Black IB, Plummer MR. Brain-derived neurotrophic factor rapidly enhances synaptic transmission in hippocampal neurons via postsynaptic tyrosine kinase receptors. *Proc Natl Acad Sci USA*. 1995;92:8074–8077.
27. Zakharenko SS, Patterson SL, Dragatsis I, et al. Presynaptic BDNF required for a presynaptic but not postsynaptic component of LTP at hippocampal CA1-CA3 synapses. *Neuron*. 2003;39:975–990.

28. McAllister AK, Katz LC, Lo DC. Neurotrophin regulation of cortical dendritic growth requires activity. *Neuron*. 1996;17:1057–1064.
29. Kang H, Schuman EM. A requirement for local protein synthesis in neurotrophin-induced hippocampal synaptic plasticity. *Science*. 1996;273:1402–1406.
30. Chen DY, Stern SA, Garcia-Osta A, et al. A critical role for IGF-II in memory consolidation and enhancement. *Nature*. 2011;469:491–497.
31. Alonso M, Vianna MRM, Depino AM, et al. BDNF-triggered events in the rat hippocampus are required for both short- and long-term memory formation. *Hippocampus*. 2002;12:551–560.
32. Ma YL, Wang HL, Wu HC, Wei CL, Lee EH. Brain-derived neurotrophic factor antisense oligonucleotide impairs memory retention and inhibits long-term potentiation in rats. *Neuroscience*. 1998;82:957–967.
33. Minichiello L, Korte M, Wolfer D, et al. Essential role for TrkB receptors in hippocampus-mediated learning. *Neuron*. 1999;24:401–414.
34. Ou L-C, Gean P-W. Transcriptional regulation of brain-derived neurotrophic factor in the amygdala during consolidation of fear memory. *Mol Pharmacol*. 2007;72:350–358.
35. Rattiner LM, Davis M, French CT, Ressler KJ. Brain-derived neurotrophic factor and tyrosine kinase receptor B involvement in amygdala-dependent fear conditioning. *J Neurosci*. 2004;24:4796–4806.
36. Chou D, Huang C-C, Hsu K-S. Brain-derived neurotrophic factor in the amygdala mediates susceptibility to fear conditioning. *Exp Neurol*. 2014;255:19–29.
37. Heldt SA, Stanek L, Chhatwal JP, Ressler KJ. Hippocampus-specific deletion of BDNF in adult mice impairs spatial memory and extinction of aversive memories. *Mol Psychiatry*. 2007;12:656–670.
38. Rosas-Vidal LE, Do-Monte FH, Sotres-Bayon F, Quirk GJ. Hippocampal—prefrontal BDNF and memory for fear extinction. *Neuropsychopharmacology*. 2014;39:2161–2169.
39. Herry C, Ferraguti F, Singewald N, Letzkus JJ, Ehrlich I, Lüthi A. Neuronal circuits of fear extinction. *Eur J Neurosci*. 2010;31:599–612.
40. Milad MR, Quirk GJ. Neurons in medial prefrontal cortex signal memory for fear extinction. *Nature*. 2002;420:70–74.
41. Peters J, Dieppa-Perea LM, Melendez LM, Quirk GJ. Induction of fear extinction with hippocampal-infralimbic BDNF. *Science*. 2010;328:1288–1290.
42. Agis-Balboa RC, Arcos-Diaz D, Wittnam J, et al. A hippocampal insulin-growth factor 2 pathway regulates the extinction of fear memories. *EMBO J*. 2011;30: 4071–4083.
43. Võikar V, Rossi J, Rauvala H, Airaksinen MS. Impaired behavioural flexibility and memory in mice lacking GDNF family receptor alpha2. *Eur J Neurosci*. 2004;20:308–312.
44. Graham BM, Zagic D, Richardson R. Low endogenous fibroblast growth factor 2 levels are associated with heightened conditioned fear expression in rats and humans. *Biol Psychiatry*. 2017;82:601–607.
45. Vecsey CG, Hawk JD, Lattal KM, et al. Histone deacetylase inhibitors enhance memory and synaptic plasticity via CREB:CBP-dependent transcriptional activation. *J Neurosci*. 2007;27:6128–6140.
46. Monsey MS, Ota KT, Akingbade IF, Hong ES, Schafe GE. Epigenetic alterations are critical for fear memory consolidation and synaptic plasticity in the lateral amygdala. Izquierdo I, ed *PLoS One*. 2011;6:e19958.
47. Lubin FD, Roth TL, Sweatt JD. Epigenetic regulation of BDNF gene transcription in the consolidation of fear memory. *J Neurosci*. 2008;28:10576–10586.
48. Fuchikami M, Yamamoto S, Morinobu S, Takei S, Yamawaki S. Epigenetic regulation of BDNF gene in response to stress. *Psychiatry Investig*. 2010;7:251–256.

49. Mizuno K, Dempster E, Mill J, Giese KP. Long-lasting regulation of hippocampal Bdnf gene transcription after contextual fear conditioning. *Genes Brain Behav.* 2012;11:651–659.
50. Roth TL, Lubin FD, Funk AJ, Sweatt JD. Lasting epigenetic influence of early-life adversity on the BDNF gene. *Biol Psychiatry.* 2009;65:760–769.
51. Roth TL, Zoladz PR, Sweatt JD, Diamond DM. Epigenetic modification of hippocampal Bdnf DNA in adult rats in an animal model of post-traumatic stress disorder. *J Psychiatr Res.* 2011;45:919–926.
52. Tsankova NM, Berton O, Renthal W, Kumar A, Neve RL, Nestler EJ. Sustained hippocampal chromatin regulation in a mouse model of depression and antidepressant action. *Nat Neurosci.* 2006;9:519–525.
53. Onishchenko N, Karpova N, Sabri F, Castrén E, Ceccatelli S. Long-lasting depression-like behavior and epigenetic changes of BDNF gene expression induced by perinatal exposure to methylmercury. *J Neurochem.* 2008;106:1378–1387.
54. Boulle F, Velthuis H, Koedam K, et al. Behavioral and neurochemical characterization of TrkB-dependent mechanisms of agomelatine in glucocorticoid receptor-impaired mice. *Eur Neuropsychopharmacol.* 2016;26:65–77.
55. Calabrese F, Guidotti G, Middelman A, Racagni G, Homberg J, Riva MA. Lack of serotonin transporter alters BDNF expression in the rat brain during early postnatal development. *Mol Neurobiol.* 2013;48:244–256.
56. Stragier E, Massart R, Salery M, et al. Ethanol-induced epigenetic regulations at the Bdnf gene in C57BL/6J mice. *Mol Psychiatry.* 2015;20:405–412.
57. Levenson JM, O'Riordan KJ, Brown KD, Trinh MA, Molfese DL, Sweatt JD. Regulation of histone acetylation during memory formation in the hippocampus. *J Biol Chem.* 2004;279:40545–40559.
58. Chwang WB, O'Riordan KJ, Levenson JM, Sweatt JD. ERK/MAPK regulates hippocampal histone phosphorylation following contextual fear conditioning. *Learn Mem.* 2006;13(3):322–328.
59. Gupta-Agarwal S, Franklin AV, Deramus T, et al. G9a/GLP histone lysine dimethyltransferase complex activity in the hippocampus and the entorhinal cortex is required for gene activation and silencing during memory consolidation. *J Neurosci.* 2012;32:5440–5453.
60. Bredy TW, Wu H, Crego C, Zellhoefer J, Sun YE, Barad M. Histone modifications around individual BDNF gene promoters in prefrontal cortex are associated with extinction of conditioned fear. *Learn Mem.* 2007;14:268–276.
61. Miller CA, Sweatt JD. Covalent modification of DNA regulates memory formation. *Neuron.* 2007;53:857–869.
62. Frankland PW, Bontempi B, Talton LE, Kaczmarek L, Silva AJ. The involvement of the anterior cingulate cortex in remote contextual fear memory. *Science.* 2004;304:881–883.
63. Baker-Andresen D, Flavell CR, Li X, Bredy TW. Activation of BDNF signaling prevents the return of fear in female mice. *Learn Mem.* 2013;20:237–240.
64. Ju L-S, Yang J-J, Lei L, et al. The combination of long-term ketamine and extinction training contributes to fear erasure by Bdnf methylation. *Front Cell Neurosci.* 2017;11:100.
65. Morris RGM, Moser EI, Riedel G, et al. Elements of a neurobiological theory of the hippocampus: the role of activity-dependent synaptic plasticity in memory. *Philos Trans R Soc Lond B Biol Sci.* 2003;358:773–786.
66. Maddox SA, Watts CS, Schafe GE. DNA methyltransferase activity is required for memory-related neural plasticity in the lateral amygdala. *Neurobiol Learn Mem.* 2014;107:93–100.

67. Miller CA, Campbell SL, Sweatt JD. DNA methylation and histone acetylation work in concert to regulate memory formation and synaptic plasticity. *Neurobiol Learn Mem.* 2008;89:599–603.
68. Maddox SA, Watts CS, Schafe GE. p300/CBP histone acetyltransferase activity is required for newly acquired and reactivated fear memories in the lateral amygdala. *Learn Mem.* 2013;20:109–119.
69. Maddox SA, Watts CS, Doyère V, Schafe GE. A naturally-occurring histone acetyltransferase inhibitor derived from Garcinia indica impairs newly acquired and reactivated fear memories. Lyons LC, ed *PLoS One.* 2013;8:e54463.
70. Wei W, Coelho CM, Li X, et al. p300/CBP-associated factor selectively regulates the extinction of conditioned fear. *J Neurosci.* 2012;32:11930–11941.
71. Lattal KM, Barrett RM, Wood MA. Systemic or intrahippocampal delivery of histone deacetylase inhibitors facilitates fear extinction. *Behav Neurosci.* 2007;121:1125–1131.
72. Stafford JM, Raybuck JD, Ryabinin AE, Lattal KM. Increasing histone acetylation in the hippocampus-infralimbic network enhances fear extinction. *Biol Psychiatry.* 2012;72:25–33.
73. Sui L, Wang Y, Ju L-H, Chen M. Epigenetic regulation of reelin and brain-derived neurotrophic factor genes in long-term potentiation in rat medial prefrontal cortex. *Neurobiol Learn Mem.* 2012;97:425–440.
74. Gupta S, Kim SY, Artis S, et al. Histone methylation regulates memory formation. *J Neurosci.* 2010;30:3589–3599.
75. Cohen-Armon M, Visochek L, Katzoff A, et al. Long-term memory requires polyADP-ribosylation. *Science.* 2004;304:1820–1822.
76. Schwartz S, Truglio M, Scott MJ, Fitzsimons HL. Long-term memory in drosophila is influenced by histone deacetylase HDAC4 interacting with SUMO-conjugating enzyme Ubc9. *Genetics.* 2016;203:1249–1264.
77. Caiafa P, Guastafierro T, Zampieri M. Epigenetics: poly(ADP-ribosyl)ation of PARP-1 regulates genomic methylation patterns. *FASEB J.* 2009;23:672–678.
78. Krishnakumar R, Kraus WL. PARP-1 regulates chromatin structure and transcription through a KDM5B-dependent pathway. *Mol Cell.* 2010;39:736–749.
79. Verdone L, La Fortezza M, Ciccarone F, Caiafa P, Zampieri M, Caserta M. Poly(ADP-ribosyl)ation affects histone acetylation and transcription. *PLoS One.* 2015;10: e0144287.
80. Kaplan GB, Vasterling JJ, Vedak PC. Brain-derived neurotrophic factor in traumatic brain injury, post-traumatic stress disorder, and their comorbid conditions: role in pathogenesis and treatment. *Behav Pharmacol.* 2010;21:427–437.
81. Quénet D, El Ramy R, Schreiber V, Dantzer F. The role of poly(ADP-ribosyl)ation in epigenetic events. *Int J Biochem Cell Biol.* 2009;41:60–65.
82. Visochek L, Steingart RA, Vulih-Shultzman I, et al. PolyADP-ribosylation is involved in neurotrophic activity. *J Neurosci.* 2005;25:7420–7428.
83. Goldberg S, Visochek L, Giladi E, Gozes I, Cohen-Armon M. PolyADP-ribosylation is required for long-term memory formation in mammals. *J Neurochem.* 2009;111:72–79.
84. Inaba H, Tsukagoshi A, Kida S. PARP-1 activity is required for the reconsolidation and extinction of contextual fear memory. *Mol Brain.* 2015;8:63.
85. Krishnan V, Han M-H, Graham DL, et al. Molecular adaptations underlying susceptibility and resistance to social defeat in brain reward regions. *Cell.* 2007;131:391–404.
86. Liu D, Diorio J, Tannenbaum B, et al. Maternal care, hippocampal glucocorticoid receptors, and hypothalamic-pituitary-adrenal responses to stress. *Science.* 1997;277:1659–1662.
87. Weaver ICG, Cervoni N, Champagne FA, et al. Epigenetic programming by maternal behavior. *Nat Neurosci.* 2004;7:847–854.

88. Elliott E, Ezra-Nevo G, Regev L, Neufeld-Cohen A, Chen A. Resilience to social stress coincides with functional DNA methylation of the Crf gene in adult mice. *Nat Neurosci*. 2010;13:1351–1353.
89. Ewald ER, Wand GS, Seifuddin F, et al. Alterations in DNA methylation of Fkbp5 as a determinant of blood-brain correlation of glucocorticoid exposure. *Psychoneuroendocrinology*. 2014;44:112–122.
90. Le François B, Soo J, Millar AM, et al. Chronic mild stress and antidepressant treatment alter 5-HT1A receptor expression by modifying DNA methylation of a conserved Sp4 site. *Neurobiol Dis*. 2015;82:332–341.
91. Covington HE, Vialou VF, LaPlant Q, Ohnishi YN, Nestler EJ. Hippocampal-dependent antidepressant-like activity of histone deacetylase inhibition. *Neurosci Lett*. 2011;493:122–126.
92. Hollis F, Duclot F, Gunjan A, Kabbaj M. Individual differences in the effect of social defeat on anhedonia and histone acetylation in the rat hippocampus. *Horm Behav*. 2011;59:331–337.
93. Patki G, Solanki N, Atrooz F, et al. Novel mechanistic insights into treadmill exercise based rescue of social defeat-induced anxiety-like behavior and memory impairment in rats. *Physiol Behav*. 2014;130:135–144.
94. Kenworthy CA, Sengupta A, Luz SM, et al. Social defeat induces changes in histone acetylation and expression of histone modifying enzymes in the ventral hippocampus, prefrontal cortex, and dorsal raphe nucleus. *Neuroscience*. 2014;264:88–98.
95. Hinwood M, Tynan RJ, Day TA, Walker FR. Repeated social defeat selectively increases δFosB expression and histone H3 acetylation in the infralimbic medial prefrontal cortex. *Cereb Cortex*. 2011;21:262–271.
96. Boulle F, Massart R, Stragier E, et al. Hippocampal and behavioral dysfunctions in a mouse model of environmental stress: normalization by agomelatine. *Transl Psychiatry* 2014;4(11):e485.
97. Suri D, Bhattacharya A, Vaidya VA. Early stress evokes temporally distinct consequences on the hippocampal transcriptome, anxiety and cognitive behaviour. *Int J Neuropsychopharmacol*. 2014;17:289–301.
98. Tran L, Schulkin J, Ligon CO, Greenwood-Van Meerveld B. Epigenetic modulation of chronic anxiety and pain by histone deacetylation. *Mol Psychiatry*. 2015;20:1219–1231.
99. Wilson CB, McLaughlin LD, Ebenezer PJ, Nair AR, Francis J. Valproic acid effects in the hippocampus and prefrontal cortex in an animal model of post-traumatic stress disorder. *Behav Brain Res*. 2014;268:72–80.
100. Uchida S, Hara K, Kobayashi A, et al. Epigenetic status of Gdnf in the ventral striatum determines susceptibility and adaptation to daily stressful events. *Neuron*. 2011;69:359–372.
101. Païzanis E, Renoir T, Lelievre V, et al. Behavioural and neuroplastic effects of the new-generation antidepressant agomelatine compared to fluoxetine in glucocorticoid receptor-impaired mice. *Int J Neuropsychopharmacol*. 2010;13:759–774.
102. Alboni S, Tascedda F, Corsini D, et al. Stress induces altered CRE/CREB pathway activity and BDNF expression in the hippocampus of glucocorticoid receptor-impaired mice. *Neuropharmacology*. 2011;60:1337–1346.
103. Ma DK, Jang M-H, Guo JU, et al. Neuronal activity-induced Gadd45b promotes epigenetic DNA demethylation and adult neurogenesis. *Science*. 2009;323:1074–1077.
104. Grassi D, Franz H, Vezzali R, et al. Neuronal activity, TGFβ-signaling and unpredictable chronic stress modulate transcription of Gadd45 family members and DNA methylation in the hippocampus. *Cereb Cortex*. 2017;27:1–16.

105. Su C-L, Su C-W, Hsiao Y-H, Gean P-W. Epigenetic regulation of BDNF in the learned helplessness-induced animal model of depression. *J Psychiatr Res*. 2016;76: 101–110.
106. Pandey SC, Zhang H, Roy A, Misra K. Central and medial amygdaloid brain-derived neurotrophic factor signaling plays a critical role in alcohol-drinking and anxiety-like behaviors. *J Neurosci*. 2006;26:8320–8331.
107. Modarresi F, Faghihi MA, Lopez-Toledano MA, et al. Inhibition of natural antisense transcripts in vivo results in gene-specific transcriptional upregulation. *Nat Biotechnol*. 2012;30:453–459.
108. Guennewig B, Cooper AA. The central role of noncoding RNA in the brain. *Int Rev Neurobiol*. 2014;116:153–194.
109. Takada S, Berezikov E, Choi YL, Yamashita Y, Mano H. Potential role of miR-29b in modulation of Dnmt3a and Dnmt3b expression in primordial germ cells of female mouse embryos. *RNA*. 2009;15:1507–1514.
110. Muiños-Gimeno M, Espinosa-Parrilla Y, Guidi M, et al. Human microRNAs miR-22, miR-138-2, miR-148a, and miR-488 are associated with panic disorder and regulate several anxiety candidate genes and related pathways. *Biol Psychiatry*. 2011;69:526–533.
111. Baudry A, Mouillet-Richard S, Schneider B, Launay J-M, Kellermann O. miR-16 targets the serotonin transporter: a new facet for adaptive responses to antidepressants. *Science*. 2010;329:1537–1541.
112. Vitali P, Basyuk E, Le Meur E, et al. ADAR2-mediated editing of RNA substrates in the nucleolus is inhibited by C/D small nucleolar RNAs. *J Cell Biol*. 2005;169:745–753.
113. Doe CM, Relkovic D, Garfield AS, et al. Loss of the imprinted snoRNA mbii-52 leads to increased 5htr2c pre-RNA editing and altered 5HT2CR-mediated behaviour. *Hum Mol Genet*. 2009;18:2140–2148.
114. Martin CBP, Hamon M, Lanfumey L, Mongeau R. Controversies on the role of 5-HT (2C) receptors in the mechanisms of action of antidepressant drugs. *Neurosci Biobehav Rev*. 2014;42:208–223.
115. Englander MT, Dulawa SC, Bhansali P, Schmauss C. How stress and fluoxetine modulate serotonin 2C receptor pre-mRNA editing. *J Neurosci*. 2005;25:648–651.
116. Bhansali P, Dunning J, Singer SE, David L, Schmauss C. Early life stress alters adult serotonin 2C receptor pre-mRNA editing and expression of the alpha subunit of the heterotrimeric G-protein G q. *J Neurosci*. 2007;27:1467–1473.
117. Hill RA, Murray SS, Halley PG, Binder MD, Martin SJ, van den Buuse M. Brain-derived neurotrophic factor expression is increased in the hippocampus of 5-HT (2C) receptor knockout mice. *Hippocampus*. 2011;21:434–445.
118. Kawahara Y, Grimberg A, Teegarden S, et al. Dysregulated editing of serotonin 2C receptor mRNAs results in energy dissipation and loss of fat mass. *J Neurosci*. 2008;28:12834–12844.
119. Mombereau C, Kawahara Y, Gundersen BB, Nishikura K, Blendy JA. Functional relevance of serotonin 2C receptor mRNA editing in antidepressant- and anxiety-like behaviors. *Neuropharmacology*. 2010;59:468–473.
120. Regue M, Poilbout C, Lanfumey L, Mongeau R. P.4.a.004 Fear extinction deficits and BDNF dysregulation in mice with full 5-HT2C receptor editing: a model for post-traumatic stress disorders? *Eur Neuropsychopharmacol*. 2015;25:S554–S555.
121. Smith JP, Book SW. Anxiety and substance use disorders: a review. *Psychiatr Times*. 2008;25:19–23.
122. Back SE, Brady KT. Anxiety disorders with comorbid substance use disorders: diagnostic and treatment considerations. *Psychiatr Ann*. 2008;38:724–729.
123. Wittchen HU, Zhao S, Kessler RC, Eaton WW. DSM-III-R generalized anxiety disorder in the National Comorbidity Survey. *Arch Gen Psychiatry*. 1994;51:355–364.

124. Kessler RC, Sonnega A, Bromet E, Hughes M, Nelson CB. Posttraumatic stress disorder in the National Comorbidity Survey. *Arch Gen Psychiatry*. 1995;52:1048–1060.
125. Reed PL, Anthony JC, Breslau N. Incidence of drug problems in young adults exposed to trauma and posttraumatic stress disorder: do early life experiences and predispositions matter? *Arch Gen Psychiatry*. 2007;64:1435–1442.
126. Kessler RC, McGonagle KA, Zhao S, et al. Lifetime and 12-month prevalence of DSM-III-R psychiatric disorders in the United States. Results from the National Comorbidity Survey. *Arch Gen Psychiatry*. 1994;51:8–19.
127. Lüthi A, Lüscher C. Pathological circuit function underlying addiction and anxiety disorders. *Nat Neurosci*. 2014;17:1635–1643.
128. Russo SJ, Mazei-Robison MS, Ables JL, Nestler EJ. Neurotrophic factors and structural plasticity in addiction. *Neuropharmacology*. 2009;56(suppl 1):73–82.
129. Logrip ML, Barak S, Warnault V, Ron D. Corticostriatal BDNF and alcohol addiction. *Brain Res*. 1628;2015:60–67.
130. Filip M, Faron-Górecka A, Kuśmider M, Gołda A, Frankowska M, Dziedzicka-Wasylewska M. Alterations in BDNF and trkB mRNAs following acute or sensitizing cocaine treatments and withdrawal. *Brain Res*. 2006;1071:218–225.
131. Lee JC, Wang LP, Tsien JZ. Dopamine rebound-excitation theory: putting brakes on PTSD. *Front Psychiatry*. 2016;7:163.
132. Lüscher C, Malenka RC. Drug-evoked synaptic plasticity in addiction: from molecular changes to circuit remodeling. *Neuron*. 2011;69:650–663.
133. Ahmadiantehrani S, Barak S, Ron D. GDNF is a novel ethanol-responsive gene in the VTA: implications for the development and persistence of excessive drinking. *Addict Biol*. 2014;19:623–633.
134. Hatami H, Hossainpour-Faizi MA, Azarfarin M, Azarfam P. Chronic ecstasy use increases neurotrophin-4 gene expression and protein levels in the rat brain. *Pharmacol Rep*. 2010;62:998–1004.
135. Graham DL, Edwards S, Bachtell RK, DiLeone RJ, Rios M, Self DW. Dynamic BDNF activity in nucleus accumbens with cocaine use increases self-administration and relapse. *Nat Neurosci*. 2007;10:1029–1037.
136. Chu N-N, Zuo Y-F, Meng L, Lee DY-W, Han J-S, Cui C-L. Peripheral electrical stimulation reversed the cell size reduction and increased BDNF level in the ventral tegmental area in chronic morphine-treated rats. *Brain Res*. 2007;1182:90–98.
137. Maze I, Russo SJ. Transcriptional mechanisms: underlying addiction-related structural plasticity. *Mol Interv*. 2010;10:219–230.
138. Numan S, Lane-Ladd SB, Zhang L, et al. Differential regulation of neurotrophin and trk receptor mRNAs in catecholaminergic nuclei during chronic opiate treatment and withdrawal. *J Neurosci*. 1998;18:10700–10708.
139. Vargas-Perez H, Ting-A Kee R, Walton CH, et al. Ventral tegmental area BDNF induces an opiate-dependent-like reward state in naive rats. *Science*. 2009;324:1732–1734.
140. Benzerouk F, Gierski F, Gorwood P, et al. Brain-derived neurotrophic factor (BDNF) Val66Met polymorphism and its implication in executive functions in adult offspring of alcohol-dependent probands. *Alcohol*. 2013;47:271–274.
141. Benzerouk F, Gierski F, Raucher-Chéné D, et al. Association study between reward dependence and a functional BDNF polymorphism in adult women offspring of alcohol-dependent probands. *Psychiatr Genet*. 2015;25:208–211.
142. Nubukpo P, Ramoz N, Girard M, Malauzat D, Gorwood P. Determinants of blood brain-derived neurotrophic factor blood levels in patients with alcohol use disorder. *Alcohol Clin Exp Res*. 2017;41:1280–1287.
143. McGough NNH, He D-Y, Logrip ML, et al. RACK1 and brain-derived neurotrophic factor: a homeostatic pathway that regulates alcohol addiction. *J Neurosci*. 2004;24:10542–10552.

144. Logrip ML, Janak PH, Ron D. Escalating ethanol intake is associated with altered corticostriatal BDNF expression. *J Neurochem*. 2009;109:1459–1468.
145. Jeanblanc J, He D-Y, Carnicella S, Kharazia V, Janak PH, Ron D. Endogenous BDNF in the dorsolateral striatum gates alcohol drinking. *J Neurosci*. 2009;29:13494–13502.
146. Miller MW. Repeated episodic exposure to ethanol affects neurotrophin content in the forebrain of the mature rat. *Exp Neurol*. 2004;189:173–181.
147. Okamoto H, Miki T, Lee K-Y, et al. Effects of chronic ethanol administration on the expression levels of neurotrophic factors in the rat hippocampus. *Okajimas Folia Anat Jpn*. 2006;83:1–6.
148. Miller R, King MA, Heaton MB, Walker DW. The effects of chronic ethanol consumption on neurotrophins and their receptors in the rat hippocampus and basal forebrain. *Brain Res*. 2002;950:137–147.
149. MacLennan AJ, Lee N, Walker DW. Chronic ethanol administration decreases brain-derived neurotrophic factor gene expression in the rat hippocampus. *Neurosci Lett*. 1995;197:105–108.
150. Hauser SR, Getachew B, Taylor RE, Tizabi Y. Alcohol induced depressive-like behavior is associated with a reduction in hippocampal BDNF. *Pharmacol Biochem Behav*. 2011;100:253–258.
151. Paizanis E, Kelaï S, Renoir T, Hamon M, Lanfumey L. Life-long hippocampal neurogenesis: environmental, pharmacological and neurochemical modulations. *Neurochem Res*. 2007;32:1762–1771.
152. Aberg E, Hofstetter CP, Olson L, Brené S. Moderate ethanol consumption increases hippocampal cell proliferation and neurogenesis in the adult mouse. *Int J Neuropsychopharmacol*. 2005;8:557–567.
153. Ducci F, Goldman D. The genetic basis of addictive disorders. *Psychiatr Clin North Am*. 2012;35:495–519.
154. Godino A, Jayanthi S, Cadet JL. Epigenetic landscape of amphetamine and methamphetamine addiction in rodents. *Epigenetics*. 2015;10:574–580.
155. Rogge GA, Wood MA. The role of histone acetylation in cocaine-induced neural plasticity and behavior. *Neuropsychopharmacology*. 2013;38:94–110.
156. Renthal W, Kumar A, Xiao G, et al. Genome-wide analysis of chromatin regulation by cocaine reveals a role for sirtuins. *Neuron*. 2009;62:335–348.
157. Massart R, Barnea R, Dikshtein Y, et al. Role of DNA methylation in the nucleus accumbens in incubation of cocaine craving. *J Neurosci*. 2015;35:8042–8058.
158. Zhou Z, Yuan Q, Mash DC, Goldman D. Substance-specific and shared transcription and epigenetic changes in the human hippocampus chronically exposed to cocaine and alcohol. *Proc Natl Acad Sci USA*. 2011;108:6626–6631.
159. Simon-O'Brien E, Alaux-Cantin S, Warnault V, Buttolo R, Naassila M, Vilpoux C. The histone deacetylase inhibitor sodium butyrate decreases excessive ethanol intake in dependent animals. *Addict Biol*. 2015;20:676–689.
160. Krishnan HR, Sakharkar AJ, Teppen TL, Berkel TDM, Pandey SC. The epigenetic landscape of alcoholism. *Int Rev Neurobiol*. 2014;115:75–116.
161. Pandey SC, Ugale R, Zhang H, Tang L, Prakash A. Brain chromatin remodeling: a novel mechanism of alcoholism. *J Neurosci*. 2008;28:3729–3737.
162. Moonat S, Sakharkar AJ, Zhang H, Tang L, Pandey SC. Aberrant histone deacetylase2-mediated histone modifications and synaptic plasticity in the amygdala predisposes to anxiety and alcoholism. *Biol Psychiatry*. 2013;73:763–773.
163. Marjonen H, Sierra A, Nyman A, et al. Early maternal alcohol consumption alters hippocampal DNA methylation, gene expression and volume in a mouse model. *PLoS One*. 2015;10:e0124931.

164. Miozzo F, Arnould H, de Thonel A, et al. Alcohol exposure promotes DNA methyltransferase DNMT3A upregulation through reactive oxygen species-dependent mechanisms. *Cell Stress Chaperones*. 2017;23:115–126. [Epub ahead of print].
165. Boulle F, van den Hove DLA, Jakob SB, et al. Epigenetic regulation of the BDNF gene: implications for psychiatric disorders. *Mol Psychiatry*. 2012;17:584–596.
166. Kumar A, Choi K-H, Renthal W, et al. Chromatin remodeling is a key mechanism underlying cocaine-induced plasticity in striatum. *Neuron*. 2005;48:303–314.
167. Wang L, Lv Z, Hu Z, et al. Chronic cocaine-induced H3 acetylation and transcriptional activation of CaMKIIalpha in the nucleus accumbens is critical for motivation for drug reinforcement. *Neuropsychopharmacology*. 2010;35:913–928.
168. Toledo-Rodriguez M, Lotfipour S, Leonard G, et al. Maternal smoking during pregnancy is associated with epigenetic modifications of the brain-derived neurotrophic factor-6 exon in adolescent offspring. *Am J Med Genet B Neuropsychiatr Genet*. 2010;153B:1350–1354.
169. Ponomarev I. Epigenetic control of gene expression in the alcoholic brain. *Alcohol Res*. 2013;35:69–76.
170. Hamid A, Wani NA, Kaur J. New perspectives on folate transport in relation to alcoholism-induced folate malabsorption—association with epigenome stability and cancer development. *FEBS J*. 2009;276:2175–2191.
171. Ponomarev I, Stelly CE, Morikawa H, Blednov YA, Mayfield RD, Harris RA. Mechanistic insights into epigenetic modulation of ethanol consumption. *Alcohol*. 2017;60:95–101.
172. Qiang M, Li JG, Denny AD, et al. Epigenetic mechanisms are involved in the regulation of ethanol consumption in mice. *Int J Neuropsychopharmacol*. 2014;18.
173. Jeanblanc J, Lemoine S, Jeanblanc V, Alaux-Cantin S, Naassila M. The Class I-Specific HDAC inhibitor MS-275 decreases motivation to consume alcohol and relapse in heavy drinking rats. *Int J Neuropsychopharmacol*. 2015;18:pyv029.
174. Al Ameri M, Al Mansouri S, Al Maamari A, Bahi A. The histone deacetylase (HDAC) inhibitor valproic acid reduces ethanol consumption and ethanol-conditioned place preference in rats. *Brain Res*. 2014;1583:122–131.
175. Book SW, Myrick H. Novel anticonvulsants in the treatment of alcoholism. *Expert Opin Investig Drugs*. 2005;14:371–376.

CHAPTER NINE

Neuroepigenetics of Schizophrenia

Giovanna Punzi*, Rahul Bharadwaj*, Gianluca Ursini*,†,1
*Lieber Institute for Brain Development, Johns Hopkins Medical Campus, Baltimore, MD, United States
†Department of Psychiatry and Behavioral Sciences, Johns Hopkins University School of Medicine, Baltimore, MD, United States
[1]Corresponding author: e-mail address: gianluca.ursini@libd.org

Contents

Abstract

Schizophrenia is a complex disorder of the brain, where genetic variants explain only a portion of risk. Neuroepigenetic mechanisms may explain the remaining share of risk, as well as the transition from susceptibility to the actual disease. Here, we discuss the most recent findings in the field of brain epigenetics applied to the study of schizophrenia.

Methylome studies have found several candidates exhibiting methylation modifications in association with the disorder, but genes affected do not always overlap. Notably, these studies converge in that genes within the schizophrenia risk *loci* or genes differentially methylated in patients affected with the disorder are dynamically regulated during early life. They also imply that schizophrenia-associated genetic variation may affect DNA methylation in fetal and adult brains. Histone modifications may help mediating the effect of genetic risk variants associated with schizophrenia, and regulating chromatin higher-order structure. The 3D-organization of chromatin in the brain

Progress in Molecular Biology and Translational Science, Volume 158
ISSN 1877-1173
https://doi.org/10.1016/bs.pmbts.2018.04.010

creates physical interactions within chromosomes, so that schizophrenia-associated genetic variants can be linked with genes distant from their *loci*; this suggests that chromatin conformation matters in the mechanism of risk for the disorder. Non-coding RNAs provide a novel and complex mechanism of gene regulation potentially significant for schizophrenia, as proposed by research on specific microRNAs and long non-coding RNAs (lncRNAs). Finally, a recent study in epitranscriptomics identifies RNA methylation as a further epigenetic mechanism active in human brain and specifically in a portion of the transcriptome associated with schizophrenia susceptibility.

These findings indicate that, as expected from the complexity of the brain and its development, several epigenetic mechanisms may intervene in the etiopathogenesis of schizophrenia. An understanding of their roles calls for research approaches integrating the investigation of different epigenetic mechanisms and of environmental and genetic risk, in the context of development.

Schizophrenia is a complex disabling disorder, affecting ~1% of the worldwide population,[1] with a dramatic impact on both the individual and society. More than 50% of patients with schizophrenia have intermittent but long-term psychiatric problems, and around 20% of them have chronic symptoms and disability.[2] Most patients are unemployed, and they have lower life expectancy compared with the general population.[2]

The pathogenesis of the disorder is not clearly established, but research into its causal mechanisms has seen substantial advances in the application of genomics, epidemiology and molecular neuroscience, encouraging ambitions for novel strategies of prevention and treatment that may take advantage from future potential insights about the etiopathogenesis of the disorder. Diagnosis of schizophrenia rests on the observation of a codified nosology of symptoms, which are usually clustered into "positive", "negative" and "cognitive" categories. *Positive* symptoms are thoughts, perceptions, and behaviors that are aberrant in content and form and that characterize the psychotic "loss of contact with reality," taking the shape of delusions, hallucinations, and disorganized speech and behavior. *Negative* symptoms characterize the so-called amotivational syndrome, and include social withdrawal, affective flattening, anhedonia, apathy, and diminished initiative and energy. Finally, *cognitive* symptoms manifest as a broad set of cognitive dysfunctions, although with individual variability.[2]

Of note, all these clinical manifestations hint at the involvement of epigenetic mechanisms in the pathophysiology of the disorder, since all the faculties disrupted in schizophrenia are faculties that humans, in normal conditions, would acquire growing up, i.e., as their brains develop. Nobody is ever born with a full set of mature cognitive functions, with an organized speech and behavior, with a sense of reality that is shared with other individuals, or with the ability of interacting with other people in a way that is socially appropriate. As these faculties are gradually acquired during development, through increasing interactions with the internal and external environment, the pathophysiology of the disorder should partly rest in the altered function of the regulatory epigenetic machinery, i.e., at the interface of the developing brain and inputs from the environment. This is consistent with the hypothesis that schizophrenia is a "developmental" brain disorder, i.e., where the early, normal maturation of the brain is disrupted.[3] Indeed, a wide range of environmental factors has been implicated in the development of schizophrenia, including maternal stress and complications during pregnancy,[4] labor/delivery and early neonatal life (i.e., obstetric complications), early-life stressors, growing up in an urban environment, migration, and cannabis use in early adolescence.[5] Exposure to these factors impacts on key periods for the development of relevant cognitive, affective, and social capabilities, which in turn parallels maturational processes of the brain[3,5]: insults in embryonic and early fetal development can affect differentiation of neural progenitors, formation of the neural tube, and neural patterning; complications in the later fetal and early postnatal periods may affect neuronal production, migration and differentiation[3]; finally, early life and adolescent unfavorable experiences could hamper myelination and regressive and progressive events involving neuronal processes and sculpting of synaptic and circuitry architectures.[6]

In this chapter, we will review the most recent evidence showing how epigenetic mechanisms in brain may have a role in the etiology and in the clinical manifestations of the disorder. After an introduction on the disorder's genetics, we will first summarize studies on brain DNA methylation in schizophrenia—the most studied mechanism on the subject, so far—and we will focus on those assessing methylation at epigenome-wide level (while we direct the reader to a recent review[7] for methylation studies on candidate genes). We will then discuss mounting evidence from research on histone modifications and chromatin looping at specific loci in schizophrenia. We will also comment on a recent study on the three-dimensional organization of chromatin in fetal brain. We will then review the evidence from brain research on non-coding RNAs in schizophrenia. Finally, we will briefly approach the emerging field of epitranscriptomics, in which regard we will highlight findings converging with the recent results from genome-wide association studies (GWAS) on schizophrenia.[8] We will then propose some future directions for epigenetic studies on schizophrenia.

1. DOES THE GENETICS OF SCHIZOPHRENIA POINT TO EPIGENETICS?

If the high heritability of schizophrenia suggests an etiologically major role for genetic variants, the same evidence is also compatible with a significant contribution of epigenetic changes since, by definition, epigenetics refer to modifications in gene expression that can be heritable and do not involve changes in the DNA sequence.[9,10] Epidemiological evidence shows that heritable genetic/epigenetic variation contributes to schizophrenia through a polygenic mechanism. To this regard, the notion of the disorder as a polygenic one is long-standing[11] and arises from the observation that, while the risk of schizophrenia aggregates in families, it does not segregate following simple Mendelian inheritance.[11,12] More specifically, the degree to which the relative risks of schizophrenia changes with different degrees of relatedness is consistent with polygenic inheritance, as well as the prevalence of sporadic cases, i.e., not linked with family history of schizophrenia.[13]

In recent years, GWAS with increasing sample sizes have been able to detect more and more risk-associated loci.[8,14,15] The state-of-the-art GWAS of schizophrenia, conducted by the Psychiatric Genetics Consortium (PGC), identified 108 risk-associated genetic loci achieving genome-wide statistical significance ($P < 5 \times 10^{-8}$), with each locus accounting for a very small fraction of genetic risk, with odds ratios generally between 0.9 and 1.1.[8] While this is highly consistent with a polygenic architecture of the disorder, it also opens to the possibility of the effect of risk alleles being modulated by epigenetic mechanisms, in addition to epistasis and ancestral heterogeneity. Indeed, for most of the 108 loci the association signal has been credibly attributed to polymorphisms in the non-coding portion of the genome, enriched for enhancers and promoter active in brain[8,16] and somewhat in cells with immune functions.[8] This fact suggests that schizophrenia-associated genetic variation may operate via epigenetic mechanisms regulating expression of the genes, which is in turn consistent with the enrichment of GWAS-significant variants in brain expression quantitative trait loci (eQTLs).[17,18] At the same time, one might speculate that the effects of schizophrenia-associated variants are modulated by further epigenetic mechanisms, acting at those loci, which would partly justify the small amount of risk explained by each genetic variant.

Interestingly, the network and pathway analysis subgroup of the PGC has developed an analysis framework to rank common pathways across schizophrenia, major depression and bipolar disorder, using summary statistics from their GWAS's. Not only histone H3K4 methylation emerges as a top pathway associated with schizophrenia,[19] but the same analysis highlights that histone methylation in general has the strongest association with GWAS-significant loci, across different psychiatric disorders. In other words, in the effort to identify the main biological mechanisms underlying the GWAS signals, this study suggests that many genes likely regulated by genetic variants associated with schizophrenia and other psychiatric disorders are involved in the regulation of histone methylation, an epigenetic mechanism sensitive to environmental factors.[19]

2. DNA METHYLATION CHANGES IN SCHIZOPHRENIA

Epigenetic mechanisms include DNA methylation and hydroxymethylation, histone modification, chromatin conformation, and non-coding RNAs-mediated gene regulation. However, the vast majority of schizophrenia epigenetic research (Table 1) has been focused on the study of DNA methylation at CpG sites, i.e., regions of DNA where a cytosine nucleotide is followed by a guanine nucleotide in the linear sequence of bases along its $5' \rightarrow 3'$ direction (Fig. 1).

The main DNA epigenetic modifications are cytosine methylation and hydroxymethylation at the carbon 5.[44] The mC5 (5-methylcytosine) and hmC5 (5-hydroxymethylcytidine) markings show a differential, but not mutually exclusive, pattern of genomic occupancy. The hmC5 marking correlates with local levels of expression[45,46] while mC5, particularly when positioned around the 5′ end of genes, seems to function mostly as negative regulator of transcription[47,48] (Fig. 1).

The first in the field to conduct a post-mortem methylome analysis, Mill et al.[20] studied tissue from the frontal brain cortex of individuals diagnosed with schizophrenia ($n = 35$) or bipolar disorder ($n = 35$) and from matched controls ($n = 35$). Via hybridization of enriched unmethylated DNA fractions to CGI-arrays,[20] they showed disease-associated DNA methylation differences in multiple loci, particularly in genes regulating glutamatergic and GABAergic neurotransmission, already believed as dysregulated in the disorder. Epigenetic dysregulation emerged also for genes involved in neuronal development, learning and motor functions. Gene-ontology analysis highlighted epigenetic disruption of loci involved in mitochondrial function, brain development, and stress response. A network approach, searching for the number of connections between nodes representing

Table 1 Shown are the Main Reported Findings of Studies that have Searched for Epigenetic Changes Associated with Schizophrenia

Study	Epigenetic Marker	Method	Tissue	Sample Size (Total *N*)	Main Findings/Genes with Differential Epigenetic Changes in Patients with Schizophrenia
Mill[20]	DNA methylation	CG-array hybridization	PFC	105	Glutamatergic, GABAergic, neurodevelopmental, stress response genes; lower connectivity in major psychosis
Chen[21]	DNA methylation	27k array	Cerebellum	118	*PIK3R1*, *NHLH1*, *SLC16A7*, *BTN3A3*, neurodevelopmental and immune response genes
Numata[22]	DNA methylation	27k array	PFC	226	*GRIA4*, *ASTN2*, *DCDC2*
Wockner[23]	DNA methylation	450k array	PFC	48	*NOS1*, *AKT1*, *DNMT1*, *SOX10*, neurodevelopmental genes
Pidsley[24]	DNA methylation	450k array	PFC and cerebellum	44	*GSDMD*, *RASA3*, *HTR5A*, *PPFIA1*, and neurodevelopmental genes; DMPs enriched for loci at which DNA methylation is dynamically altered during human fetal brain development; no schizophrenia-associated modules identified in the cerebellum
Wockner[25]	DNA methylation	450k array	PFC	92	*CERS3*, *DPPA5*, *PRDM9*, *DDX43*, *REC8*, *LY6G5C*
Hannon[26]	DNA methylation	450k array	PFC, striatum and cerebellum	166	Prenatal meQTLs fourfold enriched for schizophrenia risk variants; CpGs in the schizophrenia loci more likely to show differential methylation during the transition of prenatal to postnatal life
Jaffe[27]	DNA methylation	450k array	PFC	335	62/108 of GWAS-significant schizophrenia loci contain a meQTL within 20 kb of tag SNPs, 2104 DMPs distinct from schizophrenia risk loci-associated meQTL and linked to changes associated with the transition from prenatal to postnatal life
Gagliano[28]	DNA methylation	Restriction enzymes	PFC	208	Allele-skewed modifications (ASM) SNPs significantly overrepresented among schizophrenia SNPs, also in control brains; ASM SNPs present in sperm

Dempster[29]	DNA methylation	27k array	Blood	22 MZ twins pairs	*PUS3*, *ST6GALNAC1*
Aberg[30,31]	DNA methylation	MBD-seq	Blood	1497	Hypoxia-related genes; *FAM63B, RELN*; DMPs overlapping with schizophrenia loci
Liu[32]	DNA methylation	27k	Blood	206	Inflammatory response genes
van Eijk[33]	DNA methylation	27k	Blood	510	*CALHM1*, *HLA-C*, *PRRT1*, and *MRPL41*
Montano[34]	DNA methylation	450k	Blood	1334	923 DMPs (625 replicated, 3 in schizophrenia loci); neurodevelopmental and immune response genes
Hannon[35]	DNA methylation	450k	Blood	675	25 DMPs (one in the MHC region); DMR overlapping with schizophrenia loci; DMPs associated with schizophrenia PRS
van den Oord[36]	DNA methylation	MBD-seq	Blood	1408	7 CpG differentially methylated (one in the *IL1RAP* gene)
Tang[37]	Histone modifications	ChIP-qPCR	PFC	82	H3K9K14 hypoacetylation of *GAD67*, *HTR2C*, *TOMM70A* and *PPM1E* in young subjects with schizophrenia
Huang[38]	Histone modifications	ChIP-qPCR	PFC	100	Decreased H3K4-trimethyl. at the GAD1 promoter and increased H3K27-trimethyl., predominantly in females and in conjunction with a risk haplotype
Bharadwaj[39]	Chromatin looping	3C	PFC	17	Decreased *GAD1/GAD67* expression accompanied by a weakening of a long-range GAD1 promoter-enhancer chromatin loop

See text for comments and findings from studies on HiC, ncRNAs, and epitranscriptomics.

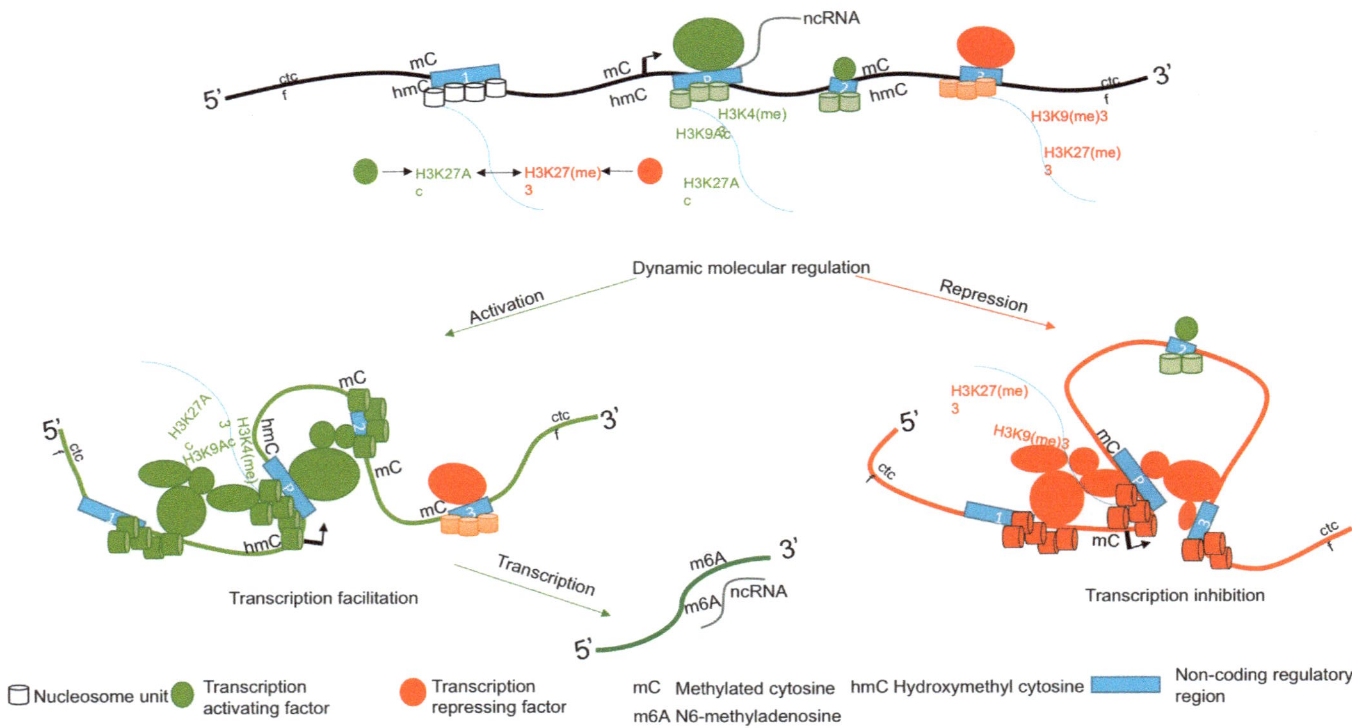

Fig. 1 See legend on opposite page.

correlated methylation, suggested a lower connectivity of different epigenomic modifications in major psychosis, pointing to a systemic epigenetic dysfunction in schizophrenia.[20] Such possibility, although intriguing, to date has not been fully explored.

Chen et al.[21] used the Illumina Infinium HumanMethylation27 Bead Chip (27k) array, assessing 27,000 CpG sites, to analyze methylation differences in another brain area, the cerebellum, of individuals diagnosed with schizophrenia ($n = 39$) or bipolar disorder ($n = 36$) and from matched controls ($n = 43$). They did not observe CpG differentially methylated in patients, after correction for multiple comparisons. However, by focusing only on CpGs correlated with gene expression, they identified four genes—*PIK3R1*, *NHLH1*, *SLC16A7*, *BTN3A3*, involved in neurotransmission, neurodevelopment, and adaptive immune responses—differentially expressed in patients, which contained CpGs differentially methylated and associated with gene expression.[21]

Fig. 1 A simplified model to illustrate how environmental signals translate to dynamic molecular regulation of chromatin and RNA to regulate gene expression. The epigenome encompasses broad mechanistic systems that operate via a variety of tools, many of which have been implicated in the pathophysiology of schizophrenia and are shown, in this figure, at the levels of DNA (mC, hmC), RNA (ncRNAs, m6A), histone octameric nucleosome units, subject to post-translational modifications (PTMs), structural proteins (insulator element binding CTCF) and regulatory complexes that activate (green) and/or repress (red) gene expression. The epigenome is largely regulatory, therefore implying a balancing role, when inducible/repressive gene expression signals are no longer operational from the environment. In so far as epigenetic regulation can occur at any single base/nucleosome unit across the genome, there are sequence-specific regulatory elements that offer platforms (regulatory regions or sequences are generally non-coding) to assemble facilitative and/or repressive protein (transcription factors, co-factors, structural proteins) complexes that in turn have both DNA and histone modifying enzymes to further enhance transcription activation or repression. These processes are highly complex and have been represented in this simplified model. Note that transcription activation leads to a different physical interaction between P (gene promoter) and regulatory elements 1 and 2 coupled with an increase in nucleosome units with transcription activating histone modifications, whereas the repressive complex at regulatory element 3 is not a part of this transcription facilitative chromatin complex. Similarly, transcription repression leads to a physical interaction between P and regulatory elements 1 (this time a repressive element) and 3 coupled with an increase in nucleosome units, with transcription repressing histone modifications, whereas the regulatory complex at element 2 is not a part of this transcription repressive chromatin complex. These alternate chromatin conformation states associated with transcription regulation is a key window to explore mechanistic and/or therapeutic control of unwanted transcription, especially in the case of pathogenic alternate transcripts that are associated with genetic risk loci.[40–43]

Using the same tool (27k array), Numata et al.[22] conducted a genome-wide methylation profiling of dorsolateral prefrontal cortex (DLPFC) of a larger cohort of 106 patients with schizophrenia and 110 controls. They identified 107 CpGs with aberrant DNA methylation in schizophrenia, at 5% Bonferroni correction; some of the genes containing these CpGs were previously implicated in schizophrenia (*GRIA4*, *ASTN2*, and *DCDC2*) and, when comparing this study with the analysis by Mill et al.,[20] the *MRPS14* gene was consistently hypermethylated.[22]

The following research on genome-wide DNA methylation modifications in schizophrenia has taken advantage from the introduction of the Illumina Infinium HumanMethylation450 Bead Chip (450k) array, containing 485,000 CpG sites. Wockner et al.[23] measured DNA methylation modifications in post-mortem frontal cortex from 24 patients with schizophrenia and 24 unaffected controls.[23] Results showed 4641 probes corresponding to 2929 unique genes as differentially methylated, including *NOS1*, *AKT1*, *DNMT1*, and *SOX10*, among other genes previously associated with schizophrenia and neurodevelopment.[23]

Pidsley et al.[24] firsts analyzed the methylation changes potentially relevant for schizophrenia in the context of neurodevelopment. They investigated epigenomic changes in post-mortem prefrontal cortex (PFC) and cerebellum from 20/21 patients with schizophrenia and 23 matched controls using the 450k array, and then assessing the disease-associated regions in prenatal samples.[24] Highly significant differentially methylated CpGs were detected in the PFC; specifically, probes in four neurodevelopmental genes were associated with schizophrenia at an FDR-corrected P value of ≤ 0.05: *GSDMD*, *RASA3*, *HTR5A*, and *PPFIA1*. Additionally, the authors performed a weighted gene co-methylation network analysis (WGCNA[49]), to identify gene networks dysregulated in the disorder. The analysis showed several modules of co-methylated loci that were associated with schizophrenia, and enriched for genes involved in neurodevelopmental processes. On the other hand, no schizophrenia-associated modules were identified in the cerebellum, consistent with region-specific DNA methylation differences in schizophrenia. Methylomic data from the prenatal samples (spanning 23–184 days post-conception) indicated that the schizophrenia-associated differentially methylated positions (DMPs) were significantly enriched for loci at which DNA methylation is dynamically altered during human fetal brain development, supporting the hypothesis that schizophrenia has an early-life component, mediated by epigenetic mechanisms.[24]

In a following study,[25] Wockner et al. sought to narrow down the results to a smaller number of differentially methylated regions (DMRs) of more functional relevance, by analyzing their previously published data,[23] together with the two datasets by Pidsley et al.[24] Even though patients had mixed profiles of methylation, the study pointed to a number of regions consistently differentially methylated in the disorder. The regions were near *CERS3*, *DPPA5*, *PRDM9* (encoding a histone methyltransferase), *DDX43*, *REC8*, *LY6G5C*, plus a region on chromosome 10.[25]

These studies, despite each based on small effects, together support the involvement of DNA methylation in the disorder. The heterogeneity of the findings may have two possible explanations: first, the methylation changes relevant for schizophrenia may be different in different populations, with different ancestry and environment; second, not all of these changes are in fact pathogenic, as many of them may merely represent an epiphenomenon of the disorder, perhaps related to medication or other factors associated with the experience of schizophrenia. In this regard, the investigation of the developmental trajectory of the schizophrenia-associated methylation modifications—as performed by Pidsley et al.[24]—together with the study of the relationship between genetic variation and epigenetic mechanisms, may help to clarify the contribution of each player, as we discuss below.

3. DNA METHYLATION, NEURODEVELOPMENT AND SCHIZOPHRENIA: A LINK WITH GENETIC RISK

Three recent studies[26–28] have analyzed the relationship between DNA methylation, neurodevelopment and genetic variation associated with schizophrenia in the most recent GWAS.[8]

Two of these studies[26,27] first focused on prenatal development to investigate genetic variants associated with methylation changes, i.e., methylation quantitative trait loci[26] (meQTLs), and early switches in CpG methylation.

Hannon et al.[26] detected ≈16,000 *cis*-meQTLs within a 1 Mb sliding window in 166 fetal brains ranging from 56 to 166 days post-conception. Using tissue from three regions of the adult brain (PFC, striatum and cerebellum), the authors found that most of the prenatal meQTLs were developmentally stable, although fetal-specific effects characterized a subset of the meQTLs. In agreement with a previous report,[50] the effect sizes of these meQTLs were small, with a median change in DNA methylation per allele of ~7%. Likewise, only a few *trans*-meQTLs ($n = 5$) of smaller effect size were found. Functionally, prenatal meQTLs fulfilled the criteria of regulatory domains,[51,52] since they overlapped with DNase I hypersensitive sites, regulatory histone marks, transcription factor binding sites, and eQTLs. Interestingly, they were strongly enriched in DNA-binding sites for the architectural protein CTCF (Fig. 1) that connects higher-order chromatin structure to lineage-specific, but also to aberrant, gene expression.[53] Importantly, prenatal meQTLs were fourfold enriched for genome-wide significant schizophrenia risk variants,[8] suggesting that they play a role in brain development. Consistently, 2903 CpGs located in the risk loci for schizophrenia[8] were more likely to show differential methylation during the transition from prenatal to postnatal life than non-schizophrenia risk loci.

Jaffe et al.[27] assessed genome-wide DNA methylation at 230,000 CpGs in post-mortem DLPFC from 335 brain donors ranging from the 14th week of gestation to 80 years of age, who were unaffected by psychiatric diseases. The analysis identified 6480 statistically significant DMRs that emerged during the transition from the second prenatal trimester to postnatal life and that were mapped to 4557 genes. Most of these genes play a role in brain development and morphogenesis,[27] suggesting that CpG relevant for schizophrenia may matter during early phases of neurodevelopment. The authors also performed a meQTL analysis in a large set of DLPFC tissue samples (191 adults diagnosed with schizophrenia and 240 matched controls), finding that 62 out of 108 genome-wide significant schizophrenia risk loci[8] contained a meQTL within 20 kb of tag SNPs and those in LD ($R^2 > 0.6$).[27] However, none of these meQTLs in the schizophrenia risk loci was specific to control or disease status. In both studies, a schizophrenia locus whose GWAS-significant variant was associated with methylation was annotated to the *AS3MT* gene. At this locus a VNTR polymorphism, linked with the variant, has been associated with promoter activity, DNA methylation within the gene, and expression of an alternative transcript up-regulated in brains of schizophrenia cases and during human stem cell differentiation toward neuronal fates.[54]

Additionally, Jaffe et al. identified 2104 CpGs in the adult brain that were differentially methylated between schizophrenia cases and controls; these CpGs were weakly but significantly enriched in the schizophrenia risk loci (40 CpGs out of 2104 CpGs).[27] On the other hand, only 97 diagnosis-associated CpGs corresponded to genome-wide significant meQTLs, suggesting that diagnosis-associated CpGs are distinct from schizophrenia risk loci-associated meQTLs, and that the epigenetic mechanisms of schizophrenia involving many methylation sites may be independent from genetic risk or may represent epiphenomena of the disorder. The authors also found that the vast majority of the schizophrenia-associated DNA methylation sites were linked to changes associated with the transition from prenatal to postnatal life, consistent with results by Hannon et al.,[26] rather than corresponding to age-related changes occurring at adolescence or young adulthood, i.e., the typical age of onset. While they speculate that both genetic and environmental risk factors for schizophrenia exert their principal effects early in development, one cannot exclude the possibility of epigenetic changes at the onset being highly heterogeneous.

The two studies[26,27] overall indicate that schizophrenia-associated loci[8] are enriched for genetic variants associated with DNA methylation during fetal and adult life and that CpG sites within schizophrenia risk loci[26] and with DNA methylation differences between patients and controls[27] are enriched for prenatal-to-postnatal epigenetic changes. This in turn indicates that epigenetic changes relevant for the disorder can be established during early life, in a period of high plasticity, further supporting the neurodevelopmental hypothesis of schizophrenia. Consistently, a recent gene expression analysis within the public resource of PFC tissue "BrainSeq" has identified widespread shifts in preferential transcript isoform use across early brain development, which were significantly enriched for genes within schizophrenia risk loci.[18]

Interestingly, Gagliano et al.[28] used a different approach to analyze whether epigenetic marks can mediate the effects of genetic variation associated with schizophrenia. After genotyping PFC samples from post-mortem control subjects ($n=76$), and patients with bipolar disorders ($n=67$), and schizophrenia ($n=65$), the authors searched for allelic differences in DNA epigenetic modifications by enriching the unmodified DNA fraction with restriction enzymes sensitive to DNA modifications. In this way, they obtained—for each SNP—a genotype call and a DNA-modification level corresponding to each allele, thus identifying SNPs with allele-skewed modifications (ASM). The results were consistent with the above described methylation studies,[26,27] in showing that brain ASM SNPs were significantly overrepresented among SNPs associated with schizophrenia; the enrichment of ASM SNPs for schizophrenia variants was much stronger compared with the enrichment for SNPs with GWAS-association with non-psychiatric disorders. The enrichment was also present in control brains, possibly suggesting that schizophrenia-related ASM effects can occur before the onset of the disorder. When the authors searched for ASM SNPs in sperm, they detected a partial overlap with brain ASM SNPs. The authors propose a model where both DNA sequence and epigenetic variation are necessary for the disorder, exerting their effects on risk synergistically at regulatory regions. Consistently, a recent multi-omic approach[55] has detected an enrichment of schizophrenia GWAS SNPs for variants associated with multiple molecular features, i.e., gene expression, DNA methylation and histone modifications.

However, while DNA methylation changes studies overall suggest that the epigenome might play a role in the etiopathogenesis of schizophrenia, this line of research presents limitations. First, all the studies so far published are based on assays that do not cover the whole DNA methylome, so that the contribution of other CpGs potentially relevant for the disorder and of non-CpG methylation[56] is still unexplored; in this regard, whole genome bisulfite sequencing studies may represent a future resource, since it can provide an assessment of methylation at single-CpG and non-CpG resolution. Second, all the main DNA methylation studies on schizophrenia do not discriminate between DNA methylation and hydroxymethylation. 5-Hydroxymethylation is considered an intermediate step in the process of active de-methylation of 5-methylcytosine[45,46]; indeed methylation and hydroxymethylation have a different relationship with gene expression.[45–48] Since DNA hydroxymethylation is particularly relevant in brain and it exhibits dynamic regulation during development,[56] the choice of techniques discriminating between the two DNA modifications may help to detect loci where dynamic regulation of DNA methylation, relevant for schizophrenia, occurs—not necessarily in early life. Third, DNA methylation studies on schizophrenia have searched for epigenetic changes "universally" associated with schizophrenia, across patients, while an assumption beyond the concept of epigenetic plasticity is that different developmental/epigenetic trajectory may lead to similar phenotypes. To this regard, single-CpG studies, performed on peripheral blood, have illustrated that the biological meaning of DNA methylation differences may be different[57] or even opposite[58] depending on the genetic background. Bioinformatics approaches are needed in order to analyze whether a complex interaction between genome and epigenome may help to explain the heterogeneous architecture of schizophrenia. Finally, given the potential confounder represented by the experience itself of schizophrenia on DNA methylation, as well as technical confounders such as cell composition, epigenome-wide studies on schizophrenia will definitely benefit from the addition of experimental systems, such as iPSCs-derived neurons and single-cell approaches.

4. THE SEARCH FOR SCHIZOPHRENIA EPIGENETIC BIOMARKERS IN BLOOD

The pathogenic processes for schizophrenia likely involve the brain. However, since brain tissue is not readily accessible in living patients, many biomarker studies on schizophrenia focus on blood. Aberg et al.[59] summarize two models explaining how methylation studies in blood can advance schizophrenia research,[59] both models do not assume that methylation in blood directly affects disease susceptibility. In the "signature" model, the factors that increase susceptibility to schizophrenia—likely acting in brain—leave a biomarker signature in blood. Thus, the methylation markings in blood represent a sort of epigenetic fossil, tracing memory of a cause of the disease, which may affect schizophrenia through processes that can be unrelated to methylation in the brain. In contrast, the "functional mirror site" model assumes a causal role of DNA methylation in the brain. When specific CpG sites in the blood mirror the methylation status of the corresponding "causative" CpG sites in the brain, we will observe associations between schizophrenia and methylation markings at the same loci in blood. Mirror sites can occur because peripheral tissues may reveal methylation markings related to the epigenetic reprogramming events affecting the germ line and embryogenesis,[60] and to environmental factors and genetic polymorphisms that can affect methylation levels in multiple tissues.[61,62] The same authors found that, when administering haloperidol to inbred mice, more than 65% of the methylation sites showed correlated changes with similar concordance rates between blood and brain, thus supporting the plausibility of a signature model.[59] However, in a more recent study on human samples, Marzi et al. detected differences in ASM between cortex, cerebellum and whole blood samples with inter-tissue correlations ranging from $r=0.42$ to 0.48.[63]

Dempster et al. performed the first genome-wide analysis of DNA methylation on peripheral blood DNA samples from 22 monozygotic (MZ) twins pairs discordant for schizophrenia and bipolar disorder, using the 27k array.[29] Despite the small sample size, numerous loci—enriched for neurodevelopmental pathways—demonstrated disease-associated DNA methylation differences between twins discordant for schizophrenia and bipolar disorders. The top differentially methylated CpG sites across the twin pairs discordant for schizophrenia was located upstream of the gene encoding PUS3, hypermethylated in affected twins, while the top differentially methylated site across all 22 psychosis-discordant MZ twin pairs was located in the promoter region of the *ST6GALNAC1* gene, hypomethylated in affected twins. An analysis in post-mortem brain samples from unrelated controls ($n = 15$) and patients with schizophrenia ($n = 15$) did not reveal significant differences in *ST6GALNAC1* methylation, although a subset of patients showed marked hypomethylation.[29]

Using methyl-CpG binding domain (MBD) protein-enriched genome sequencing (MBD-seq), Aberg et al.[30] performed a methylome-wide association study in a discovery sample of 759 patients with schizophrenia and 738 controls from Sweden. The authors identified 139 sites differentially methylated at a false discovery rate of 0.01; analyses of protein–protein interaction networks, pathway databases, and miRNA target networks, were consistent in pointing at hypoxia as a common biological denominator underlying the differentially methylated genes, compatibly with the persistence of epigenetic marks associated with early-life events. The top differentially methylated site—replicated in two independent samples—was located in the *FAM63B* gene, which is part of a network regulated by microRNAs linked with brain development. Consistent with candidate methylation studies, the authors also found hypermethylation of *RELN* in schizophrenia, although they did not control for cell composition.[30] In a following study, the same group combined these results with a GWAS meta-analysis, showing a significant overlap between the differentially methylated sites and genetic loci associated with schizophrenia.[31]

Liu et al.[32] used the 27k array to analyze blood methylation in a smaller sample of 98 patients with schizophrenia and 108 controls. They identified 20 CpG with methylation levels associated with schizophrenia, 16 of which were validated in a replication sample of 719 individuals; the 16 corresponding genes were found to be differentially expressed in an independent dataset, with 5 showing an inverse correlation between methylation marks and gene expression. The DMPs were located in genes related with inflammatory response, and 11 of them showed association with reality distortion symptoms.[32]

van Eijk et al.[33] also combined meQTL analysis with gene expression data, by analyzing DNA methylation levels (again assessed with the 27k chip) from whole blood of 260 patients with schizophrenia and 250 controls. They prioritized meQTL-CpGs that were differentially methylated and of which the corresponding genes were differentially expressed in patients compared with controls. The identified loci were strongly enriched in the loci with GWAS-significant association with schizophrenia, indicating that meQTLs derived from combined differential methylation and gene expression data can reveal susceptibility loci for schizophrenia. The discovered genes were *CALHM1*, *HLA-C*, *PRRT1*, and *MRPL41*, of which the first two have previously been associated with schizophrenia.[33]

Montano et al.[34] analyzed genome-scale DNA methylation in a large schizophrenia case–control consortium and its relationship to genetic variation, using the 450k array. After adjusting for age, sex, race/ethnicity, smoking, batch, and a measure of cell-type heterogeneity, 923 CpGs were found differentially methylated, at FDR <0.2, in the discovery dataset of 689 cases and 645 controls; 625 CpG showed changes with consistent directionality in a replication sample of 247 cases and 250 controls. 3 out of these 625 DMPs overlapped 3 loci with GWAS-significant association with schizophrenia, and in one of these loci (on chromosome 15) the schizophrenia-associated genotype was associated with methylation, without evidence of mediation effect of methylation on schizophrenia. The DMPs included genes involved in neuronal function and genes associated with schizophrenia and with immune functions; however, statistically significant differences in cell composition between patients and controls, reported by the authors, may represent a factor potentially affecting the results in this as in other studies on peripheral biomarkers.[34]

Using the same method for methylation analysis, Hannon et al.[35] performed a multistage epigenome-wide analysis, in DNA samples analyzed from whole blood of 322 controls and 353 patients with schizophrenia. They also controlled for potential confounders, such as smoking, age sex, experimental batch and cell composition. In their discovery cohort of 675 individuals, they identified 25 DMPs associated with schizophrenia (at a stringent experiment-wide significance threshold of $P < 10 \times 10^{-7}$) and located in 12 DMRs; the directionality of association with schizophrenia was overall consistent in two independent samples of 847 individuals and 96 MZ twin pairs. The top-ranked DMRs overlapped the major histocompatibility complex (MHC) on chromosome 6,[35] which is the most robustly associated locus in schizophrenia GWAS.[8] The authors calculated polygenic risk scores for schizophrenia (PRS's) in their samples, and they found DMPs associated with schizophrenia PRS's, which did not overlap with DMPs associated with schizophrenia diagnosis.[35] This was consistent with postmortem findings[26,27] potentially suggesting that the mechanism of risk of schizophrenia-associated genetic variation may imply an effect on CpG methylation, while methylation differences between patients and controls can be independent from genetic risk as identified by GWAS. Again, consistent with findings in post-mortem brain[26,27] and peripheral blood,[31] the differentially methylated loci overlapped with loci with GWAS-significant association with schizophrenia.[35] Finally, a Bayesian co-localization analysis in the discovery sample revealed that genetic variation associated with meQTLs in the schizophrenia loci, in blood[35] as in brain,[26] may be causally involved in the disease.[35]

van den Oord et al.[36] investigated the link between genetic variation, DNA methylation and schizophrenia, by focusing on point-mutated CpG sites, i.e., CpG-SNPs. Methylation of common CpG-SNPs was analyzed with MBD-seq, using DNA extracted from 712 patients with schizophrenia and 696 controls. They identified 7 CpG-SNPs that reached methylome-wide significance at a FDR <0.1; one of these CpG-SNPs, located in the *IL1RAP* gene, was differentially methylated also in post-mortem brain ($n = 66$), and was further validated in 736 independent blood samples. Notably, the authors found that 68.3% and 67.7% of CpG-SNPs were methylated, respectively, in blood and brain,[36] thus suggesting that methylation can modulate the effect of genetic variation. However, their analysis did not contemplate the possibility that the relationship between methylation and brain phenotypes may be different depending on genetic background.[57,58]

Overall, methylome-wide analyses performed in blood and brain identify large number of DMPs and corresponding annotated genes (Table 1). While there is little or no overlap in the top gene hits (derived from the most significant DMPs) between studies on schizophrenia, some genes are consistently differentially methylated in different studies. For example, as noted in a recent review by Pries et al.,[7] differential methylation of the following genes was associated with schizophrenia in at least four independent studies, across different tissue types: *PRKD2*, *NOTCH4*, *NCOR2*, *PTPRN2*, *FOXN3*, *FOXP1*, *MYT1L*, *PTGER4*, *RPP21*, *TRERF1*, *WDR20*, *ADCY6*, *AUTS2*, and *C7orf50*. In addition, some genes identified through methylome-wide analyses were also supported by candidate gene studies, such as *COMT*, *GAD1*, *RELN*, and *BDNF*.

When considering the results of pathway and gene-ontology enrichment analyses, blood-based EWAS tend to involve the immune system functioning,[30,32,35] while blood and brain-based studies point to neurodevelopmental process.[20,24,29,35] One possible explanation is that environmental risk factors associated with schizophrenia may imprint the epigenome of brain and immune cells in early life.

5. HISTONE POST-TRANSLATIONAL MODIFICATIONS AND CHROMATIN CONFORMATION STUDIES IN SCHIZOPHRENIA

Most studies on schizophrenia epigenetics have been focused on DNA methylation; however, there is no reason to expect this to be the exclusive or most relevant epigenetic mechanism for the disorder, nor a process completely independent from other epigenetic processes representing different layers of gene regulation (Fig. 1).

Histone modifications contribute to nucleosomal and facilitate supra-nucleosomal or higher-order chromatin regulation.[64] The regulation of histone modifications is more complex than the DNA methylation discussed above: far more than 100 amino acid residue-specific post-translational modifications (PTMs) are indeed thought to occur in a vertebrate cell,[65] including mono (me1), di (me2)- and tri (me3) methylation, acetylation, and crotonylation, polyadenosine triphosphate (ADP)-ribosylation and small protein (ubiquitin, small ubiquitin-like modifier—SUMO) modification of specific lysine residues, as well as arginine (R) methylation and "citrullination," serine (S) phosphorylation, tyrosine (T) hydroxylation, and several others.[65–67] Multiple combinatorial sets of histone PTMs, together with DNA methylation/hydroxymethylation, may contribute to define distinct structural and thereby functional chromatin states[68,69] (Fig. 1).

Despite the potential of such complex modifications for neurodevelopment and schizophrenia etiopathogenesis, only a few single-gene studies have focused on this process to date (Table 1), likely due to technical challenges in post-mortem tissue. Moreover, the potential role of histone genes as targets of schizophrenia-associated genetic variation remains completely unexplored. For example, the locus with the most significant association with schizophrenia—known as the MHC locus—includes a number of histone genes, yet eQTL studies are currently designed to exclude them from the analysis, as they rely on post-mitotic tissue and sometimes RNA-sequencing poly-A data (most histone genes do not have a poly-A tail, in addition to being mainly expressed during cellular replication).

Tang et al. detected significant H3K9K14 hypoacetylation at candidate genes (*GAD67*, *HTR2C*, *TOMM70A* and *PPM1E*) in young subjects with schizophrenia when compared with age-matched controls; the same marker was consistently associated with age in controls but not in patients with schizophrenia.[37]

In a study focused on the GABAergic system, Huang et al. found decreased H3K4-trimethylation in PFC of patients with schizophrenia, predominantly in females and in conjunction with a risk haplotype at the 5′ end of *GAD1* gene.[38] Subjects with schizophrenia who were biallelic for a *GAD1* promoter haplotype, showed a significant deficit in prefrontal *GAD67* transcript together with a shift in the epigenetic landscape of the surrounding chromatin, specifically including the loss of a facilitative histone methylation marking (histone H3 trimethyl-lysine 4) and excess of a repressive mark, histone H3 trimethyl-lysine 27.[38]

In a following study from the same group, Bharadwaj et al. found that these disease-associated changes in local chromatin signals at specific gene promoters were apparently accompanied by additional alterations in higher-order chromatin structures. Indeed, decreased *GAD1/GAD67* expression in schizophrenia PFC was accompanied by a weakening (i.e., lesser frequency of interaction) of a long-range promoter-enhancer chromatin loop between regulatory sequences positioned 50 kb upstream of the *GAD1* gene and the gene's transcription start site and proximal promoter.[39] The authors concluded that 3D genome architectures, including chromosomal loopings for promoter-enhancer interactions (Fig. 1) involved in the regulation of GABAergic gene expression, are subject to developmental and activity-dependent regulation, and are altered in some cases with schizophrenia.[39] The same authors detected activity-regulated long-range conserved chromatin loops bypassing up to 0.5 Mb of linear genome to modulate NMDA glutamate receptor *GRIN2B* expression in human and mouse PFC, suggesting that dynamic regulation of chromosomal conformations represents an important layer for transcriptional regulation impacting neuronal signaling and cognition, and likely psychiatric disorders.[40] While all these studies have been focused on single loci, they support the possibility that histone modifications and the three-dimensional organization of chromatin represent epigenetic mechanisms relevant for brain development and schizophrenia.

Three-dimensional physical interactions within chromosomes have been recently explored on an epigenome-wide scale by Won et al.[70] The authors generated Hi-C libraries from mid-gestation developing human cerebral cortex during the peak of neurogenesis and migration from the cortical and subcortical plate, consisting primarily of post-mitotic neurons and the germinal zone, and containing mainly mitotically active neural progenitors. In this way, they identified hundreds of genes that physically interact with enhancers gained on the human lineage. Many of these genes are under purifying selection and associated with human cognitive function.[70] They then applied the developing brain Hi-C maps to detect the potential targets of schizophrenia-associated genetic variation,[8] and identified ~500 genes that were neither adjacent to the schizophrenia-associated index SNPs, nor in LD with them, consistent with observations that linear chromosomal organization does not capture many regulatory interactions. These genes included *DRD2*, and several acetylcholine receptors, *SOX2*, *GRIN2A*, *CACNA1C*, and *FOXG1*. The authors also performed a genome editing experiment in neural progenitors to confirm the regulatory effect on gene expression of the physical interaction between a schizophrenia-associated SNP and *FOXG1*, a gene associated with brain development, located 760 kb away from the disease-associated locus. The genes physically interacting with schizophrenia-associated variants were overall enriched for biological functions related to brain development and activity, such as post-synaptic density, acetylcholine receptors, neuronal differentiation, and chromatin remodeling,[70] and they overlap genes down-regulated in the DLPFC of patients with schizophrenia in CommonMind Consortium data.[17] These results suggest that a comprehensive analysis of the genome-wide chromatin contacts may provide biological insights into gene regulatory mechanisms relevant for brain development and schizophrenia. However, caution should be used when generalizing the findings; a larger sample size or specific experiments at single loci are likely necessary to confirm if schizophrenia-associated genetic variants may effectively have a role in disrupting these physical interactions.

6. NON-CODING RNA ALTERATIONS IN SCHIZOPHRENIA

While DNA and histone modifications, and the 3D organization of the chromatin regulate protein-coding gene expression at the transcriptional level, non-coding RNAs represents one more layer of gene regulation, potentially acting at multiple levels, i.e., transcriptional, post-transcriptional and translational (Fig. 1). It has been observed that less than 2% of the transcribed human genome is protein-coding.[71] Moreover, while the protein-coding genome has undergone only modest alterations during evolution, the non-coding portion has scaled with biodiversity and organismal complexity.[72] Consistently, it has been hypothesized that the evolution of a more complex regulatory apparatus, afforded by the expansion of non-coding RNAs, has allowed the emergence of cellular complexity and abilities such as higher-order cognition, as illustrated through the rapid striking expansion of the primate brain.[73] A dysfunction of such system may hence contribute to the etiopathogenesis of complex disorders like schizophrenia, meaning that non-coding regulation may underlie neuronal complexity as much as the resulting fragility of the human brain.[74] Non-coding RNAs are a group of RNA transcripts that do not necessarily code for protein products, as they make instead potentially functional RNAs. They include small RNA (sRNA) (for example, microRNA (miRNA), small interfering RNA (siRNA) and small nucleolar RNA (snoRNA)) and lncRNA [NB: for a classification of non-coding RNAs and their functions, we refer to extensive reviews that have been published on the topic[73,75–79]].

A great number of microRNAs has been associated with schizophrenia, based on GWAS and differential expression analyses performed in post-mortem brains or peripheral tissue.[80,81] Among the most consistent associations, *miR-137* lies in the second most statistically significantly associated region in the genome, and has a known role in regulating neuronal maturation and differentiation.[8,82] Of note, several genes with confirmed *miR-137*-binding sites in their 3′ UTR belong to loci with GWAS-significant association with schizophrenia,[8] such as *TCF4* and *CACNA1C*, or are involved in synaptogenesis.[83] Another microRNAs that has been consistently linked to schizophrenia is *miR-132*, whose transcript and related targets have been shown to be altered in brain tissue of patients[84]; moreover *miR-132* has been proven to modulate synaptic plasticity and to be sensitive to NMDA receptor antagonist.[85] Involvement of microRNAs in the disorder[86] is further supported by a recent analyses of miRNA targetomes. More in details, by analyzing summary statistics from the most recent schizophrenia GWAS, Roussos et al.[86] have found that schizophrenia risk genes were more likely to be regulated by miRNAs; the strongest enrichment was detected for *miR-9-5p*, followed by *miR-485-5p* and *miR-137*.[86]

Despite lncRNAs being highly expressed in the brain, and ~40% of them being considered as actually brain-specific,[77] their role in brain disorders is mainly unexplored. In line with a role of *DISC1*[87] and *ERBB4*[88] in the pathophysiology of schizophrenia, Barry et al.[89] found that chronic down-regulation of the lncRNA *GOMAFU* in human pluripotent cell-derived neurons in vitro in fact drives splicing defects in *DISC1* and *ERRB4*.[89] These defects were consistent with those seen in patients with schizophrenia and with the lower expression of *GOMAFU* in cortex from the superior temporal gyrus of schizophrenia patients. *GOMAFU* is a lncRNA that has been shown to localize in the CA1 region of the hippocampus and in large cortical neurons,[90] and is, importantly, regulated by neuronal activity.[89]

Scarcity of evidence for a role in schizophrenia contrasts with lncRNAs' expanding involvement in brain development and plasticity,[77] suggesting that novel specific paradigms are needed to investigate them in the context of the disorder. In this regard, it seems unlikely that differential expression analysis will suffice: small amounts of lncRNAs—including novel classes such as divergent lncRNAs[91]—are likely enough to perform highly specific functions, thus posing technical challenges. For instance, at the chromatin level, a single copy of non-coding RNA per cell is sufficient to interact with DNA and transcription complexes.[92,93] Moreover, the effects of non-coding RNAs are often different depending on the cellular context where they act. Protein-coding RNAs and non-coding RNAs may act as competing endogenous RNAs, for example co-regulating each other for binding to shared microRNAs.[94] Experimental systems like iPSCs-derived neurons and bioinformatics approaches based on machine learning may be crucial to detail such RNA crosstalk in brain development and disorders like schizophrenia.

7. EPITRANSCRIPTOMICS: A NEW FRONTIER FOR NEUROEPIGENETICS

High-throughput sequencing approaches have revealed a dynamic "epitranscriptome" landscape including mRNA modifications in various organisms, such as N6-methyladenosine (m6A), N1-methyladenosine (m1A), 5-methylcytosine (m5C), 5-hydroxymethylcytosine (hm5C), and pseudouridine (ψ).[95] Among them, m6A (Fig. 1) is the most abundant internal modification in mRNAs of eukaryotic cells[96] and its functions in cell renewal and differentiation have been recently characterized.[97,98] In their recent work, Yoon et al.[99] have studied m6A in brain development. The authors have found that depletion of the methyltransferase complex involved in this epigenetic modification leads to prolonged cell-cycle progression of cortical neuronal progenitors, pointing to a role of m6A in neurogenesis.[99] By performing m6A-sequencing of day 47 human forebrain organoids and fetal human brain, they have also detected a set of m6A-tagged genes that were enriched for mental disorders and schizophrenia. Moreover, among the genes in the 108 loci with GWAS-significant association with schizophrenia,[8] 60 were m6A tagged in human and 21 genes were uniquely tagged in both human forebrain organoids and fetal brain, but not in mouse E13.5 forebrain.[99] This work opens a new avenue of research in epitranscriptomics and brain development/schizophrenia, suggesting that not only DNA but also RNA modifications may contribute to regulate how genetic risk for schizophrenia translates into different phenotypes.

8. HUMAN BRAIN SPATIOTEMPORAL CHANGES IN GENE EXPRESSION AND EPIGENETIC MARKS

The brain is heterogeneous both at a regional and at a cellular level. Many studies present findings on spatiotemporal changes in gene expression patterns, yet few have done so in a comprehensive manner,[100,101] and very few studies have analyzed epigenetic changes in brain areas other than PFC. Once again, this is partly due to the lack of large collections of post-mortem brain tissue ranging over the lifespan, coupled with a shortage of brain cell-type separation technologies. While some of these studies have addressed cellular specificity at the basic level (neuronal versus glial), detailed resolutions at the cell-type level (e.g., glutamatergic versus dopaminergic) are far and few. Moreover, the majority of these studies have focused on neocortical regions rather than deep nuclei or subcortical nuclear groups within the brain. Hence, this is still an emergent field, as the regional complexity and cellular heterogeneity of the human brain becomes more evident.

Recently, databases such as the Roadmap Epigenomics Browser (http://www.roadmapepigenomics.org/data/visualbrowser/brain) and PsychEncode (https://www.synapse.org//#!Synapse:syn5553626) have deposited both transcriptomic and epigenomic datasets generated from multiple neocortical and a few subcortical brain regions. A cursory glance at the reference browsers identifies differences namely in the levels of epigenetic signals including DNA methylation and a variety of histone modifications across the genome. It is again worth noting that these datasets are generated from bulk homogenates and not at the cell-specific level. However, these data suggest that future directions in epigenetic studies on schizophrenia may consist in looking for epigenetic regulation in specific cell types and well-defined brain areas, not limited to the PFC, which has been the focus of most post-mortem methylation studies.

To date, epigenetic data collected from different brain areas can inform about changes in gene regulation related to cell composition. For example, Lu et al.[102] have used 1796 brain samples from 1163 individuals to carry out a GWAS of two DNA methylation-based biomarkers of brain age: the epigenetic aging rate and estimated proportion of neurons. In addition to finding loci associated with aging rate across five brain regions, they also found genetic variants that predict the proportion of neurons, and overlap with schizophrenia-associated loci, specifically in PFC.[102]

We predict that, with the advent of cell-specific technologies combining epigenetic assays with gene expression (transcript and protein) levels, we will be able to generate higher resolution molecular datasets that can then be analyzed to accelerate biological and clinical understanding of schizophrenia.

9. A ROLE OF PLACENTA IN THE EARLY-LIFE EPIGENETIC CHANGES RELEVANT FOR SCHIZOPHRENIA

The heritability of schizophrenia not only indicates a major role for inherited genetic and epigenetic variants in its etiology, but also for non-inherited epigenetic modifications. Specifically, the 45%–60% concordance[103] of the disorder in MZ twins leaves plenty of room for environmental mechanisms, further pointing to epigenetic mechanisms acting early in life, since MZ twins share the same intrauterine environment. Indeed, dizygotic twins—who also share the same intrauterine environment—have a concordance rate for schizophrenia twice that of siblings, despite both groups have 50% of segregating DNA variation in common.[103] Epigenetic mechanisms acting in early life have also been invoked by several studies that have associated schizophrenia with a number of complications occurring during pregnancy and labor, at delivery, and early in neonatal life.[4]

A role of the early-life environment in risk for schizophrenia is also supported by several analyses of PFC expression of genes containing rare schizophrenia risk variants, genes affected by schizophrenia-associated CNVs, and genes within the 108 schizophrenia GWAS-significant loci. These analyses overall indicate a bias toward prenatal expression compared with postnatal expression of schizophrenia risk genes and transcripts.[3] As described above, recent DNA methylation[26,27] and Hi-C[70] studies have provided more direct evidence that epigenetic mechanisms acting during early brain development are also implicated in schizophrenia risk.

Accordingly, we have recently shown that genetic risk for schizophrenia, as measured with PRS calculated from the most GWAS-significant loci, interact with serious prenatal and perinatal complications in affecting risk for schizophrenia, so that the liability of schizophrenia of genetic risk is five time higher in individual with a history of such complications compared with its absence.[104] Consistently, we have found that the genes mapping to the schizophrenia risk loci, and interacting with those complications, are highly expressed in placenta and differentially expressed in placentae from complicated in comparison with normal pregnancies; moreover they are differentially up-regulated in placentae from male compared with female offspring. Such finding suggests a link between genetic risk for schizophrenia and placenta pathophysiology, together with a sex-biased role for the placenta in expressing genetic risk for schizophrenia.[104] These data suggest that the study of epigenetic mechanisms and gene expression in placenta, as in post-mortem fetal brain, may help to further characterize and elucidate epigenetic mechanisms relevant for schizophrenia risk during early development.

10. CONCLUSIONS

Epigenetic processes in schizophrenia are compatible with several features of the disorder, including the non-complete concordance in MZ twins, the polygenic and heterogeneous architecture of risk, the small effect of risk variants, the effect of environmental risk factors particularly acting during early life—when the epigenome is more plastic compared with adult life. In the last decade, many studies on schizophrenia have explored the involvement of neuroepigenetic mechanisms. Recent DNA methylome research stresses how epigenetic changes occurring in brain during early life may be particularly relevant for the etiopathogenesis of the disorder. They also indicate that the mechanism of risk of schizophrenia-associated genetic variation may imply an effect on CpG methylation, while methylation differences between patients and controls can also be independent from genetic risk. Studies focused on candidate loci indicate histone modifications as another epigenetic mechanism relevant for the disorder, potentially linked with genetic risk and with the three-dimensional organization of chromatin. Chromatin conformation, as emerges from 3D physical interactions within chromosomes in the fetal brain, appears to be crucial during development; consistently, the effect of schizophrenia risk variants can be mediated by physical interactions between non-adjacent chromosomal regions. Regulatory systems offered by non-coding RNAs may also contribute to the etiopathogenesis of schizophrenia, potentially affecting the expression of risk genes independently associated with the disorder. Finally, chemical modifications of RNAs may represent another level of regulation, with consequences on the disorder.

While previous studies suggested a general epigenetic abnormality in schizophrenia, related to a restrictive chromatin state, more recent findings indicate that the dysfunctional epigenetic mechanisms in schizophrenia are more complex,[105] and may also be heterogeneous in patients with schizophrenia. Such diversity in mechanisms and outcomes may be explained by a "U-inverted" shaped model, where either too little or too much epigenetic plasticity may increase the susceptibility to the disorder, the optimal condition being in the middle. The epigenetic machinery represents the interface between the developing brain and inputs from the environment, as necessary for a "normal" development of the brain, i.e., to eventually acquire those critical functions that are altered in schizophrenia. A diminished epigenetic plasticity may obviously prevent a normal development of the brain, thus increasing risk for the disorder. On the other hand, in responding to stressful factors known to increase risk, such as prenatal and perinatal complications and postnatal harmful events, an exasperated epigenetic plasticity may equally interfere with brain development. Therefore in this U-inverted shaped model both a defect and an excess of epigenetic response increase risk for the disorder. Such model would fit in with the enrichment of schizophrenia risk loci for regions under epigenetic regulation shown by many studies, with the heterogeneity of the epigenetic findings, and with the interaction between genetic risk and prenatal and perinatal complications.

While the emerging scenario is complex, future studies will likely leverage novel techniques to unveil further epigenetic mechanisms behind schizophrenia; also, they will likely expand the study of epigenetics to different brain areas, tissues, and specific cell types. Whether these data will, by themselves, clarify the etiopathogenesis of the disorder is uncertain. However, integrating the study of different epigenetic mechanisms with the analysis of environmental and genetic risk has the potential to inform a better understanding of the disorder's biological origins. To this purpose, novel research paradigms are needed, since traditional designs based on case–control comparisons need to give way to more complex strategies that take into account human individual trajectories of development. For example, the integration of genotype and environmental exposure data, with epigenetic markers, detailed phenotyping, and machine learning approaches, may help to understand how the same genotypes may be translated into different phenotypes, and how similar phenotypes may be the results of developmental trajectories originating from different genotypes, consistent with the concept of epigenetic plasticity. The use of cellular and experimental models may help to validate the different developmental trajectories of risk.

ACKNOWLEDGMENTS

We are grateful to the Lieber and Maltz families for their visionary support that funded the work of this project. This work is also partially supported by P50 MH094268 grant to G.U.

REFERENCES

1. Saha S, Chant D, Welham J, McGrath J. A systematic review of the prevalence of schizophrenia. *PLoS Med.* 2005;2(5):e141.
2. Owen MJ, Sawa A, Mortensen PB. Schizophrenia. *Lancet.* 2016;388(10039):86–97.
3. Birnbaum R, Weinberger DR. Genetic insights into the neurodevelopmental origins of schizophrenia. *Nat Rev Neurosci.* 2017;18(12):727–740.
4. Cannon M, Jones PB, Murray RM. Obstetric complications and schizophrenia: historical and meta-analytic review. *Am J Psychiatry.* 2002;159(7):1080–1092.
5. van Os J, Kenis G, Rutten BP. The environment and schizophrenia. *Nature.* 2010;468(7321):203–212.
6. Insel TR. Rethinking schizophrenia. *Nature.* 2010;468(7321):187–193.
7. Pries LK, Guloksuz S, Kenis G. DNA methylation in schizophrenia. *Adv Exp Med Biol.* 2017;978:211–236.
8. Schizophrenia Working Group of the Psychiatric Genomics Consortium. Biological insights from 108 schizophrenia-associated genetic loci. *Nature.* 2014;511(7510):421–427.
9. Dupont C, Armant DR, Brenner CA. Epigenetics: definition, mechanisms and clinical perspective. *Semin Reprod Med.* 2009;27(5):351–357.
10. Berger SL, Kouzarides T, Shiekhattar R, Shilatifard A. An operational definition of epigenetics. *Genes Dev.* 2009;23(7):781–783.
11. Gottesman II, Shields J. A polygenic theory of schizophrenia. *Proc Natl Acad Sci USA.* 1967;58(1):199–205.
12. McGue M, Gottesman II, Rao DC. Resolving genetic models for the transmission of schizophrenia. *Genet Epidemiol.* 1985;2(1):99–110.
13. Yang J, Visscher PM, Wray NR. Sporadic cases are the norm for complex disease. *Eur J Hum Genet.* 2010;18(9):1039–1043.
14. International Schizophrenia Consortium, Purcell SM, Wray NR, et al. Common polygenic variation contributes to risk of schizophrenia and bipolar disorder. *Nature.* 2009;460(7256):748–752.
15. Schizophrenia Psychiatric Genome-Wide Association Study (GWAS) Consortium. Genome-wide association study identifies five new schizophrenia loci. *Nat Genet.* 2011;43(10):969–976.
16. Li Y, Kellis M. Joint Bayesian inference of risk variants and tissue-specific epigenomic enrichments across multiple complex human diseases. *Nucleic Acids Res.* 2016;44(18):e144.
17. Fromer M, Roussos P, Sieberts SK, et al. Gene expression elucidates functional impact of polygenic risk for schizophrenia. *Nat Neurosci.* 2016;19(11):1442–1453.
18. Jaffe AE, Straub RE, Shin JH, et al. Developmental and genetic regulation of the human cortex transcriptome in schizophrenia. *bioRxiv.* 2017;.
19. Network and Pathway Analysis Subgroup of Psychiatric Genomics Consortium. Psychiatric genome-wide association study analyses implicate neuronal, immune and histone pathways. *Nat Neurosci.* 2015;18(2):199–209.
20. Mill J, Tang T, Kaminsky Z, et al. Epigenomic profiling reveals DNA-methylation changes associated with major psychosis. *Am J Hum Genet.* 2008;82(3):696–711.

21. Chen C, Zhang C, Cheng L, et al. Correlation between DNA methylation and gene expression in the brains of patients with bipolar disorder and schizophrenia. *Bipolar Disord*. 2014;16(8):790–799.
22. Numata S, Ye T, Herman M, Lipska BK. DNA methylation changes in the postmortem dorsolateral prefrontal cortex of patients with schizophrenia. *Front Genet*. 2014;5:280.
23. Wockner LF, Noble EP, Lawford BR, et al. Genome-wide DNA methylation analysis of human brain tissue from schizophrenia patients. *Transl Psychiatry*. 2014;4:e339.
24. Pidsley R, Viana J, Hannon E, et al. Methylomic profiling of human brain tissue supports a neurodevelopmental origin for schizophrenia. *Genome Biol*. 2014;15(10):483.
25. Wockner LF, Morris CP, Noble EP, et al. Brain-specific epigenetic markers of schizophrenia. *Transl Psychiatry*. 2015;5:e680.
26. Hannon E, Spiers H, Viana J, et al. Methylation QTLs in the developing brain and their enrichment in schizophrenia risk loci. *Nat Neurosci*. 2016;19(1):48–54.
27. Jaffe AE, Gao Y, Deep-Soboslay A, et al. Mapping DNA methylation across development, genotype and schizophrenia in the human frontal cortex. *Nat Neurosci*. 2016;19(1):40–47.
28. Gagliano SA, Ptak C, Mak DYF, et al. Allele-skewed DNA modification in the brain: relevance to a schizophrenia GWAS. *Am J Hum Genet*. 2016;98(5):956–962.
29. Dempster EL, Pidsley R, Schalkwyk LC, et al. Disease-associated epigenetic changes in monozygotic twins discordant for schizophrenia and bipolar disorder. *Hum Mol Genet*. 2011;20(24):4786–4796.
30. Aberg KA, McClay JL, Nerella S, et al. Methylome-wide association study of schizophrenia: identifying blood biomarker signatures of environmental insults. *JAMA Psychiatry*. 2014;71(3):255–264.
31. Kumar G, Clark SL, McClay JL, et al. Refinement of schizophrenia GWAS loci using methylome-wide association data. *Hum Genet*. 2015;134(1):77–87.
32. Liu J, Chen J, Ehrlich S, et al. Methylation patterns in whole blood correlate with symptoms in schizophrenia patients. *Schizophr Bull*. 2014;40(4):769–776.
33. van Eijk KR, de Jong S, Strengman E, et al. Identification of schizophrenia-associated loci by combining DNA methylation and gene expression data from whole blood. *Eur J Hum Genet*. 2015;23(8):1106–1110.
34. Montano C, Taub MA, Jaffe A, et al. Association of DNA methylation differences with schizophrenia in an epigenome-wide association study. *JAMA Psychiatry*. 2016;73(5):506–514.
35. Hannon E, Dempster E, Viana J, et al. An integrated genetic-epigenetic analysis of schizophrenia: evidence for co-localization of genetic associations and differential DNA methylation. *Genome Biol*. 2016;17(1):176.
36. van den Oord EJ, Clark SL, Xie LY, et al. A whole methylome CpG-SNP association study of psychosis in blood and brain tissue. *Schizophr Bull*. 2016;42(4):1018–1026.
37. Tang B, Dean B, Thomas EA. Disease- and age-related changes in histone acetylation at gene promoters in psychiatric disorders. *Transl Psychiatry*. 2011;1:e64.
38. Huang HS, Matevossian A, Whittle C, et al. Prefrontal dysfunction in schizophrenia involves mixed-lineage leukemia 1-regulated histone methylation at GABAergic gene promoters. *J Neurosci*. 2007;27(42):11254–11262.
39. Bharadwaj R, Jiang Y, Mao W, et al. Conserved chromosome 2q31 conformations are associated with transcriptional regulation of GAD1 GABA synthesis enzyme and altered in prefrontal cortex of subjects with schizophrenia. *J Neurosci*. 2013;33(29):11839–11851.
40. Bharadwaj R, Peter CJ, Jiang Y, et al. Conserved higher-order chromatin regulates NMDA receptor gene expression and cognition. *Neuron*. 2014;84(5):997–1008.

41. Yang R, Kerschner JL, Gosalia N, et al. Differential contribution of cis-regulatory elements to higher order chromatin structure and expression of the CFTR locus. *Nucleic Acids Res*. 2016;44(7):3082–3094.
42. Morgan SL, Mariano NC, Bermudez A, et al. Manipulation of nuclear architecture through CRISPR-mediated chromosomal looping. *Nat Commun*. 2017;8:15993.
43. Gheldof N, Smith EM, Tabuchi TM, et al. Cell-type-specific long-range looping interactions identify distant regulatory elements of the CFTR gene. *Nucleic Acids Res*. 2010;38(13):4325–4336.
44. Kriaucionis S, Heintz N. The nuclear DNA base 5-hydroxymethylcytosine is present in Purkinje neurons and the brain. *Science*. 2009;324(5929):929–930.
45. Jin SG, Wu X, Li AX, Pfeifer GP. Genomic mapping of 5-hydroxymethylcytosine in the human brain. *Nucleic Acids Res*. 2011;39(12):5015–5024.
46. Song CX, Szulwach KE, Fu Y, et al. Selective chemical labeling reveals the genome-wide distribution of 5-hydroxymethylcytosine. *Nat Biotechnol*. 2011;29(1):68–72.
47. Maunakea AK, Nagarajan RP, Bilenky M, et al. Conserved role of intragenic DNA methylation in regulating alternative promoters. *Nature*. 2010;466(7303):253–257.
48. Sharma RP, Grayson DR, Guidotti A, Costa E. Chromatin, DNA methylation and neuron gene regulation—the purpose of the package. *J Psychiatry Neurosci*. 2005;30(4):257–263.
49. Langfelder P, Horvath S. WGCNA: an R package for weighted correlation network analysis. *BMC Bioinformatics*. 2008;9:559.
50. Gibbs JR, van der Brug MP, Hernandez DG, et al. Abundant quantitative trait loci exist for DNA methylation and gene expression in human brain. *PLoS Genet*. 2010;6(5): e1000952.
51. Banovich NE, Lan X, McVicker G, et al. Methylation QTLs are associated with coordinated changes in transcription factor binding, histone modifications, and gene expression levels. *PLoS Genet*. 2014;10(9):e1004663.
52. Gutierrez-Arcelus M, Ongen H, Lappalainen T, et al. Tissue-specific effects of genetic and epigenetic variation on gene regulation and splicing. *PLoS Genet*. 2015;11(1): e1004958.
53. Flavahan WA, Drier Y, Liau BB, et al. Insulator dysfunction and oncogene activation in IDH mutant gliomas. *Nature*. 2016;529(7584):110–114.
54. Li M, Jaffe AE, Straub RE, et al. A human-specific AS3MT isoform and BORCS7 are molecular risk factors in the 10q24.32 schizophrenia-associated locus. *Nat Med*. 2016;22(6):649–656.
55. Ng B, White CC, Klein HU, et al. An xQTL map integrates the genetic architecture of the human brain's transcriptome and epigenome. *Nat Neurosci*. 2017;20(10): 1418–1426.
56. Lister R, Mukamel EA, Nery JR, et al. Global epigenomic reconfiguration during mammalian brain development. *Science*. 2013;341(6146):1237905.
57. Ursini G, Bollati V, Fazio L, et al. Stress-related methylation of the catechol-O-methyltransferase Val 158 allele predicts human prefrontal cognition and activity. *J Neurosci*. 2011;31(18):6692–6698.
58. Ursini G, Cavalleri T, Fazio L, et al. BDNF rs6265 methylation and genotype interact on risk for schizophrenia. *Epigenetics*. 2016;11(1):11–23.
59. Aberg KA, Xie LY, McClay JL, et al. Testing two models describing how methylome-wide studies in blood are informative for psychiatric conditions. *Epigenomics*. 2013;5(4):367–377.
60. Efstratiadis A. Parental imprinting of autosomal mammalian genes. *Curr Opin Genet Dev*. 1994;4(2):265–280.
61. Sutherland JE, Costa M. Epigenetics and the environment. *Ann NY Acad Sci*. 2003;983:151–160.

62. Kerkel K, Spadola A, Yuan E, et al. Genomic surveys by methylation-sensitive SNP analysis identify sequence-dependent allele-specific DNA methylation. *Nat Genet.* 2008;40(7):904–908.
63. Marzi SJ, Meaburn EL, Dempster EL, et al. Tissue-specific patterns of allelically-skewed DNA methylation. *Epigenetics.* 2016;11(1):24–35.
64. Akbarian S. Epigenetic mechanisms in schizophrenia. *Dialogues Clin Neurosci.* 2014;16(3):405–417.
65. Tan M, Luo H, Lee S, et al. Identification of 67 histone marks and histone lysine crotonylation as a new type of histone modification. *Cell.* 2011;146(6):1016–1028.
66. Kouzarides T. Chromatin modifications and their function. *Cell.* 2007;128(4): 693–705.
67. Taverna SD, Li H, Ruthenburg AJ, Allis CD, Patel DJ. How chromatin-binding modules interpret histone modifications: lessons from professional pocket pickers. *Nat Struct Mol Biol.* 2007;14(11):1025–1040.
68. Zhou VW, Goren A, Bernstein BE. Charting histone modifications and the functional organization of mammalian genomes. *Nat Rev Genet.* 2011;12(1):7–18.
69. Roadmap Epigenomics Consortium, Kundaje A, Meuleman W, et al. Integrative analysis of 111 reference human epigenomes. *Nature.* 2015;518(7539):317–330.
70. Won H, de la Torre-Ubieta L, Stein JL, et al. Chromosome conformation elucidates regulatory relationships in developing human brain. *Nature.* 2016;538(7626):523–527.
71. ENCODE Project Consortium. An integrated encyclopedia of DNA elements in the human genome. *Nature.* 2012;489(7414):57–74.
72. Liu G, Mattick JS, Taft RJ. A meta-analysis of the genomic and transcriptomic composition of complex life. *Cell Cycle.* 2013;12(13):2061–2072.
73. Barry G, Mattick JS. The role of regulatory RNA in cognitive evolution. *Trends Cogn Sci.* 2012;16(10):497–503.
74. Barry G. Integrating the roles of long and small non-coding RNA in brain function and disease. *Mol Psychiatry.* 2014;19(4):410–416.
75. Iwakawa HO, Tomari Y. The functions of microRNAs: mRNA decay and translational repression. *Trends Cell Biol.* 2015;25(11):651–665.
76. O'Connor RM, Gururajan A, Dinan TG, Kenny PJ, Cryan JF. All roads lead to the miRNome: miRNAs have a central role in the molecular pathophysiology of psychiatric disorders. *Trends Pharmacol Sci.* 2016;37(12):1029–1044.
77. Briggs JA, Wolvetang EJ, Mattick JS, Rinn JL, Barry G. Mechanisms of long non-coding RNAs in mammalian nervous system development, plasticity, disease, and evolution. *Neuron.* 2015;88(5):861–877.
78. Engreitz JM, Haines JE, Perez EM, et al. Local regulation of gene expression by lncRNA promoters, transcription and splicing. *Nature.* 2016;539(7629):452–455.
79. Pelechano V, Steinmetz LM. Gene regulation by antisense transcription. *Nat Rev Genet.* 2013;14(12):880–893.
80. Beveridge NJ, Cairns MJ. MicroRNA dysregulation in schizophrenia. *Neurobiol Dis.* 2012;46(2):263–271.
81. Mellios N, Sur M. The emerging role of microRNAs in schizophrenia and autism spectrum disorders. *Front Psychiatry.* 2012;3:39.
82. Wright C, Turner JA, Calhoun VD, Perrone-Bizzozero N. Potential impact of miR-137 and its targets in schizophrenia. *Front Genet.* 2013;4:58.
83. Strazisar M, Cammaerts S, van der Ven K, et al. MIR137 variants identified in psychiatric patients affect synaptogenesis and neuronal transmission gene sets. *Mol Psychiatry.* 2015;20(4):472–481.
84. Miller BH, Zeier Z, Xi L, et al. MicroRNA-132 dysregulation in schizophrenia has implications for both neurodevelopment and adult brain function. *Proc Natl Acad Sci USA.* 2012;109(8):3125–3130.

85. Wayman GA, Davare M, Ando H, et al. An activity-regulated microRNA controls dendritic plasticity by down-regulating p250GAP. *Proc Natl Acad Sci USA*. 2008;105(26):9093–9098.
86. Hauberg ME, Roussos P, Grove J, Borglum AD, Mattheisen M, Schizophrenia Working Group of the Psychiatric Genomics Consortium. Analyzing the role of microRNAs in schizophrenia in the context of common genetic risk variants. *JAMA Psychiatry*. 2016;73(4):369–377.
87. Brandon NJ, Sawa A. Linking neurodevelopmental and synaptic theories of mental illness through DISC1. *Nat Rev Neurosci*. 2011;12(12):707–722.
88. Law AJ, Kleinman JE, Weinberger DR, Weickert CS. Disease-associated intronic variants in the ErbB4 gene are related to altered ErbB4 splice-variant expression in the brain in schizophrenia. *Hum Mol Genet*. 2007;16(2):129–141.
89. Barry G, Briggs JA, Vanichkina DP, et al. The long non-coding RNA Gomafu is acutely regulated in response to neuronal activation and involved in schizophrenia-associated alternative splicing. *Mol Psychiatry*. 2014;19(4):486–494.
90. Sone M, Hayashi T, Tarui H, Agata K, Takeichi M, Nakagawa S. The mRNA-like noncoding RNA Gomafu constitutes a novel nuclear domain in a subset of neurons. *J Cell Sci*. 2007;120(pt 15):2498–2506.
91. Luo S, Lu JY, Liu L, et al. Divergent lncRNAs regulate gene expression and lineage differentiation in pluripotent cells. *Cell Stem Cell*. 2016;18(5):637–652.
92. Magistri M, Faghihi MA, St Laurent 3rd G, Wahlestedt C. Regulation of chromatin structure by long noncoding RNAs: focus on natural antisense transcripts. *Trends Genet*. 2012;28(8):389–396.
93. Holoch D, Moazed D. RNA-mediated epigenetic regulation of gene expression. *Nat Rev Genet*. 2015;16(2):71–84.
94. Tay Y, Rinn J, Pandolfi PP. The multilayered complexity of ceRNA crosstalk and competition. *Nature*. 2014;505(7483):344–352.
95. Li X, Xiong X, Yi C. Epitranscriptome sequencing technologies: decoding RNA modifications. *Nat Methods*. 2016;14(1):23–31.
96. Desrosiers RC, Friderici KH, Rottman FM. Characterization of Novikoff hepatoma mRNA methylation and heterogeneity in the methylated 5' terminus. *Biochemistry*. 1975;14(20):4367–4374.
97. Wang Y, Li Y, Toth JI, Petroski MD, Zhang Z, Zhao JC. N6-methyladenosine modification destabilizes developmental regulators in embryonic stem cells. *Nat Cell Biol*. 2014;16(2):191–198.
98. Batista PJ, Molinie B, Wang J, et al. m(6)A RNA modification controls cell fate transition in mammalian embryonic stem cells. *Cell Stem Cell*. 2014;15(6):707–719.
99. Yoon KJ, Ringeling FR, Vissers C, et al. Temporal control of mammalian cortical neurogenesis by m(6)A methylation. *Cell*. 2017;171(4):877–889.e817.
100. Kang HJ, Kawasawa YI, Cheng F, et al. Spatio-temporal transcriptome of the human brain. *Nature*. 2011;478(7370):483–489.
101. Hawrylycz M, Miller JA, Menon V, et al. Canonical genetic signatures of the adult human brain. *Nat Neurosci*. 2015;18(12):1832–1844.
102. Lu AT, Hannon E, Levine ME, et al. Genetic architecture of epigenetic and neuronal ageing rates in human brain regions. *Nat Commun*. 2017;8:15353.
103. Gottesman II, Wolfgram DL. *Schizophrenia Genesis: The Origins of Madness*. New York: Freeman; 1991.
104. G. Ursini, G. Punzi, Q. Chen, et al., Convergence of placenta biology and genetic risk for schizophrenia. Nat Med, in press. https://doi.org/10.1038/s41591-018-0021-y.
105. Gavin DP, Floreani C. Epigenetics of schizophrenia: an open and shut case. *Int Rev Neurobiol*. 2014;115:155–201.

CHAPTER TEN

Neuroepigenetics of Post-Traumatic Stress Disorder

Grace S. Kim[*,†], Alicia K. Smith[‡,§], Caroline M. Nievergelt[¶], Monica Uddin[*,||,#,1]

[*]Neuroscience Program, University of Illinois at Urbana–Champaign, Urbana, IL, United States
[†]Medical Scholars Program, University of Illinois College of Medicine, Urbana, IL, United States
[‡]Department of Gynecology and Obstetrics, Emory University, Atlanta, GA, United States
[§]Department of Psychiatry and Behavioral Sciences, Emory University, Atlanta, GA, United States
[¶]Department of Psychiatry, University of California, San Diego, La Jolla, CA, United States
[||]Department of Psychology, University of Illinois at Urbana–Champaign, Urbana, IL, United States
[#]Carl R. Woese Institute for Genomic Biology, University of Illinois at Urbana–Champaign, Urbana, IL, United States
[1]Corresponding author: e-mail address: muddin@illinois.edu

Contents

Abstract

While diagnosis of PTSD is based on behavioral symptom clusters that are most directly associated with brain function, epigenetic studies of PTSD in humans to date have been limited to peripheral tissues. Animal models of PTSD have been key for understanding the epigenetic alterations in the brain most directly relevant to endophenotypes of PTSD, in particular those pertaining to fear memory and stress response. This chapter provides an overview of neuroepigenetic studies based on animal models of PTSD, with an emphasis on the effect of stress on fear memory. Where relevant, we also describe human-based studies with relevance to neuroepigenetic insights gleaned from animal work and suggest promising directions for future studies of PTSD neuroepigenetics in living humans that combine peripheral epigenetic measures with measures of central nervous system activity, structure and function.

Progress in Molecular Biology and Translational Science, Volume 158
ISSN 1877-1173
https://doi.org/10.1016/bs.pmbts.2018.04.001

1. INTRODUCTION

Post-traumatic stress disorder (PTSD) is a psychiatric disorder that presents with intrusive and persistent re-experiencing of traumatic events, avoidance of distressing trauma-related stimuli, negative alterations in cognition and mood, and alterations in arousal/reactivity after exposure to a traumatic event.[1] While most people experience a traumatic event at some point in their lives (~50%–90% of US population,[2–4] ~70% world-wide[5,6]), only a small subset develop PTSD over a lifetime (~7% and ~4% lifetime prevalence of PTSD in the USA[7,8] and across countries,[5] respectively). This suggests that exposure to trauma is necessary, but not sufficient, for the development of PTSD, and that pre-existing factors, including genetic background and early life experiences, interact with trauma exposure, via epigenetic mechanisms, to elicit a spectrum of individual responses to trauma.

Epigenetic mechanisms biochemically encode individual experiences and environmental exposures into the genomic architecture of each cell to modulate gene expression, and, by extension, cell functioning and phenotype. In this way, the epigenome provides the genome with constant regulatory feedback from environmental cues and serves as a link between pre-existing factors and response to trauma exposure. Since PTSD requires exposure to a traumatic event as a diagnostic criterion, it serves as an ideal model for distinguishing between predisposing vulnerability/resilience factors and response to precipitating trauma exposure, which may be encoded as short-lived or persistent epigenetic changes that propagate post-traumatic physiological changes and symptoms.

While diagnosis of PTSD is based on behavioral symptom clusters that are most directly associated with brain function, epigenetic studies of PTSD in humans (detailed in these reviews[9,10]) have been limited to peripheral tissues, such as blood, buccal tissue, and saliva. Very few studies have been conducted in postmortem human brain tissues of individuals with PTSD,[11–14] and none have explored epigenetic mechanisms. Animal models of PTSD (reviewed here[15–17]), predominantly in rodents, have been key for understanding the epigenetic alterations in the brain most directly relevant to endophenotypes of PTSD, which can broadly be divided into dysregulation of fear memory and stress response. In this chapter, we will discuss neuroepigenetic studies based on animal models of PTSD, with an emphasis on the effect of stress on fear memory. Where possible, we will relate these findings to those from the human PTSD literature, conducted in peripheral tissue, and suggest promising directions for future studies of PTSD neuroepigenetics in humans.

2. NEUROEPIGENETIC REGULATION OF FEAR LEARNING AND MEMORY

One of the hallmark symptoms of PTSD is the persistence of intense fear memories that underlie intrusive re-experiencing of traumatic events. Recall of these traumatic memories may be triggered by non-threatening stimuli under safe conditions due to inappropriate generalization of trauma-associated stimuli,[18,19] and is thought to reflect impaired fear extinction learning and overactive fear memory consolidation and reconsolidation.[20] Animal models of PTSD examining the neural underpinnings of fear learning and memory have utilized fear conditioning paradigms,[20,21] and have been instrumental for investigating the global and gene-specific neuroepigenetic mechanisms underlying each stage of fear memory processing. These studies have focused on key brain regions dysregulated in PTSD, notably the amygdala, hippocampus, and medial prefrontal cortex (mPFC),[22,23] and have provided many insights into the dynamic regulation of participating epigenetic processes, which themselves direct the transcriptional and translational activities involved in synaptic plasticity, fear memory processing, and behavioral response.

While the fear conditioning paradigm and fear circuitry (Fig. 1) have been reviewed elsewhere,[20,21,24–26] an outline of the different stages of fear memory sets the framework for the current discussion. Memory formation involves stabilization of the acquired memory (i.e., memory consolidation) and requires de novo protein synthesis, with contextual processing and associations highly dependent on the hippocampus and cue associations dependent on the amygdala.[25,27] Recently consolidated contextual fear memories in the hippocampus are transferred to the cortex for long-term, remote memory storage and maintenance (i.e., systems-level memory consolidation),[28] while long-term storage of cued fear associations remain localized in the amygdala.[29] Memory retrieval destabilizes previously consolidated memories and requires reconsolidation to re-stabilize, and possibly update, these labile memory traces.[30,31] By presenting fear-associated conditioned stimuli without the fear-inducing unconditioned stimulus, new extinction memories can be learned on top of the original fear associations. Taken together, memories are dynamic and can be transformed over time via strengthening/weakening of memory traces or via

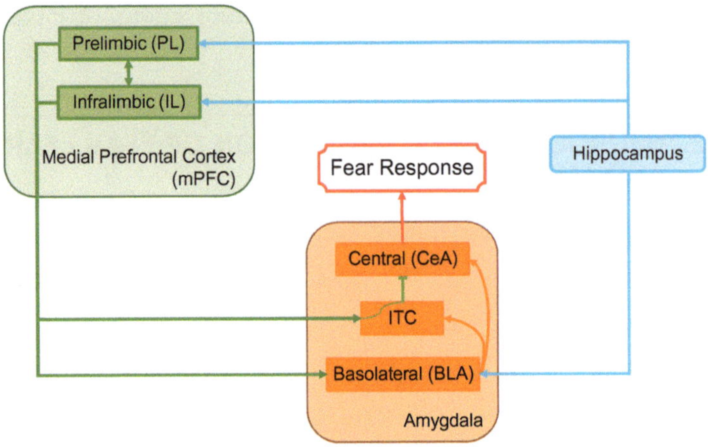

Fig. 1 Schematic diagram of neural circuitry involved in fear learning. The hippocampus, which is involved in contextual processing, has projections to the basolateral amygdala (BLA) and medial prefrontal cortex (mPFC), namely the prelimbic (PL) and infralimbic (IL) cortices. These projections are necessary for contextual fear memory retrieval and for modulating renewal and suppression of fear expression after extinction via PL and IL projections, respectively. The IL region of the mPFC is important for fear extinction learning and sends projections to the intercalated neurons (ITC), directly and indirectly, which inhibit excitation of the central amygdala (CeA). The PL region projects to the BLA to signal excitation of the CeA. Outputs from the CeA drive expression of the fear response. Refer to reviews for more detail.[21,24,25]

updating/overriding of traces by integration of new experiences.[30] Understanding these processes will enable development of preventative interventions and treatments for PTSD that combine medication and cognitive–behavioral therapies and inform effective delivery of interventions.

2.1 Molecular Mechanisms of Fear Memory Reconsolidation and Extinction

The persistent and intrusive recall of traumatic memories in PTSD is most directly associated with dysregulated enhancement of fear memory reconsolidation and impairment of fear memory extinction.[20] Therefore, PTSD-relevant interventions and preclinical evaluations have focused on interference of reconsolidation, to weaken or erase the original traumatic memory (i.e., forgetting), and on strengthening of extinction learning, to learn "disassociation" of fear-inducing component of memory while leaving the rest intact.[32,33]

While a distinct process, reconsolidation shares many key features with consolidation, such as activation of extracellular signal-regulated kinase/mitogen-activated protein kinase (ERK/MAPK) signaling,[34,35] which regulates histone modifications, and interaction of different neuroepigenetic mechanisms.[36] For example, retrieval of fear memory was found to promote global increase in histone H3 acetylation and phosphorylation in the CA1 region of the hippocampus,[37] and in histone H3 acetylation in the lateral amygdala (LA; phosphorylation was not examined in the latter study),[38] thus inducing increased transcriptional activity in these brain regions. Additionally, histone hyperacetylation in the LA, induced via histone deacetylase (HDAC) inhibitors shortly after retrieval of fear memory, was shown to enhance reconsolidation of cued fear memories; in contrast, histone hypoacetylation, induced via infusion of either broad spectrum or p300/CBP-specific histone acetyltransferase (HAT) inhibitors showed the opposite effect.[39,40] Since improvement of memory consolidation with HDAC inhibitors has been associated with enhanced induction of long-term potentiation (LTP)[41,42] and dendritic sprouting[43] similar mechanisms may be involved in fear memory reconsolidation. Similarly, inhibition of DNA methyltransferase (DNMT) shortly after retrieval impairs long-term reconsolidation of cued fear memory in the LA.[44] As observed in consolidation, pre-treatment with an HDAC inhibitor was able to rescue deficits due to DNMT inhibition,[38] illustrating a key role for the interaction between epigenetic processes in the regulation of fear memory and synaptic plasticity.

Fear extinction learning is also associated with increased levels of histone acetylation[45]: while fear extinction formation and consolidation is highly dependent on the infralimbic region of the ventromedial prefrontal cortex (ILPFC),[46] systemic, intrahippocampal and intra-ILPFC infusion of general HDAC inhibitors, before and after contextual fear memory retrieval, have all been shown to enhance extinction learning.[45,47] Additionally, HDAC inhibitor treatments have been shown to increase histone acetylation in association with fear extinction enacted with weaker extinction paradigms, which otherwise would not have been successful,[45,48] with corresponding effects on mRNA expression.[48] Together with the reconsolidation-based findings described above, these fear extinction-related findings demonstrate that retrieval and long-term consolidation of fear memories, as well as fear extinction learning, involve multiple, potentially interacting epigenetic processes that can impact downstream gene function.

Another factor with effects that are epigenetically regulated is time since initial fear memory formation. While retrieval of recent memories (i.e., 1 day) induces a limited period of hippocampal neuroplasticity, such plasticity is absent or reduced in retrieval of remote memories (i.e., 1 month), and corresponds to differences in fear extinction training success and epigenetic patterns. This time-related difference in memory lability is particularly relevant to PTSD, as the disorder requires persistence of symptoms for at least 1 month for a diagnosis. More specifically, relative to remote memory retrieval, recent memory retrieval induces more abundant H3K9/14 acetylation and lacks HDAC2 binding at the promoter of *Fos*, an immediate early gene (IEG) crucial for neuronal activation, and is more successful in fear extinction learning (i.e., reconsolidation updating).[49] This difference in HDAC2 binding was determined to be mediated by an increase in HDAC2 nitrosylation (which releases HDAC2 from chromatin[50]) after recent, but not remote, memory retrieval.[49] Importantly, remote memories were persistently attenuated (i.e., successfully extinguished) following application of an HDAC2-targeting inhibitor, which was able to "epigenetically prime" the expression of neuroplasticity-related genes[49]—suggesting potential capability for inducing neuroplasticity in remote memories. These animal-based findings resolve the effects of various factors, such as type of epigenetic-modifying adjuvant, mainly HDAC inhibitors, specific paradigm for extinction-based exposure therapy, timing of adjuvant delivery during therapy, and time since original trauma memory, and have translational implications for informing treatment planning that is optimized and tailored for individual cases.

Dynamic regulation of DNA cytosine modifications is also implicated in fear extinction learning, and studies suggest that hydroxylation of 5-methylcytosine to 5-hydroxymethylcytosine (5hmC) in the ILPFC is important for successful extinction.[51,52] Li et al.[52] revealed that fear extinction leads to rapid Tet3-mediated, global 5hmC redistribution that may drive selective accumulation of 5hmC for promoting a "primed" transcriptionally active epigenetic state in the ILPFC. Similarly, Rudenko et al.[51] demonstrated the importance of Tet1 for fear memory extinction using Tet1 knockout (KO) mice. Impairment of extinction learning in these mice was associated with hypermethylation of the *Npas4* promoter region after extinction training, which has implications for the regulation of many neuronal activity-regulated genes, including *Fos*.[51] A recent study examining how epigenetic marks coordinate regulation after memory retrieval reported an important link between H3K4me3 and 5hmC marks in the hippocampal CA1 region and anterior cingulate cortex (ACC): Retrieval of a recent, contextual fear memory increased H3K4me3 and 5hmC levels, both globally and at a CpG-enriched coding region of *Npas4*—but not *Fos*—in the hippocampal CA1 region.[53] In contrast, retrieval of a remote memory increased H3K4me3 and 5hmC levels at CpG-enriched coding region of *Fos*—but not *Npas4*—in the ACC.[53] These findings reveal important insights regarding how epigenetic marks coordinate differential targeting of important genes based on brain region and time since initial fear memory formation.

A key gene that was previously noted to play a pivotal role in fear memory formation has also been found to be critical for fear extinction, with supporting evidence in humans. Specifically, Soliman et al.[54] revealed that the brain-derived neurotrophic factor (*BDNF*) Val66Met genotype ($BDNF_{Met}$ variant) was associated with impaired fear extinction learning in both humans and an inbred genetic knock-in mouse strain expressing a variant *BDNF* allele. This is further supported in the human literature by analyses revealing that *BDNF* Val66Met genotype predicts response to exposure therapy in PTSD,[55] and in the animal literature by similar reports of impairment in extinction learning using a conditioned taste aversion paradigm on the *BDNF* variant knock-in mouse strain.[56] In the epigenetic literature, fear extinction was shown to increase histone H4 acetylation at *Bdnf* promoter 4 and mRNA levels of *Bdnf* exon I and IV in the PFC.[48] Of note, BDNF signaling may be involved in the sex differences observed in fear extinction learning. Baker-Andresen et al. found that female mice exhibited impaired retention of fear extinction and greater DNA methylation increase, with corresponding mRNA level decrease, of *Bdnf* exon IV in the medial PFC, compared to male mice.[57] They further explored targeted activation of BDNF signaling in female mice by administering a TrkB agonist before memory retrieval and extinction training and found that this sexually dimorphic impairment in fear extinction learning and fear response was blocked.[57] (However, it was not tested in male mice.) This finding has translational implications for potential sex-specific interventions in PTSD.

In addition to DNA cytosine and histone modifications, non-coding RNAs, especially microRNAs (miRNA), are emerging as key epigenetic players. While the role of miRNA in reconsolidation has not yet been reported, a number of studies implicate miRNA regulation in fear extinction. Fear extinction training leads to an activity-dependent increase in miR-128b expression in the ILPFC, and is proposed to facilitate extinction by negatively regulating genes associated with plasticity and retrieval of original fear memory, such as *Rcs*.[58] Additionally, overexpression of miR-144-3p in the BLA has been shown to rescue fear extinction memory in extinction-impaired mice and is thought to target plasticity-associated genes.[59] Of note, *Pten*, which regulates plasticity-associated signaling cascades, colocalizes with miR-144-3p in BLA and shows functional downregulation following successful fear extinction.[59] In humans, blood *DICER1* expression was found to be significantly lower in PTSD with comorbid depression, implicating dysregulation of miRNA processing.[60] Consistent with this, lower blood *DICER1* expression was also significantly associated with higher amygdalar activation to fearful stimuli—a neural intermediate phenotype for PTSD[60] and depletion of *DICER1* in the central amygdala was demonstrated to increase anxiety-like behavior in mice via inactivation of miRNA processing.[61] These studies suggest that miRNAs serve an important role in the regulation of fear extinction and stress response.

3. EFFECT OF STRESS ON FEAR LEARNING AND MEMORY

Traumatic events imprint particularly strong memories compared to other life events and may also influence future memory formation of related events. This is reflected in investigations that have revealed that stress exposure prior to fear conditioning enhances conditional fear responding and impairs fear extinction (as outlined in Refs. 62–64; Fig. 2). This stress-induced sensitization of fear response is largely mediated by the hypothalamic–pituitary–adrenal (HPA) axis, which signals systemic release of glucocorticoids (GCs) and glucocorticoid receptor (GR) activation, for regulation of GC-responsive genes and other transcription factors. GRs, encoded by the *NR3C1* gene, are found throughout the brain and are particularly abundant in the hippocampus, where they influence contextual memory formation and play a major role in coordinating behavioral, autonomic, and neuroendocrine facets of stress response.[64,67]

Stress-induced GC binding to GRs has been shown to strengthen memory consolidation of stressful events and induce epigenetic alterations, globally[68] and specifically to activate expression of neuroplasticity-related genes.[69] This has been demonstrated both in studies that utilize fear conditioning paradigms, which are inherently stressful, as well as various stress paradigms. Psychological stressors have been shown to significantly increase the number of granule neurons stained for histone phosphorylation and acetylation (i.e., H3S10p-K14ac) in the dentate gyrus (DG) of the hippocampus, and this increase can be inhibited by GR antagonists.[70] Additionally, these histone marks were found to specifically associate with promoters of *Fos* and

A

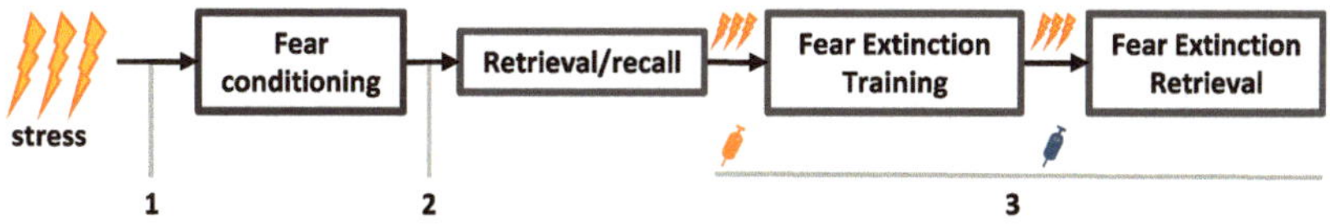

B

1. Stressor-induced

Brain region	Changes	Ref
DG	↑ H3S10p-K14ac → ↑ Fos, Egr1 expression	[70,72]
Dorsal DG	↑ DNAm *Bdnf* IV	[74]
Dorsal CA1	↑ DNAm *Bdnf* IV → ↓ Bdnf mRNA	[74,75]
Ventral CA3	↓ DNAm *Bdnf* IV	[74]
BLA	– DNAm *Bdnf* IV	[74]
mPFC	– DNAm *Bdnf* IV	[74]

2. Stress-enhanced fear learning (SEFL; after fear conditioning)

Brain region	Changes	Ref
Hippocampus	↑ H3/H4ac *Bdnf* I & IV → ↑ Bdnf mRNA/protein → ↑ contextual freezing (↑ TrkB protein)	[81]
Hippocampus	Distinct lncRNA expression profile, associated w/ *Fos* & *Camk2a*	[83]
Amygdala	↑ DNAm *Ntrs1* → ↓ Ntrs1 mRNA → ↑ freezing	[82]

3. Manipulation of fear extinction in SEFL model

Brain region	Manipulation	Changes	Ref
Hippocampus	HDAC inhibitor (vorinostat)	↑ H3/H4ac (global) → ↑ NR2B & CaMKIIα/β proteins → ↑ contextual fear extinction	[66]
Amygdala	Dexamethasone (HPA axis suppression)	↑ DNAm *Fkbp5* (intron 5 CpG4; GRE) → ↓ Fkbp5 mRNA ↔ ↑ fear extinction learning/retention (Δ Dnmt & Tet mRNA)	[65]

Fig. 2 (A) Schematic representation of general experimental paradigm used to determine the effect of stress on fear learning and memory. Stress exposure prior to fear conditioning enhances conditional fear responding and impairs fear extinction learning (i.e., extinction acquisition and/or extinction retrieval), which corresponds to increased fear behavior. (B) Tables summarizing discussed epigenetic changes at time points noted in (A), by brain region (i.e., hippocampus [blue], amygdala [orange], and prefrontal cortex [green]). Timing of manipulation is indicated by syringe (orange[65]; blue[66]).

Egr1/Zif268/Ngfia, another IEG crucial for fear learning and behavior-activated neuronal activity,[71,72] thereby facilitating a rapid induction of gene expression. Signaling pathway studies have revealed that stress-induced GR activation can facilitate rapid NMDAR-mediated ERK-MAPK signaling to downstream nuclear kinases MSK1/2 and Elk-1, for induction of these transcriptionally activating histone marks in the DG.[64] These findings underscore the GR-dependent epigenetic mechanisms required to mediate gene expression within the DG following a psychologically stressful challenge, and shed light on the molecular substrates for inducing long-term changes in neuronal function linked to memory consolidation of stressful experiences.[64]

From a locus-specific perspective, *Bdnf* has received considerable interest as a mediator of stress-sensitized fear response, as it has with the fear extinction studies described above. *Bdnf* is transcriptionally regulated by GR in an exon- and tissue-specific manner and is differentially expressed in response to stress.[73] In addition to fear conditioning studies, which revealed alterations in regulation of *Bdnf* expression via various histone modifications and DNA methylation marks, Roth et al.[74] used a complex psychosocial stress paradigm (involving restraint stress, exposure to a live predator, and social housing instability) to induce hypermethylation of the *Bdnf* exon IV promoter in the dorsal DG and, more prominently, in the dorsal CA1 region of the hippocampus, with corresponding decrease in *Bdnf* mRNA levels in these regions. Interestingly, they also reported hypomethylation of *Bdnf* exon IV in the ventral CA3 and no methylation changes in other hippocampal subregions, suggesting that psychosocial stressors selectively drive DNA methylation dynamics, particularly on the *Bdnf* gene, in key dorsal hippocampal regions and may underlie the persistent deficits in memory and stress regulation observed in PTSD.[74] The CA1 region-specific downregulation of *Bdnf* was also noted in another frequently used model of PTSD,[15] which exposed rats to a predator scent stressor (PSS) and differentiated individuals that exhibited a PTSD-like "extreme behavioral response" from those with "minimal behavioral response."[75] The consistency of findings across these different models of PTSD provides robust evidence for the involvement of hippocampus-specific *Bdnf* dysregulation in the pathophysiology of PTSD. This region-specific epigenetic difference may also underlie dysregulation of the fear circuitry by producing an imbalance of BDNF production between the hippocampus and amygdala.[76]

While classical fear conditioning provides the foundation for understanding fear memories and is inherently stressful, animal models of PTSD that expose the animal to more intense trauma/stressor(s) are better able to reproduce core PTSD symptoms.[15,16] A variety of environmental, physical, and psychological stressors exist for animal models of PTSD, and some well-validated models include single prolonged stress (SPS)[77] and its adaptations,[78] PSS,[15] and predator-based psychosocial stress paradigms.[76] Animals exposed to these stressors exhibit characteristics that reflect those generally observed in PTSD, such as increased anxiety-like behaviors, exaggerated startle, and abnormal GC response (i.e., reduced basal GC levels and enhanced GC negative feedback), indicating that they may be appropriate for investigating psychological, trauma-induced cognitive/behavioral and neuroendocrine alterations.[17,79,80] In particular, animal models of PTSD that expose animals to trauma/stressor(s) before fear conditioning show greater face, construct, and predictive validity for PTSD than fear conditioning alone,[62,77] and can be used to determine the epigenetic effect of stress on subsequent fear learning.[63] In conjunction with our understanding from studies that investigated neuroepigenetics of stress and fear memory independently, epigenetic investigations using paradigms such as stress-enhanced fear learning (SEFL) should elucidate distinct mechanisms underlying the interaction between these two major facets of PTSD and have special promise for identifying mechanisms underlying resistance to extinction learning and the role of the BDNF pathway, which is widely implicated in both stress and fear memory.[63]

While only a few epigenetic studies have been based on the SEFL model so far, their findings are generally consistent with and contribute to our current understanding of the neuroepigenetics behind stress and fear learning. Takei et al.[81] used a SEFL paradigm with a SPS before contextual fear conditioning to show that SPS rats had higher levels of contextual freezing than sham-treated rats. SPS rats also showed increased acetylation of histone H3 and H4 at the exon I and IV *Bdnf* promoter regions that corresponded with increased exon I and IV *Bdnf* mRNA and total BDNF protein levels in the hippocampus after fear learning, relative to sham-treated rats.[81] Additionally, protein levels of BDNF receptor TrkB were significantly higher in hippocampus of SPS rats compared to controls, suggesting enhanced hippocampal BDNF/TrkB signaling in response to fear conditioning in this animal model of PTSD.[81] Furthermore, subsequent work showed that treatment with an HDAC inhibitor using this SPS SEFL paradigm facilitated contextual fear extinction in the hippocampus, and was accompanied by an increase in acetylated histones H3 and H4 in this brain region.[66]

Other studies have revealed significant SEFL-induced alterations in DNA methylation levels on key anxiety- and stress-related genes, including the anxiety-related neurotensin 1 receptor gene, *Ntsr1*, and local GR regulator, *Fkbp5*, in the amygdala. Toda et al.[82] applied maternal separation as a stressor to find increased DNA methylation levels in the promoter region of *Ntsr1*, which corresponded with decreased *Ntsr1* mRNA levels.[82] Sawamura et al.[65] used the restraint/immobilization stressor to demonstrate that dexamethasone treatment (i.e., HPA axis suppression) prior to extinction training enhances stress-sensitized fear extinction learning and corresponds with alterations in *Dnmt/Tet* and *Fkbp5* DNA methylation levels that regulate expression of these genes in the amygdala.[65] These findings suggest that dexamethasone treatment enhances fear extinction learning by inducing dynamic regulation of DNA methylation to alter *Fkbp5*-mediated GC sensitivity.

In addition to epigenetic regulation of specific genes of interest, Qingzhen et al.[83] found distinct long non-coding RNA (lncRNA) expression profiles associated with SEFL-PTSD-like syndrome in the hippocampus, and while our understanding of lncRNAs is still scarce, this study suggests a possible role for lncRNA in PTSD dysregulation, a finding that echoes at least one genome-based study of PTSD in humans.[84] When looking at specific lncRNA and pathways, they reported alterations in lncRNA associated with regulating *Fos* and *Camk2a*,[83] both of which have been implicated in learning and memory processes in previous works.[85,86] Another expression study using the SEFL paradigm found that PTSD-like mice exhibit a general transcriptional attenuation profile that is associated with upregulation of HDAC5 (but no changes in DNMTs or HDAC2) in the bed nucleus of the stria terminalis (BNST), an important brain region involved in integration of stress/fear response, suggesting a role for this specific HDAC enzyme in regulating PTSD-associated expression in the BNST.[87] Taken together, these studies demonstrate that SEFL phenotypes are likely regulated by multiple epigenetic mechanisms, and suggest that interventions targeting these mechanisms hold promise for treatment of PTSD.[66]

4. EPIGENETIC EFFECTS OF STRESS/TRAUMA RESPONSE IN THE BRAIN AND PERIPHERY

PTSD is most frequently an acute stress response to trauma exposure that fails to return to homeostasis and further progresses after onset, into continued acute stress response to recurring aversive memories of the traumatic event(s), and chronic or repetitive stress response stemming from PTSD symptoms themselves.[88] Physiologically, this distress can be characterized in stages of systemic dysregulation,[89] including anatomical sites outside of the brain, which may be partially coordinated by maladaptive alterations of HPA axis and sympathetic nervous system sensitivity/responsivity, as well as neuroimmune dynamics.[88,90] Although an increasing number of human-based

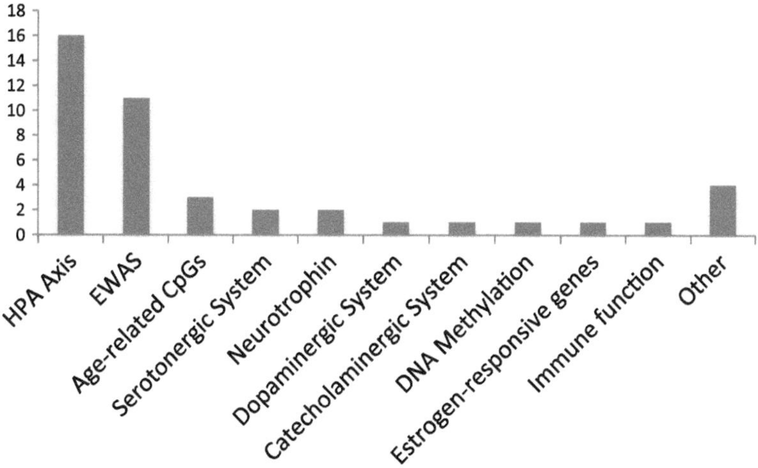

Fig. 3 Summary of 43 human-based epigenetic studies of PTSD published as of July 2017, enumerated and categorized according to relevant biological system or epigenome-wide association study (EWAS) approach: Hypothalamic–Pituitary–Adrenal (HPA) Axis[99–114]; EWAS[91–96,115–119]; Age-related CpGs[120–122]; Serotonergic system[123,124]; Neurotrophin[125,126]; Dopaminergic System[127]; Catecholaminergic system[128]; DNA methylation machinery[129]; Estrogen-responsive genes[130]; Immune Function[131]; Other.[132–135]

epigenetic studies of PTSD have adopted an epigenome-wide association study (EWAS) approach (e.g., Refs. 91–98), the bulk of studies published to date have focused on candidate genes within these and related systems (Fig. 3). A number of comprehensive reviews have been published in recent years, and the reader is referred to this work for more detailed information on these studies.[10,136–138] Here, we focus our discussion on human-based studies with relevance to the neuroepigenetic insights gleaned from animal work.

5. HUMAN-BASED EPIGENETIC STUDIES INVOLVING FEAR LEARNING AND MEMORY

As outlined in the earlier sections of this chapter, consolidation/reconsolidation and extinction of fear memories play a key role in the initial etiology of PTSD, and animal models have contributed much to enhance our understanding of the neuroepigenetic mechanisms involved in these processes. To date, however, there have been only a handful of studies in human clinical populations that parallel in some way the findings from the more mechanistic animal work; importantly, unlike the animal work, these human-based studies have drawn exclusively from living individuals and thus have relied upon measurements of epigenetic factors in peripheral tissues, predominantly whole blood and saliva.

The earliest work to evaluate fear-related behavior in relation to epigenetic factors involved assessing fear inhibition, an intermediate phenotype highly relevant to PTSD risk. Specifically, Norrholm and colleagues[128] sought to test the potential interaction between molecular variation (both genetic and epigenetic) in the catechol-*O*-methyltransferase (*COMT*) locus and PTSD diagnosis on fear-potentiated startle responses during fear conditioning and extinction in a highly traumatized sample drawn from the Grady Trauma Project.[139] Using a well-established model of fear-potentiated startle that measures subjects' ability to discriminate between danger and safety signals,[140] the authors found that individuals with higher blood-derived DNA methylation levels at two CpG sites in the promoter region of *COMT* showed higher levels of fear-potentiated startle when presented with the safety signal, compared to individuals with lower DNA methylation at these same two sites. Of note, although the authors reported an association between homozygous Met158 *COMT* genotype and impaired fear inhibition, as well as an association between the same genotype and increased DNA methylation levels at four promoter-region CpG sites, the two CpGs mentioned above were independently associated with impaired fear inhibition, suggesting that epigenetic regulation of this locus may contribute to fear processing in humans.

Additional work evaluating epigenetic associations with fear memory-related phenotypes has involved survivors of the Rwandan genocide.[99] In this study, the authors focused specifically on the gene encoding the glucocorticoid receptor (GR), *NR3C1*. The GR binds to cortisol and thus plays a key role in modulating the body's response to stress; in addition, evidence from animal and human studies indicates that elevated GCs, such as cortisol, both enhance consolidation of emotional memories/fear extinction processes and reduce memory retrieval,[141] rendering the study of *NR3C1* particularly salient for understanding PTSD-related epigenetic regulation. Using saliva-derived DNA, the authors found that DNA methylation in a CpG overlapping a transcription-factor binding site was negatively correlated with severity of symptoms related to intrusive memories in male, but not female, survivors of the Rwandan genocide; no associations were found in relation to avoidance and hyperarousal symptoms. Moreover, higher DNA methylation levels were associated with a lower lifetime PTSD risk in men, but not women. Extending these findings with an fMRI study in healthy Swiss subjects, the authors found that increased DNA methylation at the same *NR3C1* CpG site in men was associated with reduced recognition of previously viewed pictures; and that this higher methylation was associated with increased activation in the ventrolateral prefrontal cortex, a brain region implicated in memory retrieval.[142] Intriguingly, in a small study of male combat veterans with PTSD, Yehuda and colleagues found that DNA methylation at the *NR3C1* promoter region, measured in blood, predicted response to prolonged exposure therapy—a therapy grounded in confronting and working through traumatic memories—such that subjects with higher DNA methylation pre-treatment predicted lower self-reported PTSD symptoms at follow-up.[100] Taken together, these findings suggest that higher DNA methylation at functionally relevant CpG sites within the *NR3C1* promoter region may confer a protective effect against vulnerability to PTSD, at least among males. More broadly, these findings suggest that DNA methylation in the GR may index memory-related phenotypes with strong relevance for PTSD, even when measured in peripheral tissues.

Most recently, the first longitudinal EWAS has identified two replicable differentially methylated regions in *ZFP57* and *RNF39* that become hypomethylated in blood as a response to combat trauma in soldiers who develop PTSD symptoms.[94] These genes have previously been implicated in behavioral susceptibility to stress in animal models. *ZFP57* is involved in transcriptional regulation of genomic imprinting[143] and may play a role in modulating responses to stress in hippocampal neurons through interactions with other transcriptional regulators.[144] *RNF39* is thought to be involved in the early phase of synaptic plasticity and its expression is upregulated after hippocampal LTP induction.[145,146] It is also strongly upregulated in the lateral amygdala after Pavlovian fear conditioning[146] and in the DG of the hippocampus in response to corticosterone challenge.[147] While the relationship between DNA hypomethylation of these genes in blood and regulation of these genes in the brain is not yet clear, blood-based DNA methylation levels in *ZFP57* and *RNF39* may serve as brain-relevant peripheral biomarkers indicative of trauma-induced susceptibility for PTSD.

In contrast to these male-specific findings, a recent study set out to test the hypothesis that individual variation in response to estrogen levels contributes to fear regulation and PTSD risk in women[130]; this hypothesis is especially notable given the twofold increased risk of PTSD in women compared to men.[3] Drawing again on participants from the Grady Trauma Project, and focusing on genes with demonstrated expression in blood, the investigators identified a single CpG site that associated with PTSD following Bonferroni correction, which fell within the histone deacetylase 4 (*HDAC4*) gene: methylation was higher in PTSD cases than controls; and higher methylation at this site was associated with lower estradiol levels. Because HDAC4 had been linked in previous studies to learning and memory-related processes, the authors further sought to determine the relation between molecular variation at this locus and fear-related traits using both human and animal approaches. Available GWAS data indicated that subjects with the CC genotype at rs7570903, which associated with *HDAC4* methylation and expression levels, showed enhanced fear expression and that it interacted with PTSD diagnosis to predict increased fear conditioning and higher fear-potentiated startle during the early extinction phase; moreover, this risk genotype was associated with increased resting-state functional connectivity in fear-memory-related regions of the brain, specifically the amygdala and cingulate cortex. Follow-up animal experiments specifically targeting the amygdala suggested that estrogen levels mediate the effects of tone-shock exposure on alterations in *HDAC4* mRNA expression, with higher estrogen apparently buffering against tone-shock related regulation of this gene. Taken together, these results indicate that estrogen levels (at time of tone-shock exposure) influence susceptibility to fear-related phenotypes in a manner that may mediate risk for PTSD in females, and that these PTSD-associated epigenetic effects are also apparent in peripheral blood.

6. TOWARD INVESTIGATING NEUROEPIGENETICS OF PTSD IN HUMANS

While the diagnosis of PTSD is most directly associated with brain function, epigenetic studies of PTSD in humans have been limited to peripheral tissues. Insights on epigenetic alterations in the brain have relied on animal models of PTSD, which are limited in translatability and still require validation in humans. In other stress-related psychiatric conditions, postmortem human brains have been investigated to validate and complement neuroepigenetic insights from animal studies (e.g., Ref. 148). Although postmortem brain studies may be used to understand the neuroepigenetics of PTSD in the future, these studies cannot be longitudinal in design, and therefore will not be able to inform whether any epigenetic differences exist prior to onset of PTSD (indicative of pre-existing susceptibility factor), develop in response to trauma exposure, or develop during the course of PTSD. A current approach that is used to associate structural differences and functional activity in the brain with peripheral epigenetic profiles (e.g., Refs. 101,102,125,149) can utilize a longitudinal design in living individuals to identify the most robust and brain-relevant biomarkers[150] and distinguish whether the biomarker is indicative of risk for developing PTSD, response to trauma, pathophysiology, or treatment efficacy. In fact, since epigenetic biomarkers need to come from accessible, peripheral tissue for clinical utility, this approach may be more relevant for biomarker discovery.

An exciting direction for studying the neuroepigenetics of PTSD relies on the development of imaging technologies that may measure and localize activity of epigenetic processes in living humans. Development of radiotracers that penetrate the blood–brain barrier and bind to key epigenetic enzymes will enable such measurements using PET imaging. In fact, [^{11}C]Martinostat has been characterized and used to quantify the distribution of class I HDAC expression in various tissues including the brain, due to its efficient blood–brain barrier penetrance.[151,152] This radiotracer was also used in the first effort to acquire in vivo measurements of neuroepigenetic regulation in living humans.[153] However, an important confound to note is that [^{11}C]Martinostat, which selectively binds HDAC isoforms most implicated in regulation of neuroplasticity and cognitive function, can induce epigenetic alterations itself. In neural progenitor cell culture, [^{11}C]Martinostat induced changes in expression of genes associated with synaptic plasticity (i.e., *BDNF* and *synaptophysin*) and neurodegeneration (i.e., progranulin), as well as increased histone acetylation.[153] Additionally, differential expression of HDAC subtypes have been noted in postmortem brain of subjects with other psychiatric disorders, and were shown to be disorder, HDAC subtype, and brain region specific.[154] However, [^{11}C]Martinostat is unable to distinguish between these HDAC subtypes (i.e., isoforms 1,2,3). Therefore, postmortem human brain studies of dysregulated HDAC isoform expression in PTSD will be needed to benchmark HDAC isoform-specific density and distribution in different regions of the brain. Further developments stemming from this tracer,[155] as well as synthesis and preclinical testing of other promising epigenetic radiotracers, such as [18F]-TFAHA (selective for HDAC IIa),[156] have been reported and may improve and expand our capabilities for in vivo epigenetic measurements.

PET imaging of neuroepigenetic regulation can also be conducted at the same time as structural and functional MRI imaging using an MR-PET scanner to acquire a variety of other measurements that inform brain structure and function. For example, white tract matter integrity, measured using diffusion tensor imaging (DTI), has been found to correlate with epigenetic aging, calculated from DNA methylation profiles of peripheral blood, and to share common genetic influences with this peripherally-derived measure.[157] This is relevant to PTSD, which is associated with increased risk for a number of age-related comorbidities and hypothesized to accelerate cellular aging.[158]

In fact, lifetime PTSD severity has been found to be associated with accelerated aging[120,159] based on one DNA methylation-based calculation of cellular age,[160] as well as reduced neural integrity (in the genu of the corpus callosum), as indexed by DTI.[120] Since cumulative life stress has been shown to be strongly predictive of accelerated epigenetic aging,[121] the persistent symptoms and residual effects of PTSD may themselves be chronic stressors that promote accelerated epigenetic aging. This is in contrast to current PTSD diagnosis or severity, which has not been found to associate with accelerated epigenetic aging.[159] As a matter of fact, a longitudinal study of military personnel reported an apparent reversal of epigenetic aging during development of PTSD symptoms,[122] which suggests an initial attempt at compensation before loss of control and accelerated epigenetic aging.

Additionally, life stressors later in life may influence epigenetic aging more strongly than those earlier in life, possibly as a result of cumulative "wear and tear" on the epigenome that render older chronically stressed individuals more susceptible to epigenetic aging than younger individuals.[121] It is also worth noting that earlier stressors (e.g., childhood traumas) may moderate susceptibility to later life stressors in complex ways (i.e., either enhance or suppress susceptibility) such that the cumulative effects that contribute to epigenetic remodeling over a lifespan are not linear and depend on the stage/context of each contributing life stressor.[161] Future longitudinal studies that integrate neuroimaging markers with epigenome-based metrics (e.g., epigenetic aging measures from immune cells) can help advance our understanding of how life stressors build on each other to shape an individual's immunological age and neural integrity over time. This approach would be well-suited for following at-risk children who are chronically exposed to violence[162] and assessing whether interventions (e.g., supportive family environments[163]) reverse accelerated epigenetic aging and buffer against detrimental effects of life stressors.

In summary, collection of multidimensional neuroimaging data with epigenomic profiling of peripheral tissue holds particular promise for characterizing the neuroepigenetic basis of PTSD from a life course perspective. Specifically, this relatively non-invasive approach can be used at multiple timepoints (preferably using a prospective design to collect before and after trauma exposure) to discover the trajectory of PTSD-associated epigenetic changes in different brain regions and determine how they deviate from trauma-exposed non-PTSD and non-trauma exposed controls. Moreover, comprehensive history and phenotype information including lifetime exposures to stressors and traumas (especially during childhood), type of trauma, environmental conditions (e.g., socioeconomic status), social stressors and support, and comorbid conditions, are critical to include in the multidimensional dataset. Given the heterogeneous presentation of PTSD across different symptom clusters, this phenotype information will be essential for differential diagnosis and identification of subtypes of PTSD based on underlying pathophysiology. The RDoC initiative provides a framework by which this multidimensional data can be leveraged to understand individual post-traumatic conditions based on these neurobiological and neuroepigenetic metrics and symptoms.[164,165] Through these continuing advances, future research should be able to better unpack the neuroepigenetic basis of PTSD in living individuals, thereby improving prospects for treatment and prevention.

ACKNOWLEDGMENTS

This work was supported by NIH grants R01MH108826 and R01MH106595, and the University of Illinois' CompGen Initiative.

REFERENCES

1. American Psychiatric Association. *Diagnostic and Statistical Manual of Mental Disorders: Fifth Edition (DSM-5)*. Arlington, VA: American Psychiatric Publishing; 2013.
2. Breslau N. The epidemiology of trauma, PTSD, and other posttrauma disorders. *Trauma Violence Abuse*. 2009;10(3):198–210.
3. Kessler RC, Sonnega A, Bromet E, Hughes M, Nelson CB. Posttraumatic stress disorder in the National Comorbidity Survey. *Arch Gen Psychiatry*. 1995;52(12): 1048–1060.
4. Kilpatrick DG, Resnick HS, Milanak ME, Miller MW, Keyes KM, Friedman MJ. National estimates of exposure to traumatic events and PTSD prevalence using DSM-IV and DSM-5 criteria. *J Trauma Stress*. 2013;26(5):537–547.
5. Liu H, Petukhova MV, Sampson NA, et al. Association of DSM-IV posttraumatic stress disorder with traumatic experience type and history in the World Health Organization World Mental Health Surveys. *JAMA Psychiatry*. 2017;74(3):270–281.
6. Benjet C, Bromet E, Karam EG, et al. The epidemiology of traumatic event exposure worldwide: results from the World Mental Health Survey Consortium. *Psychol Med*. 2016;46(2):327–343.
7. Kessler RC, Berglund P, Demler O, Jin R, Merikangas KR, Walters EE. Lifetime prevalence and age-of-onset distributions of DSM-IV disorders in the National Comorbidity Survey Replication. *Arch Gen Psychiatry*. 2005;62(6):593–602.
8. Kessler RC, Wang PS. The descriptive epidemiology of commonly occurring mental disorders in the United States. *Annu Rev Public Health*. 2008;29:115–129.

9. Sheerin CM, Lind MJ, Bountress K, Nugent NR, Amstadter AB. The genetics and epigenetics of PTSD: overview, recent advances, and future directions. *Curr Opin Psychol.* 2017;14:5–11.
10. Zannas AS, Provencal N, Binder EB. Epigenetics of posttraumatic stress disorder: current evidence, challenges, and future directions. *Biol Psychiatry.* 2015;78(5): 327–335.
11. Bracha HS, Garcia-Rill E, Mrak RE, Skinner R. Postmortem locus coeruleus neuron count in three American veterans with probable or possible war-related PTSD. *J Neuropsychiatry Clin Neurosci.* 2005;17(4):503–509.
12. Licznerski P, Duric V, Banasr M, et al. Decreased SGK1 expression and function contributes to behavioral deficits induced by traumatic stress. *PLoS Biol.* 2015;13(10): e1002282.
13. Su YA, Wu J, Zhang L, et al. Dysregulated mitochondrial genes and networks with drug targets in postmortem brain of patients with posttraumatic stress disorder (PTSD) revealed by human mitochondria-focused cDNA microarrays. *Int J Biol Sci.* 2008;4(4):223–235.
14. Young KA, Thompson PM, Cruz DA, Williamson DE, Selemon LD. BA11 FKBP5 expression levels correlate with dendritic spine density in postmortem PTSD and controls. *Neurobiol Stress.* 2015;2:67–72.
15. Cohen H, Kozlovsky N, Alona C, Matar MA, Joseph Z. Animal model for PTSD: from clinical concept to translational research. *Neuropharmacology.* 2012;62(2):715–724.
16. Flandreau EI, Toth M. Animal Models of PTSD: A Critical Review. In: *Current Topics in Behavioral Neurosciences.* Berlin, Heidelberg: Springer; 2017. https://doi.org/10.1007/7854_2016_65.
17. Goswami S, Rodriguez-Sierra O, Cascardi M, Pare D. Animal models of post-traumatic stress disorder: face validity. *Front Neurosci.* 2013;7:89.
18. Acheson DT, Geyer MA, Baker DG, et al. Conditioned fear and extinction learning performance and its association with psychiatric symptoms in active duty Marines. *Psychoneuroendocrinology.* 2015;51:495–505.
19. Morey RA, Dunsmoor JE, Haswell CC, et al. Fear learning circuitry is biased toward generalization of fear associations in posttraumatic stress disorder. *Transl Psychiatry.* 2015;5:e700.
20. Careaga MB, Girardi CE, Suchecki D. Understanding posttraumatic stress disorder through fear conditioning, extinction and reconsolidation. *Neurosci Biobehav Rev.* 2016;71:48–57.
21. Johnson LR, McGuire J, Lazarus R, Palmer AA. Pavlovian fear memory circuits and phenotype models of PTSD. *Neuropharmacology.* 2012;62(2):638–646.
22. Kwapis JL, Wood MA. Epigenetic mechanisms in fear conditioning: implications for treating post-traumatic stress disorder. *Trends Neurosci.* 2014;37(12):706–720.
23. Zovkic IB, Sweatt JD. Epigenetic mechanisms in learned fear: implications for PTSD. *Neuropsychopharmacology.* 2013;38(1):77–93.
24. Orsini CA, Maren S. Neural and cellular mechanisms of fear and extinction memory formation. *Neurosci Biobehav Rev.* 2012;36(7):1773–1802.
25. Maren S, Phan KL, Liberzon I. The contextual brain: implications for fear conditioning, extinction and psychopathology. *Nat Rev Neurosci.* 2013;14(6):417–428.
26. Shin LM, Liberzon I. The neurocircuitry of fear, stress, and anxiety disorders. *Neuropsychopharmacology.* 2010;35(1):169–191.
27. Maren S. Neurobiology of Pavlovian fear conditioning. *Annu Rev Neurosci.* 2001;24: 897–931.
28. Frankland PW, Bontempi B, Talton LE, Kaczmarek L, Silva AJ. The involvement of the anterior cingulate cortex in remote contextual fear memory. *Science.* 2004; 304(5672):881–883.

29. Bergstrom HC. The neurocircuitry of remote cued fear memory. *Neurosci Biobehav Rev.* 2016;71:409–417.
30. Nadel L, Hupbach A, Gomez R, Newman-Smith K. Memory formation, consolidation and transformation. *Neurosci Biobehav Rev.* 2012;36(7):1640–1645.
31. Jarome TJ, Lubin FD. Epigenetic mechanisms of memory formation and reconsolidation. *Neurobiol Learn Mem.* 2014;115:116–127.
32. Zovkic IB, Meadows JP, Kaas GA, Sweatt JD. Interindividual variability in stress susceptibility: a role for epigenetic mechanisms in PTSD. *Front Psychiatry.* 2013;4:60.
33. Lattal KM, Wood MA. Epigenetics and persistent memory: implications for reconsolidation and silent extinction beyond the zero. *Nat Neurosci.* 2013;16(2):124–129.
34. Cestari V, Rossi-Arnaud C, Saraulli D, Costanzi M. The MAP(K) of fear: from memory consolidation to memory extinction. *Brain Res Bull.* 2014;105:8–16.
35. Duvarci S, Nader K, LeDoux JE. Activation of extracellular signal-regulated kinase–mitogen-activated protein kinase cascade in the amygdala is required for memory reconsolidation of auditory fear conditioning. *Eur J Neurosci.* 2005;21(1):283–289.
36. Schafe GE. Epigenetics and memory. In: Giese KP, Radwanska K, eds. *Novel Mechanisms of Memory.* Switzerland: Springer International Publishing; 2016: 141–160.
37. Lubin FD, Sweatt JD. The IkappaB kinase regulates chromatin structure during reconsolidation of conditioned fear memories. *Neuron.* 2007;55(6):942–957.
38. Maddox SA, Schafe GE. Epigenetic alterations in the lateral amygdala are required for reconsolidation of a Pavlovian fear memory. *Learn Mem.* 2011;18(9):579–593.
39. Maddox SA, Watts CS, Doyere V, Schafe GE. A naturally-occurring histone acetyltransferase inhibitor derived from Garcinia indica impairs newly acquired and reactivated fear memories. *PLoS One.* 2013;8(1):e54463.
40. Maddox SA, Watts CS, Schafe GE. p300/CBP histone acetyltransferase activity is required for newly acquired and reactivated fear memories in the lateral amygdala. *Learn Mem.* 2013;20(2):109–119.
41. Levenson JM, O'Riordan KJ, Brown KD, Trinh MA, Molfese DL, Sweatt JD. Regulation of histone acetylation during memory formation in the hippocampus. *J Biol Chem.* 2004;279(39):40545–40559.
42. Monsey MS, Ota KT, Akingbade IF, Hong ES, Schafe GE. Epigenetic alterations are critical for fear memory consolidation and synaptic plasticity in the lateral amygdala. *PLoS One.* 2011;6(5):e19958.
43. Fischer A, Sananbenesi F, Wang X, Dobbin M, Tsai LH. Recovery of learning and memory is associated with chromatin remodelling. *Nature.* 2007;447(7141):178–182.
44. Maddox SA, Watts CS, Schafe GE. DNA methyltransferase activity is required for memory-related neural plasticity in the lateral amygdala. *Neurobiol Learn Mem.* 2014;107:93–100.
45. Stafford JM, Raybuck JD, Ryabinin AE, Lattal KM. Increasing histone acetylation in the hippocampus-infralimbic network enhances fear extinction. *Biol Psychiatry.* 2012;72(1):25–33.
46. Milad MR, Quirk GJ. Neurons in medial prefrontal cortex signal memory for fear extinction. *Nature.* 2002;420(6911):70–74.
47. Lattal KM, Barrett RM, Wood MA. Systemic or intrahippocampal delivery of histone deacetylase inhibitors facilitates fear extinction. *Behav Neurosci.* 2007;121(5): 1125–1131.
48. Bredy TW, Wu H, Crego C, Zellhoefer J, Sun YE, Barad M. Histone modifications around individual BDNF gene promoters in prefrontal cortex are associated with extinction of conditioned fear. *Learn Mem.* 2007;14(4):268–276.
49. Graff J, Joseph NF, Horn ME, et al. Epigenetic priming of memory updating during reconsolidation to attenuate remote fear memories. *Cell.* 2014;156(1–2):261–276.

50. Nott A, Watson PM, Robinson JD, Crepaldi L, Riccio A. S-Nitrosylation of histone deacetylase 2 induces chromatin remodelling in neurons. *Nature*. 2008;455(7211): 411–415.
51. Rudenko A, Dawlaty MM, Seo J, et al. Tet1 is critical for neuronal activity-regulated gene expression and memory extinction. *Neuron*. 2013;79(6):1109–1122.
52. Li X, Wei W, Zhao QY, et al. Neocortical Tet3-mediated accumulation of 5-hydroxymethylcytosine promotes rapid behavioral adaptation. *Proc Natl Acad Sci USA*. 2014;111(19):7120–7125.
53. Webb WM, Sanchez RG, Perez G, et al. Dynamic association of epigenetic H3K4me3 and DNA 5hmC marks in the dorsal hippocampus and anterior cingulate cortex following reactivation of a fear memory. *Neurobiol Learn Mem*. 2017;142(pt A):66–78.
54. Soliman F, Glatt CE, Bath KG, et al. A genetic variant BDNF polymorphism alters extinction learning in both mouse and human. *Science*. 2010;327(5967):863–866.
55. Felmingham KL, Dobson-Stone C, Schofield PR, Quirk GJ, Bryant RA. The brain-derived neurotrophic factor Val66Met polymorphism predicts response to exposure therapy in posttraumatic stress disorder. *Biol Psychiatry*. 2013;73(11):1059–1063.
56. Yu H, Wang Y, Pattwell S, et al. Variant BDNF Val66Met polymorphism affects extinction of conditioned aversive memory. *J Neurosci*. 2009;29(13):4056–4064.
57. Baker-Andresen D, Flavell CR, Li X, Bredy TW. Activation of BDNF signaling prevents the return of fear in female mice. *Learn Mem*. 2013;20(5):237–240.
58. Lin Q, Wei W, Coelho CM, et al. The brain-specific microRNA miR-128b regulates the formation of fear-extinction memory. *Nat Neurosci*. 2011;14(9):1115–1117.
59. Murphy CP, Li X, Maurer V, et al. MicroRNA-mediated rescue of fear extinction memory by miR-144-3p in extinction-impaired mice. *Biol Psychiatry*. 2017;81(12): 979–989.
60. Wingo AP, Almli LM, Stevens JS, et al. DICER1 and microRNA regulation in posttraumatic stress disorder with comorbid depression. *Nat Commun*. 2015;6:10106.
61. Haramati S, Navon I, Issler O, et al. MicroRNA as repressors of stress-induced anxiety: the case of amygdalar miR-34. *J Neurosci*. 2011;31(40):14191–14203.
62. Rau V, DeCola JP, Fanselow MS. Stress-induced enhancement of fear learning: an animal model of posttraumatic stress disorder. *Neurosci Biobehav Rev*. 2005;29(8): 1207–1223.
63. Blouin AM, Sillivan SE, Joseph NF, Miller CA. The potential of epigenetics in stress-enhanced fear learning models of PTSD. *Learn Mem*. 2016;23(10):576–586.
64. Reul JM. Making memories of stressful events: a journey along epigenetic, gene transcription, and signaling pathways. *Front Psychiatry*. 2014;5:5.
65. Sawamura T, Klengel T, Armario A, et al. Dexamethasone treatment leads to enhanced fear extinction and dynamic Fkbp5 regulation in amygdala. *Neuropsychopharmacology*. 2016;41(3):832–846.
66. Matsumoto Y, Morinobu S, Yamamoto S, et al. Vorinostat ameliorates impaired fear extinction possibly via the hippocampal NMDA-CaMKII pathway in an animal model of posttraumatic stress disorder. *Psychopharmacology (Berl)*. 2013;229(1):51–62.
67. Lupien SJ, McEwen BS, Gunnar MR, Heim C. Effects of stress throughout the lifespan on the brain, behaviour and cognition. *Nat Rev Neurosci*. 2009;10(6):434–445.
68. Bilang-Bleuel A, Ulbricht S, Chandramohan Y, De Carli S, Droste SK, Reul JM. Psychological stress increases histone H3 phosphorylation in adult dentate gyrus granule neurons: involvement in a glucocorticoid receptor-dependent behavioural response. *Eur J Neurosci*. 2005;22(7):1691–1700.
69. Trollope AF, Gutierrez-Mecinas M, Mifsud KR, Collins A, Saunderson EA, Reul JM. Stress, epigenetic control of gene expression and memory formation. *Exp Neurol*. 2012;233(1):3–11.
70. Chandramohan Y, Droste SK, Reul JM. Novelty stress induces phospho-acetylation of histone H3 in rat dentate gyrus granule neurons through coincident signalling via the

N-methyl-D-aspartate receptor and the glucocorticoid receptor: relevance for c-fos induction. *J Neurochem*. 2007;101(3):815–828.
71. Clayton AL, Rose S, Barratt MJ, Mahadevan LC. Phosphoacetylation of histone H3 on c-fos- and c-jun-associated nucleosomes upon gene activation. *EMBO J*. 2000;19(14): 3714–3726.
72. Gutierrez-Mecinas M, Trollope AF, Collins A, et al. Long-lasting behavioral responses to stress involve a direct interaction of glucocorticoid receptors with ERK1/2-MSK1-Elk-1 signaling. *Proc Natl Acad Sci USA*. 2011;108(33):13806–13811.
73. Bennett MR, Lagopoulos J. Stress and trauma: BDNF control of dendritic-spine formation and regression. *Prog Neurobiol*. 2014;112:80–99.
74. Roth TL, Zoladz PR, Sweatt JD, Diamond DM. Epigenetic modification of hippocampal Bdnf DNA in adult rats in an animal model of post-traumatic stress disorder. *J Psychiatr Res*. 2011;45(7):919–926.
75. Kozlovsky N, Matar MA, Kaplan Z, Kotler M, Zohar J, Cohen H. Long-term down-regulation of BDNF mRNA in rat hippocampal CA1 subregion correlates with PTSD-like behavioural stress response. *Int J Neuropsychopharmacol*. 2007;10(6):741–758.
76. Zoladz PR, Diamond DM. Predator-based psychosocial stress animal model of PTSD: preclinical assessment of traumatic stress at cognitive, hormonal, pharmacological, cardiovascular and epigenetic levels of analysis. *Exp Neurol*. 2016;284(pt B):211–219.
77. Yamamoto S, Morinobu S, Takei S, et al. Single prolonged stress: toward an animal model of posttraumatic stress disorder. *Depress Anxiety*. 2009;26(12):1110–1117.
78. Wang W, Liu Y, Zheng H, et al. A modified single-prolonged stress model for post-traumatic stress disorder. *Neurosci Lett*. 2008;441(2):237–241.
79. Liberzon I, Krstov M, Young EA. Stress-restress: effects on ACTH and fast feedback. *Psychoneuroendocrinology*. 1997;22(6):443–453.
80. Belda X, Fuentes S, Nadal R, Armario A. A single exposure to immobilization causes long-lasting pituitary-adrenal and behavioral sensitization to mild stressors. *Horm Behav*. 2008;54(5):654–661.
81. Takei S, Morinobu S, Yamamoto S, Fuchikami M, Matsumoto T, Yamawaki S. Enhanced hippocampal BDNF/TrkB signaling in response to fear conditioning in an animal model of posttraumatic stress disorder. *J Psychiatr Res*. 2011;45(4):460–468.
82. Toda H, Boku S, Nakagawa S, et al. Maternal separation enhances conditioned fear and decreases the mRNA levels of the neurotensin receptor 1 gene with hypermethylation of this gene in the rat amygdala. *PLoS One*. 2014;9(5):e97421.
83. Qingzhen L, Jiehua M, Zhiyang Y, Hongjun L, Chunlong C, Weiyan L. Distinct hippocampal expression profiles of lncRNAs in rats exhibiting a PTSD-like syndrome. *Mol Neurobiol*. 2016;53(4):2161–2168.
84. Guffanti G, Galea S, Yan L, et al. Genome-wide association study implicates a novel RNA gene, the lincRNA AC068718.1, as a risk factor for post-traumatic stress disorder in women. *Psychoneuroendocrinology*. 2013;38(12):3029–3038.
85. Minatohara K, Akiyoshi M, Okuno H. Role of immediate-early genes in synaptic plasticity and neuronal ensembles underlying the memory trace. *Front Mol Neurosci*. 2015;8:78.
86. Easton AC, Lourdusamy A, Loth E, et al. CAMK2A polymorphisms predict working memory performance in humans. *Mol Psychiatry*. 2013;18(8):850–852.
87. Lebow M, Neufeld-Cohen A, Kuperman Y, Tsoory M, Gil S, Chen A. Susceptibility to PTSD-like behavior is mediated by corticotropin-releasing factor receptor type 2 levels in the bed nucleus of the stria terminalis. *J Neurosci*. 2012;32(20):6906–6916.
88. Yehuda R, Hoge CW, McFarlane AC, et al. Post-traumatic stress disorder. *Nat Rev Dis Primers*. 2015;1:15057.
89. McFarlane AC, Lawrence-Wood E, Van Hooff M, Malhi GS, Yehuda R. The need to take a staging approach to the biological mechanisms of PTSD and its treatment. *Curr Psychiatry Rep*. 2017;19(2):10.

90. Daskalakis NP, Cohen H, Nievergelt CM, et al. New translational perspectives for blood-based biomarkers of PTSD: from glucocorticoid to immune mediators of stress susceptibility. *Exp Neurol.* 2016;284(pt B):133–140.
91. Chen Y, Li X, Kobayashi I, Tsao D, Mellman TA. Expression and methylation in post-traumatic stress disorder and resilience; evidence of a role for odorant receptors. *Psychiatry Res.* 2016;245:36–44.
92. Kuan PF, Waszczuk MA, Kotov R, et al. An epigenome-wide DNA methylation study of PTSD and depression in World Trade Center responders. *Transl Psychiatry.* 2017;7(6):e1158.
93. Mehta D, Klengel T, Conneely KN, et al. Childhood maltreatment is associated with distinct genomic and epigenetic profiles in posttraumatic stress disorder. *Proc Natl Acad Sci USA.* 2013;110(20):8302–8307.
94. Rutten BPF, Vermetten E, Vinkers CH, et al. Longitudinal analyses of the DNA methylome in deployed military servicemen identify susceptibility loci for post-traumatic stress disorder. *Mol Psychiatry.* 2017.
95. Smith AK, Conneely KN, Kilaru V, et al. Differential immune system DNA methylation and cytokine regulation in post-traumatic stress disorder. *Am J Med Genet B Neuropsychiatr Genet.* 2011;156B(6):700–708.
96. Uddin M, Aiello AE, Wildman DE, et al. Epigenetic and immune function profiles associated with posttraumatic stress disorder. *Proc Natl Acad Sci USA.* 2010;107(20): 9470–9475.
97. Mehta D, Bruenig D, Carrillo-Roa T, et al. Genomewide DNA methylation analysis in combat veterans reveals a novel locus for PTSD. *Acta Psychiatr Scand.* 2017;136(5): 493–505.
98. Hammamieh R, Chakraborty N, Gautam A, et al. Whole-genome DNA methylation status associated with clinical PTSD measures of OIF/OEF veterans. *Transl Psychiatry.* 2017;7(7):e1169.
99. Vukojevic V, Kolassa IT, Fastenrath M, et al. Epigenetic modification of the glucocorticoid receptor gene is linked to traumatic memory and post-traumatic stress disorder risk in genocide survivors. *J Neurosci.* 2014;34(31):10274–10284.
100. Yehuda R, Daskalakis NP, Desarnaud F, et al. Epigenetic biomarkers as predictors and correlates of symptom improvement following psychotherapy in combat veterans with PTSD. *Front Psychiatry.* 2013;4:118.
101. Klengel T, Mehta D, Anacker C, et al. Allele-specific FKBP5 DNA demethylation mediates gene-childhood trauma interactions. *Nat Neurosci.* 2013;16(1):33–41.
102. Sadeh N, Spielberg JM, Logue MW, et al. SKA2 methylation is associated with decreased prefrontal cortical thickness and greater PTSD severity among trauma-exposed veterans. *Mol Psychiatry.* 2016;21(3):357–363.
103. Perroud N, Paoloni-Giacobino A, Prada P, et al. Increased methylation of glucocorticoid receptor gene (NR3C1) in adults with a history of childhood maltreatment: a link with the severity and type of trauma. *Transl Psychiatry.* 2011;1:e59.
104. Perroud N, Rutembesa E, Paoloni-Giacobino A, et al. The Tutsi genocide and transgenerational transmission of maternal stress: epigenetics and biology of the HPA axis. *World J Biol Psychiatry.* 2014;15(4):334–345.
105. Ressler KJ, Mercer KB, Bradley B, et al. Post-traumatic stress disorder is associated with PACAP and the PAC1 receptor. *Nature.* 2011;470(7335):492–497.
106. Yehuda R, Daskalakis NP, Bierer LM, et al. Holocaust exposure induced intergenerational effects on FKBP5 methylation. *Biol Psychiatry.* 2016;80(5):372–380. https://doi.org/10.1016/j.biopsych.2015.08.005.
107. Yehuda R, Daskalakis NP, Lehrner A, et al. Influences of maternal and paternal PTSD on epigenetic regulation of the glucocorticoid receptor gene in Holocaust survivor offspring. *Am J Psychiatry.* 2014;171(8):872–880.

108. Yehuda R, Flory JD, Bierer LM, et al. Lower methylation of glucocorticoid receptor gene promoter 1F in peripheral blood of veterans with posttraumatic stress disorder. *Biol Psychiatry*. 2015;77(4):356–364.
109. Labonte B, Azoulay N, Yerko V, Turecki G, Brunet A. Epigenetic modulation of glucocorticoid receptors in posttraumatic stress disorder. *Transl Psychiatry*. 2014;4:e368.
110. Schechter DS, Moser DA, Paoloni-Giacobino A, et al. Methylation of NR3C1 is related to maternal PTSD, parenting stress and maternal medial prefrontal cortical activity in response to child separation among mothers with histories of violence exposure. *Front Psychol*. 2015;6:690.
111. Boks MP, Rutten BP, Geuze E, et al. SKA2 methylation is involved in cortisol stress reactivity and predicts the development of post-traumatic stress disorder (PTSD) after military deployment. *Neuropsychopharmacology*. 2016;41(5):1350–1356.
112. Kaminsky Z, Wilcox HC, Eaton WW, et al. Epigenetic and genetic variation at SKA2 predict suicidal behavior and post-traumatic stress disorder. *Transl Psychiatry*. 2015;5: e627.
113. Sadeh N, Wolf EJ, Logue MW, et al. Epigenetic variation at Ska2 predicts suicide phenotypes and internalizing psychopathology. *Depress Anxiety*. 2016;33(4): 308–315.
114. Clive ML, Boks MP, Vinkers CH, et al. Discovery and replication of a peripheral tissue DNA methylation biosignature to augment a suicide prediction model. *Clin Epigenetics*. 2016;8:113.
115. Uddin M, Galea S, Chang SC, et al. Epigenetic signatures may explain the relationship between socioeconomic position and risk of mental illness: preliminary findings from an urban community-based sample. *Biodemography Soc Biol*. 2013;59(1):68–84.
116. Bam M, Yang X, Zhou J, et al. Evidence for epigenetic regulation of pro-inflammatory cytokines, interleukin-12 and interferon gamma, in peripheral blood mononuclear cells from PTSD patients. *J Neuroimmune Pharmacol*. 2016;11(1):168–181.
117. Bam M, Yang X, Zumbrun EE, et al. Dysregulated immune system networks in war veterans with PTSD is an outcome of altered miRNA expression and DNA methylation. *Sci Rep*. 2016;6:31209.
118. Powers A, Almli L, Smith A, et al. A genome-wide association study of emotion dysregulation: evidence for interleukin 2 receptor alpha. *J Psychiatr Res*. 2016;83: 195–202.
119. Zhou J, Nagarkatti P, Zhong Y, et al. Dysregulation in microRNA expression is associated with alterations in immune functions in combat veterans with post-traumatic stress disorder. *PLoS One*. 2014;9(4):e94075.
120. Wolf EJ, Logue MW, Hayes JP, et al. Accelerated DNA methylation age: associations with PTSD and neural integrity. *Psychoneuroendocrinology*. 2016;63:155–162.
121. Zannas AS, Arloth J, Carrillo-Roa T, et al. Lifetime stress accelerates epigenetic aging in an urban, African American cohort: relevance of glucocorticoid signaling. *Genome Biol*. 2015;16:266.
122. Boks MP, van Mierlo HC, Rutten BP, et al. Longitudinal changes of telomere length and epigenetic age related to traumatic stress and post-traumatic stress disorder. *Psychoneuroendocrinology*. 2015;51:506–512.
123. Koenen KC, Uddin M, Chang SC, et al. SLC6A4 methylation modifies the effect of the number of traumatic events on risk for posttraumatic stress disorder. *Depress Anxiety*. 2011;28(8):639–647.
124. Schechter DS, Moser DA, Pointet VC, et al. The association of serotonin receptor 3A methylation with maternal violence exposure, neural activity, and child aggression. *Behav Brain Res*. 2017;325(Pt B):268–277.
125. Moser DA, Paoloni-Giacobino A, Stenz L, et al. BDNF methylation and maternal brain activity in a violence-related sample. *PLoS One*. 2015;10(12):e0143427.

126. Kim TY, Kim SJ, Chung HG, Choi JH, Kim SH, Kang JI. Epigenetic alterations of the BDNF gene in combat-related post-traumatic stress disorder. *Acta Psychiatr Scand.* 2017;135(2):170–179.
127. Chang SC, Koenen KC, Galea S, et al. Molecular variation at the SLC6A3 locus predicts lifetime risk of PTSD in the Detroit Neighborhood Health Study. *PLoS One.* 2012;7(6):e39184.
128. Norrholm SD, Jovanovic T, Smith AK, et al. Differential genetic and epigenetic regulation of catechol-O-methyltransferase is associated with impaired fear inhibition in posttraumatic stress disorder. *Front Behav Neurosci.* 2013;7:30.
129. Sipahi L, Wildman DE, Aiello AE, et al. Longitudinal epigenetic variation of DNA methyltransferase genes is associated with vulnerability to post-traumatic stress disorder. *Psychol Med.* 2014;44(15):3165–3179 .
130. Maddox SA, Kilaru V, Shin J, et al. Estrogen-dependent association of HDAC4 with fear in female mice and women with PTSD. *Mol Psychiatry.* 2017.
131. Rusiecki JA, Byrne C, Galdzicki Z, et al. PTSD and DNA methylation in select immune function gene promoter regions: a repeated measures case-control study of U.S. military service members. *Front Psychiatry.* 2013;4:56.
132. Almli LM, Stevens JS, Smith AK, et al. A genome-wide identified risk variant for PTSD is a methylation quantitative trait locus and confers decreased cortical activation to fearful faces. *Am J Med Genet B Neuropsychiatr Genet.* 2015;168B(5):327–336.
133. Sipahi L, Uddin M, Hou ZC, et al. Ancient evolutionary origins of epigenetic regulation associated with posttraumatic stress disorder. *Front Hum Neurosci.* 2014; 8:284.
134. Uddin M, Galea S, Chang SC, et al. Gene expression and methylation signatures of MAN2C1 are associated with PTSD. *Dis Markers.* 2011;30(2–3):111–121.
135. Rusiecki JA, Chen L, Srikantan V, et al. DNA methylation in repetitive elements and post-traumatic stress disorder: a case-control study of US military service members. *Epigenomics.* 2012;4(1):29–40.
136. Rampp C, Binder EB, Provencal N. Epigenetics in posttraumatic stress disorder. *Prog Mol Biol Transl Sci.* 2014;128:29–50.
137. Ryan J, Chaudieu I, Ancelin ML, Saffery R. Biological underpinnings of trauma and post-traumatic stress disorder: focusing on genetics and epigenetics. *Epigenomics.* 2016;8(11):1553–1569.
138. Matosin N, Cruceanu C, Binder EB. Preclinical and clinical evidence of DNA methylation changes in response to trauma and chronic stress. *Chronic Stress.* 2017;1:1–20.
139. Gillespie CF, Bradley B, Mercer K, et al. Trauma exposure and stress-related disorders in inner city primary care patients. *Gen Hosp Psychiatry.* 2009;31(6):505–514.
140. Norrholm SD, Jovanovic T, Olin IW, et al. Fear extinction in traumatized civilians with posttraumatic stress disorder: relation to symptom severity. *Biol Psychiatry.* 2011;69(6):556–563.
141. de Quervain DJ, Aerni A, Schelling G, Roozendaal B. Glucocorticoids and the regulation of memory in health and disease. *Front Neuroendocrinol.* 2009;30(3): 358–370.
142. Taylor SF, Welsh RC, Wager TD, Phan KL, Fitzgerald KD, Gehring WJ. A functional neuroimaging study of motivation and executive function. *Neuroimage.* 2004;21(3): 1045–1054.
143. Li X, Ito M, Zhou F, et al. A maternal-zygotic effect gene, Zfp57, maintains both maternal and paternal imprints. *Dev Cell.* 2008;15(4):547–557.
144. Jakobsson J, Cordero MI, Bisaz R, et al. KAP1-mediated epigenetic repression in the forebrain modulates behavioral vulnerability to stress. *Neuron.* 2008;60(5):818–831.
145. Matsuo R, Asada A, Fujitani K, Inokuchi K. LIRF, a gene induced during hippocampal long-term potentiation as an immediate-early gene, encodes a novel RING finger protein. *Biochem Biophys Res Commun.* 2001;289(2):479–484.

146. Ploski JE, Park KW, Ping J, Monsey MS, Schafe GE. Identification of plasticity-associated genes regulated by Pavlovian fear conditioning in the lateral amygdala. *J Neurochem*. 2010;112(3):636–650.
147. Datson NA, van den Oever JM, Korobko OB, Magarinos AM, de Kloet ER, McEwen BS. Previous history of chronic stress changes the transcriptional response to glucocorticoid challenge in the dentate gyrus region of the male rat hippocampus. *Endocrinology*. 2013;154(9):3261–3272.
148. McGowan PO, Sasaki A, D'Alessio AC, et al. Epigenetic regulation of the glucocorticoid receptor in human brain associates with childhood abuse. *Nat Neurosci*. 2009;12(3):342–348.
149. McNerney MW, Sheng T, Nechvatal JM, et al. Integration of neural and epigenetic contributions to posttraumatic stress symptoms: the role of hippocampal volume and glucocorticoid receptor gene methylation. *PLoS One*. 2018;13(2):e0192222.
150. Nikolova YS, Hariri AR. Can we observe epigenetic effects on human brain function? *Trends Cogn Sci*. 2015;19(7):366–373.
151. Wang C, Schroeder FA, Wey HY, et al. In vivo imaging of histone deacetylases (HDACs) in the central nervous system and major peripheral organs. *J Med Chem*. 2014;57(19):7999–8009.
152. Wang C, Schroeder FA, Hooker JM. Visualizing epigenetics: current advances and advantages in HDAC PET imaging techniques. *Neuroscience*. 2014;264:186–197.
153. Wey HY, Gilbert TM, Zurcher NR, et al. Insights into neuroepigenetics through human histone deacetylase PET imaging. *Sci Transl Med*. 2016;8(351):351ra106.
154. Schroeder FA, Gilbert TM, Feng N, et al. Expression of HDAC2 but not HDAC1 transcript is reduced in dorsolateral prefrontal cortex of patients with schizophrenia. *ACS Chem Neurosci*. 2017;8(3):662–668.
155. Strebl MG, Wang C, Schroeder FA, et al. Development of a fluorinated class-I HDAC radiotracer reveals key chemical determinants of brain penetrance. *ACS Chem Neurosci*. 2016;7(5):528–533.
156. Bonomi R, Mukhopadhyay U, Shavrin A, et al. Novel histone deacetylase class IIa selective substrate radiotracers for PET imaging of epigenetic regulation in the brain. *PLoS One*. 2015;10(8):e0133512.
157. Hodgson K, Carless MA, Kulkarni H, et al. Epigenetic age acceleration assessed with human white-matter images. *J Neurosci*. 2017;37(18):4735–4743.
158. Miller MW, Sadeh N. Traumatic stress, oxidative stress and post-traumatic stress disorder: neurodegeneration and the accelerated-aging hypothesis. *Mol Psychiatry*. 2014;19(11):1156–1162.
159. Wolf EJ, Maniates H, Nugent N, et al. Traumatic stress and accelerated DNA methylation age: a meta-analysis. *Psychoneuroendocrinology*. 2017. https://doi.org/10.1016/j.psyneuen.2017.12.007.
160. Hannum G, Guinney J, Zhao L, et al. Genome-wide methylation profiles reveal quantitative views of human aging rates. *Mol Cell*. 2013;49(2):359–367.
161. Zannas AS, Chrousos GP. Epigenetic programming by stress and glucocorticoids along the human lifespan. *Mol Psychiatry*. 2017;22(5):640–646.
162. Jovanovic T, Vance LA, Cross D, et al. Exposure to violence accelerates epigenetic aging in children. *Sci Rep*. 2017;7(1):8962.
163. Brody GH, Yu T, Chen E, Beach SR, Miller GE. Family-centered prevention ameliorates the longitudinal association between risky family processes and epigenetic aging. *J Child Psychol Psychiatry*. 2016;57(5):566–574.
164. Cuthbert BN, Insel TR. Toward the future of psychiatric diagnosis: the seven pillars of RDoC. *BMC Med*. 2013;11:126.
165. Michopoulos V, Norrholm SD, Jovanovic T. Diagnostic biomarkers for posttraumatic stress disorder: promising horizons from translational neuroscience research. *Biol Psychiatry*. 2015;78(5):344–353.

CHAPTER ELEVEN

Using Epigenetic Tools to Investigate Antidepressant Response

Laura M. Fiori, Rixing Lin, Chelsey Ju, Raoul Belzeaux, Gustavo Turecki[1]
McGill Group for Suicide Studies, Douglas Mental Health University Institute, Department of Psychiatry, McGill University, Montreal, QC, Canada
[1]Corresponding author: e-mail address: gustavo.turecki@mcgill.ca

Contents

Abstract

Major depressive disorder is a chronic and debilitating illness. It is most commonly treated with antidepressant drugs, however, as the majority of patients do not respond on their first trial or following several adequate trials, there is great interest in identifying biological factors that may help select the most appropriate treatment for each patient and in understanding biological processes that mediate treatment response. Epigenetic factors, such as non-coding RNAs (ncRNAs), hold potential as biomarkers of antidepressant response. In this chapter, we review key methodological considerations when investigating ncRNA biomarkers, including biological samples and technologies which have been used in these studies. Secondly, we summarize findings from studies investigating ncRNAs in antidepressant treatment response. Finally, we discuss some of the future directions which will be necessary for the development of clinically relevant epigenetic tools.

Progress in Molecular Biology and Translational Science, Volume 158
ISSN 1877-1173
https://doi.org/10.1016/bs.pmbts.2018.04.004

1. INTRODUCTION

1.1 Major Depressive Disorder and Antidepressant Treatment

Major depressive disorder (MDD) is a serious and debilitating illness with a lifetime prevalence of 16.2%.[1] Antidepressants are the most common treatment for MDD, and although they can be effective, the majority of patients do not achieve remission with their first prescribed antidepressant,[2] and typically 30%–40% of patients do not respond following several trials.[3] MDD symptom presentation and antidepressant treatment responses are highly heterogeneous and currently unpredictable. Although this variability results from individual differences in the interplay between various sociodemographic factors and biological pathways (gene × environment interactions),[4] the exact mechanisms are not yet understood. To address these issues, there is great interest in identifying biological factors (biomarkers) that may help us better understand the underlying pathophysiology of MDD, the mechanisms that underlie treatment response, and importantly, how to select the most appropriate treatment for each patient.

1.2 Biomarkers for MDD and Antidepressant Response

Biomarkers are defined as "a characteristic that is objectively measured and evaluated as an indicator of normal biological processes, pathogenic processes, or pharmacologic responses to a therapeutic intervention."[5] With respect to studies of MDD and antidepressant response, relevant biomarker categories include diagnostic, prognostic, monitoring, predictive, and response biomarkers.[6] Given the heterogeneous presentation of MDD and its comorbidity with other psychiatric disorders, diagnostic biomarkers would be invaluable, as they could provide greater diagnostic sensitivity and accuracy than symptom-based rating scales. Of similar benefit, prognostic biomarkers may be used to assess the likelihood of depression recurrence after treatment. One of the focuses of this chapter is antidepressant response, for which predictive and response biomarkers are the most relevant, as they would allow a clinician to select the most effective treatment (predictive), and to determine when adequate antidepressant response has been achieved. Finally, monitoring biomarkers may be assessed throughout diagnosis, treatment, and response, in order to evaluate improvement or worsening of MDD symptoms.

1.3 Epigenetic Regulators of Gene Expression

Although many early studies in psychiatry attempted to relate pathological states and antidepressant response to genetic variance, it became clear that genetic differences between individuals explain only a small proportion of disease risk and drug response.[7] At the same time, epigenetic regulation, which mediates the interaction between the environment and the genome, became increasingly recognized as an important source of information.[8,9] Epigenetic modifications are a collection of chemical and physical alterations in the genome that regulate the activity of genes.[10] More broadly, epigenetic regulators can be defined as molecules which directly or indirectly regulate gene expression. Although protein-coding genes and their respective messenger RNAs (mRNAs) are functional effectors of epigenetic differences, how epigenetic differences are associated with MDD pathology and/or influenced by antidepressant treatment can be investigated to understand underlying pathophysiological changes. This may provide potential new targets for therapeutic intervention. DNA methylation and post-translational histone modifications are the most well-known epigenetic regulators, and have been investigated extensively in certain psychiatric contexts. However, there is little research evaluating their role in antidepressant response in humans. We have thus focused this review on small non-coding RNAs (ncRNAs), which have been investigated specifically in the context of treatment response.

Although numerous species of small ncRNAs exist, the majority of studies to date have focused on microRNAs (miRNAs). miRNAs are single-stranded RNA that are 17- to 22-nt in length, and play essential roles in regulating gene expression through their involvement in the RNA-induced silencing complex.[11] This complex typically induces translational silencing and degradation of target mRNA molecules through interactions with the 3′ untranslated region (UTR) of the target mRNA and a complementary region of the miRNA (seed region). One miRNA can regulate the expression of multiple genes,[12] and as there are relatively few miRNA (approximately 2000 in humans), miRNA-focused studies have much greater statistical power than those focusing on genetic variance or mRNA expression. Furthermore, they are of particular interest as biomarkers as they are known to be released into circulation and stably transported in the blood.[13]

In this chapter, we review several key methodological considerations when investigating ncRNA biomarkers, including the types of biological samples (Section 2.1) and technologies (Section 2.2) which have been used in these studies. Secondly, we summarize studies that explored ncRNAs in antidepressant treatment response (Section 3). Finally, we discuss future directions which will be necessary for the development of clinically relevant epigenetic tools (Section 4).

2. METHODOLOGICAL CONSIDERATIONS

The importance of ncRNAs in antidepressant response has only begun to be recognized, and optimal sources of sample material, and most suitable quantification technologies have yet to be determined. Methodologies have thus varied greatly between studies, and must be taken into account when comparing their results. In this section, we first describe types of biological samples which have been used in these investigations, along with how this impacts the interpretation and functional relevance of findings. Secondly, we provide an overview of the technologies which have been used to investigate small ncRNAs.

2.1 Biological Samples

Blood is one of the most accessible tissues, and thus, the most commonly-used source of biological material for biomarker studies. Blood is comprised of a number of components, which may be analyzed as a whole, or separated into various fractions. Although other peripheral samples are available, including saliva, cerebrospinal fluid, and urine, these are less popular due to issues such as difficulties in collection, and the quantity and quality of RNA and DNA they provide. Accordingly, this section focuses exclusively on blood-derived samples, and describe some of their key characteristics, and how their selection may impact the interpretation of results.

2.1.1 Whole Blood

Collection and analysis of whole blood is the most straightforward use of peripheral blood samples. Whole blood is a heterogeneous mixture of erythrocytes (red blood cells), leukocytes (white blood cells), platelets, and plasma. Depending upon the downstream application, specialized blood collection tubes are used in order to address protein coagulation and/or preserve RNA integrity. In addition to its simplicity, another advantage to analyzing whole blood is that disease- or treatment-associated epigenetic effects from any specific blood component can be detected simultaneously. However, individual differences in cellular composition can skew results, thus blood cellular profiles should be assessed and bioinformatically corrected for if necessary.[14] Furthermore, erythrocytes are present in substantially higher numbers than other cells, diluting the effects of epigenetic alterations specific to other components to below levels detectable using current technologies. Finally, globin mRNA is highly expressed in erythrocytes, and represents approximately two-thirds of whole blood RNAs.[15] As such, molecular techniques often must be employed for gene expression studies in order to deplete globin mRNAs prior to analysis.

2.1.2 Leukocytes

Leukocytes are nucleated blood cells which play essential roles in immune response. Although they comprise only a small fraction of cells in the blood, they contribute the majority of DNA, making them the primary source of methylation and histone information. The use of leukocytes offers two advantages: firstly, it removes the issue of globin mRNA and its influence on sample integrity, and secondly, it reduces heterogeneity between individuals and is thus less influenced by disease-related differences in blood composition.

2.1.3 Plasma and Serum

Although both whole blood and leukocyte samples have been investigated extensively in psychiatry, MDD is viewed largely as a brain disorder, and the relationship between epigenetic effects in blood cells and pathological changes in the brain are unclear. Unlike blood-derived cells, cell-free components of blood (plasma or serum) harbor nucleic acids and other components originating from all tissues of the body, including the brain. Cell-free circulating nucleic acids (cfCNAs) are DNA and RNA molecules released by cells into circulation and transported in the plasma within microvesicles, exosomes, or bound to HDL.[16]

While cell-free samples have not been well-studied in psychiatry, they have been analyzed extensively in other pathological conditions, and have shown great promise for identifying biomarkers of disease and treatment response. This has been particularly true in the field of oncology, as nucleic acids released by tumor cells are highly enriched in plasma, and display tumor-specific epigenetic profiles. The methylation patterns at certain genomic regions in circulating DNA have been proposed as biomarkers for cancer diagnosis.[17,18] Additionally, plasma miRNAs have been used for diagnosis,[19] monitoring treatment response,[20] and prediction of recurrence.[21] The utility of cfCNAs has also been demonstrated in diabetes,[22,23] trauma,[24] and a number of brain pathologies.[23,25] This relationship between pathological states in the brain and cfCNAs indicates that they can cross the blood–brain barrier and thus can act as functionally relevant biomarkers for drug response.

2.2 Technology

Broadly, epigenetic studies can be divided into either targeted, or large-scale investigations, with the choice between these two options being largely driven by cost and if the researcher has an a priori hypothesis regarding a particular epigenetic modification. Numerous technologies have been developed for both approaches, and have been applied to the investigation of peripheral samples. While a full analysis of these approaches, and how data from these investigations may be combined, is beyond the scope of this review, more in depth reviews are available elsewhere.[26,27]

Gene expression profiling has been performed extensively in the investigation of MDD and antidepressant response. High-throughput approaches allow simultaneous measurement of thousands of transcripts, which can then be related to clinical characteristics or compared across timepoints to investigate treatment-related alterations. Although there are numerous technologies used for high-throughput studies, they can be divided into two classes: affinity-based technologies which interrogate large panels of predetermined transcripts, and next-generation sequencing (NGS) which can allow the investigation of all RNA species within a sample. The microarray is most well-used affinity-based approach. While microarrays are less costly, and their bioinformatic processing steps relatively standardized and straightforward, their utility is contingent upon the predetermined sequences they interrogate. Additionally, as the cost of sequencing technologies continues to decrease, affinity-based microarray technologies are being phased out. NGS methods are more amenable to investigation of ncRNAs as that they do not require the preselection of targets. However, ncRNAs are often expressed at much lower levels than mRNAs and may still be difficult to quantify accurately. As with all high-throughput investigations, the use of robust bioinformatics pipelines, statistical correction for multiple testing, and validation of findings with other technological platforms and/or additional cohorts is required.

Targeted investigations are more cost-effective and require less material, which is a particularly important consideration when using plasma or serum samples. A variety of technologies are available, including quantitative real-time polymerase chain reaction (qRT-PCR), and low-throughput affinity-based methods.

3. CURRENT FINDINGS

Investigation of biological factors related to antidepressant response, and research assessing the epigenetics of psychiatric disorders, have been ongoing for over a decade, yet studies combining these two fields are still relatively uncommon. Nonetheless, many interesting findings have emerged, demonstrating the potential utility of ncRNAs as biomarkers for antidepressant response. As mentioned above, the majority of published studies exploring ncRNAs in treatment response have investigated miRNAs. These have been performed in peripheral tissues in humans, and rodent models of MDD, and are comprised of both high-throughput and targeted approaches.

Summaries of studies investigating miRNAs in treatment response and MDD are summarized in Tables 1 and 2. We have grouped this section by the biochemical and signaling pathways which were reported to be the targets of response-associated miRNAs. However, as mentioned above, each miRNA can regulate the expression of numerous genes, with diverse

Table 1 miRNAs Associated with Antidepressant Treatment Response

miRNA	Treatment	Results	Reference
miR-1202	Citalopram	Lower expression shows predictive value for response before treatment, with increase in expression associated with response after treatment	28–30
miR-146a-5p miR-146b-5p miR-24-3p miR-425-3p	Duloxetine Escitalopram Nortriptyline	Higher expression in post-mortem brains of MDD subjects. Decreased expression associated with treatment response	31
miR-124	Naturalistic study	Decreased level of expression associated with treatment response	32
miR-135a	Cognitive behavioral therapy Escitalopram	Trend for higher expression after CBT when compared to escitalopram	33
miR-145 miR-20b	Naturalistic study	Increased expression associated with treatment response	34

Table 2 miRNAs Associated with MDD

miRNA	Expression in MDD	Tissue	Reference
miR-20a	Downregulated	Prefrontal cortex	35
let-7a-5p	Downregulated	Peripheral blood	36
let-7d-5p	Downregulated	Peripheral blood	36
let-7f-5p	Downregulated	Peripheral blood	36
miR-124a	Upregulated	Brodmann area 44	37
miR-146a-5p	Upregulated	Ventrolateral prefrontal cortex	31
miR-146b-5p	Upregulated	Ventrolateral prefrontal cortex	31
miR-148b	Downregulated	Prefrontal cortex	35
miR-152-3p	Downregulated	Prefrontal cortex	38
miR-1915-3p	Downregulated	Peripheral blood	36
miR-1972	Downregulated	Peripheral blood	39
miR-199a-5p	Upregulated	Peripheral blood	36
miR-20b	Downregulated	Prefrontal cortex	35
miR-24-3p	Upregulated	Ventrolateral prefrontal cortex	31
	Upregulated	Peripheral blood	36
miR-26b	Downregulated	Peripheral blood	39
miR-34a	Upregulated	Brodmann area 44	37
	Downregulated	Prefrontal cortex	35
miR-34b	Downregulated	Prefrontal cortex	35
miR-34c-5p	Upregulated	Brodmann area 44	37
miR-425-3p	Upregulated	Ventrolateral prefrontal cortex	31
	Upregulated	Peripheral blood	36
miR-4485	Downregulated	Peripheral blood	39
miR-4498	Downregulated	Peripheral blood	39
miR-497	Upregulated	Brodmann area 44	37
miR-508-3p	Downregulated	Prefrontal cortex	38
miR-873	Upregulated	Brodmann area 44	37
miR-4743	Downregulated	Peripheral blood	39

biological functions. Consequently, differential miRNA expression can have far-reaching effects, and identifying the most phenotypically-relevant target genes and pathways can be challenging.

3.1 miR-16 and miR-135a (Serotonin)

Serotonin (5-HT) is produced by neurons in the midbrain raphe nuclei (RN), which projects extensively throughout the brain. The most popular antidepressants are selective serotonin reuptake inhibitors (SSRIs) whose antidepressant effects involve the inhibition of presynaptic 5-HT recapture and modulation of the expression of 5-HT receptor subtypes. miR-16 is predicted to target the 3′UTR of the 5-HT transporter (SERT) and has been investigated in several studies.[40,41] Using 1C11 neuroectodermal cells, which can differentiate into serotonergic or noradrenergic neurons, one study demonstrated that miR-16 levels were elevated in noradrenergic cells, whereas miR-16 levels remained steady in serotonergic cells.[40] In mice, miR-16 expression was lower in the RN compared to the locus coeruleus (LC), a noradrenergic neuron-rich region.[40] This suggested that miR-16 may play a role in silencing SERT transcripts during noradrenergic differentiation. Additionally, SERT protein expression in 1C11 noradrenergic cells was restored after miR-16 knock-down but noradrenergic-associated function was unaffected. Similarly, miR-16 knock-down showed higher SERT levels in rat RN compared to negative control rat RN.[41] miR-16 knock-down also induced noradrenergic cells to synthesize, store and degrade 5-HT.[40] Fluoxetine infusion into the mouse RN increased miR-16 levels by 2.5-fold, suggesting that SSRIs may regulate SERT expression through miR-16 in the RN. Direct infusion of fluoxetine into the LC failed to induce any change in miR-16 expression; however, infusion into the RN reduced miR-16 expression in the LC by 30% and increased SERT expression and tryptophan hydroxylase (rate limiting enzyme in 5-HT synthesis) activity.[40] This effect may be a result of the bilateral connection between the RN and the LC. Overall, this study demonstrated that miR-16 may have distal effects on brain regions receiving serotonergic projections, and can be targeted by SSRIs to regulate SERT expression. miR-16 was also profiled in relation to serotonin in cerebral spinal fluid (CSF). Serotonin and miR-16 are both significantly lower in CSF of MDD patients when compared to healthy controls.[41] Moreover CSF levels of miR-16 were negatively correlated with depression severity. To replicate these human findings, anti-miR-16 was injected intracerebroventricularly into rats. This resulted in a decrease in CSF miR-16, with healthy injected rats showing depressive-like behaviors. Injected rats also displayed lower CSF serotonin levels, similar to observations in depressed humans.[41]

miR-135 has also been studied in relation to serotonergic neuron activity, and shown to regulate SERT and Htr1a (5-HT1A autoreceptor) in the mouse RN. Reporter gene assays revealed that miR-135 targeted both the SERT 3′UTR and Htr1a 3′UTR, and that it could reduce translation of these genes by 30%–40%.[33] Further analysis revealed the existence of three highly conserved variants of miR-135: miR-135a-1, miR-135a-2, and miR-135b. Chronic social defeat stress in mice did not alter miR-135a levels in the RN. However, chronic or acute administration of imipramine (non-specific 5-HT-reuptake inhibitor), in both stressed and non-stressed mice, significantly increased miR-135a expression.[33] Moreover, administration of fluoxetine (SSRI) resulted in elevated miR-135a levels, whereas administration of reboxetine (norepinephrine reuptake inhibitor) had no effect on miR-135a expression. miR-135a levels were also significantly lower in human RN of suicide completers compared to controls, giving additional support for a specific role of miR-135a in the serotonin system.[33] miR-135a overexpression specifically in 5-HT neurons was found to reduce anxiety-like and MDD-like behaviors in mice, while miR-135a knock-down in the RN increased anxiety-like behaviors and decreased the response to antidepressant treatment.[33] In the same study, miR-135a was measured in humans and found to be decreased in whole blood in MDD patients compared to controls. Interestingly, miR-135a blood levels showed a statistically significant increase after cognitive behavioral therapy but not after SSRI antidepressant treatment. These results demonstrated that miR-135a is responsive to antidepressants, has a regulatory role in serotonergic neurons and could also act as a biomarker for a depressive state and antidepressant response.

To evaluate the biomarker potential of these two miRNA, we recently characterized the expression of both miR-16 and miR-135a in whole blood collected from two cohorts of MDD patients both before and after 8 weeks of antidepressant treatment.[28] Interestingly, although there were significant differences between responders and non-responders in the levels of both miRNAs, the effects varied between cohorts. In our first cohort, patients were treated with either the SSRI escitalopram or the serotonin–norepinephrine reuptake inhibitor (SNRI) desvenlafaxine, where we found significantly decreased miR-16 levels at both baseline and after treatment in responders, but no differences in miR-135a levels. However, in the second cohort, where patients were treated with duloxetine (SNRI), we observed no differences in baseline miR-16, but it was increased in responders after treatment. Additionally, miR-135a was decreased in responders at baseline and subsequently increased after treatment. The variability between cohorts suggested that additional variables may influence these effects. Indeed, a number of studies investigating the relationship between these miRNAs and depression- and anxiety-related phenotypes found them to be either significantly upregulated or downregulated depending upon the phenotype and tissue being examined.[41–46]

3.2 miR-9 and miR-326 (Dopamine)

The mesolimbic dopamine system plays an important role in regulation of motivation, reward, attention, and mood. Dopamine (DA) is the most abundant catecholaminergic neurotransmitter in the mammalian brain, and is involved in incentive-motivational pathways as well as in response to aversive stimuli.[47] Lowered DA neurotransmission has been linked to MDD, and possible explanations include diminished DA release from presynaptic neurons, changes in receptor number/function, and intracellular signal processing dysfunction.[48] Among dopamine-related genes, the dopamine receptor D2 (DRD2) has been the most extensively associated with MDD.[48] Alterations to DRD2 have been identified in several rodent models of MDD. Increased binding of DRD2 in the nucleus accumbens (NAc) and striatum, and increased striatal DRD2 mRNA levels were detected in adult rats subjected to maternal deprivation (MD). Additionally, increased accumbal DRD2 protein levels were detected in adult rats subjected to chronic unpredicted stress (CUS).[49]

It has been proposed that increases in DRD2 expression may be the result of dysregulated miRNA expression in the NAc and striatum. Accordingly, miR-9 was found to be downregulated in the NAc of CUS rats and in the striatum of MD rats, suggesting that miR-9 in the NAc is disrupted by stress in adulthood (modeled by CUS rats), whereas miR-9 in the striatum is disrupted by early-life adversity (modeled by MD rats).[49] Furthermore, when rats were exposed to MD as well as CUS, miR-9 was more significantly downregulated in both the NAc and striatum. Treatment with escitalopram did not reverse miR-9 downregulation, indicating that miR-9 may be specific to stress susceptibility and not antidepressant response.[49] To show that miR-9 directly targets and interacts with DRD2, miR-9 mimics were transfected into 5H-SY5Y cells, which resulted in significantly reduced DRD2 protein expression. Additionally, a luciferase assay with a DRD2 3′UTR reporter and miR-9 mimics significantly reduced luciferase reporter (DRD2 3′UTR) activity.[49] miR-9 levels were also found to be negatively correlated with DRD2 protein levels,[49] providing further evidence that miR-9 downregulation plays a role in the disrupted expression of DRD2 in the NAc and striatum.

The same study also reported miR-326 as being upregulated in the NAc and downregulated in the striatum of CUS rats, and upregulated in the striatum of MD rats.[49] miR-326 expression was found to be positively correlated with DRD2 protein levels in the NAc and the striatum although not correlated with DRD2 mRNA levels. Interestingly, the disrupted expression profile of miR-326 was reversed after 4 weeks of escitalopram treatment, suggesting that miR-326 activity is not a direct regulation of DRD2 gene expression, but may be involved in escitalopram response.[49]

3.3 miR-1202 and miR-335 (Glutamate)

The expression of a number of glutamate receptors, including glutamate metabotropic receptor 4 (GRM4), have been found to be dysregulated in depressed suicide completers,[50] and shown to be altered by antidepressant treatment.[51] GRM4 is a group III glutamate receptor and has been implicated in the regulation of anxiety-related behaviors.[52,53] To date, two miRNAs which are predicted to target GRM4 have been implicated in antidepressant response, miR-1202 and miR-335.

miR-1202 is primate-specific, enriched in brains, and has been shown to target GRM4.[29] miR-1202 was found to be downregulated in the brains of depressed suicide completers and appears to mediate antidepressant responses, as chronic treatment of citalopram (SSRI) or imipramine (TCA) in neural progenitor cells resulted in an upregulation of miR-1202 and a decrease in GRM4 mRNA.[29] This effect seems to be specific to antidepressant drugs, as levels were unchanged by treatment with non-serotonergic drugs. In addition, in MDD patients, those who responded to antidepressant treatment showed a decrease in peripheral blood levels of miR-1202 at baseline compared to non-responders and controls. Furthermore, after 8 weeks of antidepressant treatment the responder group showed an upregulation of miR-1202.[29] More recently, we characterized miR-1202 levels in two additional antidepressant-treated cohorts,[28] where it displayed decreased baseline levels in responders, and increased levels as a function of treatment response. As these findings were found for two antidepressant classes (SSRIs and SNRIs), these results imply that lower miR-1202 levels may represent a biomarker for clinical response to antidepressant monotherapy in general. Further support for the relationship between miR-1202 and GRM4 was found in a recent study which identified a variant in the 3′UTR of GRM4 predicted to disrupt miR-1202 binding, which was associated with MDD.[54]

miR-335 has also been proposed to target GRM4 and similar results to those reported for miR-1202 were observed after treatment with citalopram (SSRI); miR-335 expression was upregulated and GRM4 expression was decreased in peripheral blood.[55]

3.4 miR-24, miR-146a, miR-146b, and miR-425 (MAPK/Wnt Signaling)

We recently profiled miRNAs in antidepressant response using small RNA-seq and found miR-24-3p, miR-146a-5p, miR-146b-5p, and miR-425-3p to be significantly downregulated in responders.[31] Results were validated using qPCR and replicated in two additional external cohorts. The observed differential expression of miR-146a-5p, miR-146b-5p, miR-24-3p, and miR-425-3p were specific to response, as no significant differences were observed in treatment non-responders. These miRNAs were further investigated in the frontal cortex of depressed suicide completers, where they were significantly upregulated compared to non-suicide controls. These findings resemble those of miR-1202, which displays an inverse relationship between levels in the brain and levels associated with improvements of MDD symptoms.[31] Interestingly, both miR-24-3p and miR-425-3p have previously been shown to be upregulated in peripheral blood samples of depressed patients relative to healthy controls.[34,36]

In our study, findings were also replicated in mice which exhibited depressive-like behavior following chronic social defeat stress. These mice were treated with imipramine or saline controls for 14 days, at which point they were classified as responders or non-responders. Importantly, a significant downregulation of miR-146b-5p, miR-24-3p and miR-425 (miR-146a-5p downregulated but not significant) were exclusively observed in responders.[31]

In silico target prediction and pathway analysis suggested that these four miRNAs regulate genes involved in the MAPK/Wnt signaling pathway. Whole-genome mRNA levels were quantified in blood, and these four miRNAs were found to be strongly correlated with genes in the MAPK/Wnt signaling pathway, with more significant correlations observed in treatment responders. These interactions were further assessed in human embryonic kidney cells (HEK293). Canonical gene expression in HEK293 cells show high levels of MAPK/Wnt and low levels of the candidate miRNAs. After treatment with miRNA mimics, a downregulation of genes involved in MAPK/Wnt signaling was observed.[31]

This study was the largest investigation of miRNAs in antidepressant response to date. Most importantly, dysregulated expression of these four miRNAs was observed in whole blood, leukocytes, and plasma. This strongly supports their potential utility as biomarkers of antidepressant response.

4. CONCLUSIONS AND FUTURE PERSPECTIVES

ncRNAs represent a promising source for peripheral biomarkers of antidepressant response. In this chapter, we highlighted a number of miRNAs that appear to modulate signaling pathways of relevance to MDD and treatment response. Furthermore, their levels have been shown to be responsive to treatment, demonstrating that in addition to diagnosis, they may also hold utility as predictive and response biomarkers. With that said, considerable work is necessary prior to clinical application.

Firstly, although promising, results have been replicated across only a few studies; thus, additional trials in larger and better-characterized cohorts is necessary. This is particularly important in the context of epigenetic modifications, which are highly susceptible to the environment, and easily confounded by clinical and sociodemographic factors such as gender or early-life adversity.[56,57]

Secondly, it is unclear which peripheral sample is optimal for biomarker studies. The studies described herein performed measures in either whole blood or leukocytes. However, the neurobiological relevance of epigenetic alterations in blood cells is still unclear. As technologies improve, future biomarker studies should move toward investigations of cfCNAs from plasma or serum, as they can provide greater insight into pathological processing occurring across the body, including the brain.

Finally, although other epigenetic marks, such as DNA methylation and post-translational histone modifications, have been investigated in other psychiatric contexts, few studies have examined their relationship with antidepressant response. Among these, there is evidence of antidepressant-associated effects at the promoter regions of SERT[58,59] and brain-derived neurotrophic factor,[60–62] as well as treatment-associated expression differences of the modifier histone deacetylase 5.[63] Future studies should explore other epigenetic modifications, while including gene expression measures, to obtain a broader picture of the relationship between antidepressant treatment and the epigenomic landscape. Ultimately, it is likely that combinations, not individual biomarkers, will have the greatest utility to diagnose, predict and monitor MDD and antidepressant treatment. In the future, researchers should assess and integrate multiple types of information (i.e., genetic, epigenetic, and gene expression), to identify the most meaningful panel of biomarkers.

REFERENCES

1. Kupfer DJ, Frank E, Phillips ML. Major depressive disorder: new clinical, neurobiological, and treatment perspectives. *Lancet.* 2012;379(9820):1045–1055.
2. Rush AJ, Trivedi MH, Wisniewski SR, et al. Acute and longer-term outcomes in depressed outpatients requiring one or several treatment steps: a STAR*D report. *Am J Psychiatry.* 2006;163(11):1905–1917.
3. Trivedi MH, Rush AJ, Wisniewski SR, et al. Evaluation of outcomes with citalopram for depression using measurement-based care in STAR*D: implications for clinical practice. *Am J Psychiatry.* 2006;163(1):28–40.
4. El-Hage W, Leman S, Camus V, Belzung C. Mechanisms of antidepressant resistance. *Front Pharmacol.* 2013;4:146.
5. Biomarkers Definitions Working Group. Biomarkers and surrogate endpoints: preferred definitions and conceptual framework. *Clin Pharmacol Ther.* 2001;69(3):89–95.
6. FDA-NIH. *BEST (Biomarkers, EndpointS, and other Tools) Resource.* Maryland. 2016.
7. Lin E, Lane HY. Genome-wide association studies in pharmacogenomics of antidepressants. *Pharmacogenomics.* 2015;16(5):555–566.
8. McGowan PO, Sasaki A, D'Alessio AC, et al. Epigenetic regulation of the glucocorticoid receptor in human brain associates with childhood abuse. *Nat Neurosci.* 2009;12(3):342–348.
9. Weaver IC, Cervoni N, Champagne FA, et al. Epigenetic programming by maternal behavior. *Nat Neurosci.* 2004;7(8):847–854.
10. Tsankova N, Renthal W, Kumar A, Nestler EJ. Epigenetic regulation in psychiatric disorders. *Nat Rev Neurosci.* 2007;8(5):355–367.
11. Wahid F, Shehzad A, Khan T, Kim YY. MicroRNAs: synthesis, mechanism, function, and recent clinical trials. *Biochim Biophys Acta.* 2010;1803(11):1231–1243.
12. Issler O, Chen A. Determining the role of microRNAs in psychiatric disorders. *Nat Rev Neurosci.* 2015;16(4):201–212.
13. Valadi H, Ekstrom K, Bossios A, Sjostrand M, Lee JJ, Lotvall JO. Exosome-mediated transfer of mRNAs and microRNAs is a novel mechanism of genetic exchange between cells. *Nat Cell Biol.* 2007;9(6):654–659.
14. Shannon CP, Yang CX, Tebbutt SJ. A bloody primer: analysis of RNA-seq from tissue admixtures. *Methods Mol Biol.* 2018;1712:175–201.
15. Mastrokolias A, den Dunnen JT, van Ommen GB, 't Hoen PA, van Roon-Mom WM. Increased sensitivity of next generation sequencing-based expression profiling after globin reduction in human blood RNA. *BMC Genomics.* 2012;13:28.
16. Creemers EE, Tijsen AJ, Pinto YM. Circulating microRNAs: novel biomarkers and extracellular communicators in cardiovascular disease? *Circ Res.* 2012;110(3):483–495.
17. Margolin G, Petrykowska HM, Jameel N, Bell DW, Young AC, Elnitski L. Robust detection of DNA hypermethylation of ZNF154 as a pan-cancer locus with in silico modeling for blood-based diagnostic development. *J Mol Diagn.* 2016;18(2):283–298.
18. Warton K, Samimi G. Methylation of cell-free circulating DNA in the diagnosis of cancer. *Front Mol Biosci.* 2015;2:13.

19. Zhou X, Wen W, Zhu J, et al. A six-microRNA signature in plasma was identified as a potential biomarker in diagnosis of esophageal squamous cell carcinoma. *Oncotarget.* 2017;8:34468–34480.
20. Qu L, Li L, Zheng X, et al. Circulating plasma microRNAs as potential markers to identify EGFR mutation status and to monitor epidermal growth factor receptor-tyrosine kinase inhibitor treatment in patients with advanced non-small cell lung cancer. *Oncotarget.* 2017;8:45807–45824.
21. Dejima H, Iinuma H, Kanaoka R, Matsutani N, Kawamura M. Exosomal microRNA in plasma as a non-invasive biomarker for the recurrence of non-small cell lung cancer. *Oncol Lett.* 2017;13(3):1256–1263.
22. Zhang K, Lin G, Han Y, Xie J, Li J. Circulating unmethylated insulin DNA as a potential non-invasive biomarker of beta cell death in type 1 diabetes: a review and future prospect. *Clin Epigenetics.* 2017;9:44.
23. Lehmann-Werman R, Neiman D, Zemmour H, et al. Identification of tissue-specific cell death using methylation patterns of circulating DNA. *Proc Natl Acad Sci USA.* 2016;113(13):E1826–E1834.
24. Gogenur M, Burcharth J, Gogenur I. The role of total cell-free DNA in predicting outcomes among trauma patients in the intensive care unit: a systematic review. *Crit Care.* 2017;21(1):14.
25. Holdenrieder S, Stieber P. Clinical use of circulating nucleosomes. *Crit Rev Clin Lab Sci.* 2009;46(1):1–24.
26. Lopez JP, Diallo A, Cruceanu C, et al. Biomarker discovery: quantification of microRNAs and other small non-coding RNAs using next generation sequencing. *BMC Med Genomics.* 2015;8:35.
27. Maze I, Shen L, Zhang B, et al. Analytical tools and current challenges in the modern era of neuroepigenomics. *Nat Neurosci.* 2014;17(11):1476–1490.
28. Fiori LM, Lopez JP, Richard-Devantoy S, et al. Investigation of miR-1202, miR-135a, and miR-16 in major depressive disorder and antidepressant response. *Int J Neuropsychopharmacol.* 2017;20:619–623.
29. Lopez JP, Lim R, Cruceanu C, et al. miR-1202 is a primate-specific and brain-enriched microRNA involved in major depression and antidepressant treatment. *Nat Med.* 2014;20(7):764–768.
30. Lopez JP, Pereira F, Richard-Devantoy S, et al. Co-variation of peripheral levels of miR-1202 and brain activity and connectivity during antidepressant treatment. *Neuropsychopharmacology.* 2017;42:2043–2051.
31. Lopez JP, Fiori LM, Cruceanu C, et al. MicroRNAs 146a/b-5 and 425-3p and 24-3p are markers of antidepressant response and regulate MAPK/Wnt-system genes. *Nat Commun.* 2017;8:15497.
32. He S, Liu X, Jiang K, et al. Alterations of microRNA-124 expression in peripheral blood mononuclear cells in pre- and post-treatment patients with major depressive disorder. *J Psychiatr Res.* 2016;78:65–71.
33. Issler O, Haramati S, Paul ED, et al. MicroRNA 135 is essential for chronic stress resiliency, antidepressant efficacy, and intact serotonergic activity. *Neuron.* 2014;83(2):344–360.
34. Belzeaux R, Bergon A, Jeanjean V, et al. Responder and nonresponder patients exhibit different peripheral transcriptional signatures during major depressive episode. *Transl Psychiatry.* 2012;2:e185.
35. Smalheiser NR, Lugli G, Rizavi HS, Torvik VI, Turecki G, Dwivedi Y. MicroRNA expression is down-regulated and reorganized in prefrontal cortex of depressed suicide subjects. *PLoS One.* 2012;7(3):e33201.
36. Maffioletti E, Cattaneo A, Rosso G, et al. Peripheral whole blood microRNA alterations in major depression and bipolar disorder. *J Affect Disord.* 2016;200:250–258.

37. Lopez JP, Fiori LM, Gross JA, et al. Regulatory role of miRNAs in polyamine gene expression in the prefrontal cortex of depressed suicide completers. *Int J Neuropsychopharmacol.* 2014;17(1):23–32.
38. Smalheiser NR, Lugli G, Zhang H, Rizavi H, Cook EH, Dwivedi Y. Expression of microRNAs and other small RNAs in prefrontal cortex in schizophrenia, bipolar disorder and depressed subjects. *PLoS One.* 2014;9(1):e86469.
39. Fan HM, Sun XY, Guo W, et al. Differential expression of microRNA in peripheral blood mononuclear cells as specific biomarker for major depressive disorder patients. *J Psychiatr Res.* 2014;59:45–52.
40. Baudry A, Mouillet-Richard S, Schneider B, Launay JM, Kellermann O. miR-16 targets the serotonin transporter: a new facet for adaptive responses to antidepressants. *Science.* 2010;329(5998):1537–1541.
41. Song MF, Dong JZ, Wang YW, et al. CSF miR-16 is decreased in major depression patients and its neutralization in rats induces depression-like behaviors via a serotonin transmitter system. *J Affect Disord.* 2015;178:25–31.
42. Honda M, Kuwano Y, Katsuura-Kamano S, et al. Chronic academic stress increases a group of microRNAs in peripheral blood. *PLoS One.* 2013;8(10):e75960.
43. Bai M, Zhu X, Zhang Y, et al. Abnormal hippocampal BDNF and miR-16 expression is associated with depression-like behaviors induced by stress during early life. *PLoS One.* 2012;7(10):e46921.
44. Zurawek D, Kusmider M, Faron-Gorecka A, et al. Time-dependent miR-16 serum fluctuations together with reciprocal changes in the expression level of miR-16 in mesocortical circuit contribute to stress resilient phenotype in chronic mild stress—an animal model of depression. *Eur Neuropsychopharmacol.* 2016;26(1):23–36.
45. Manniromi C, Camon J, De Vito F, et al. Acute stress alters amygdala microRNA miR-135a and miR-124 expression: inferences for corticosteroid dependent stress response. *PLoS One.* 2013;8(9):e73385.
46. Zurawek D, Kusmider M, Faron-Gorecka A, et al. Reciprocal microRNA expression in mesocortical circuit and its interplay with serotonin transporter define resilient rats in the chronic mild stress. *Mol Neurobiol.* 2016;54:5741–5751.
47. He M, Yan H, Duan ZX, et al. Genetic distribution and association analysis of DRD2 gene polymorphisms with major depressive disorder in the Chinese Han population. *Int J Clin Exp Pathol.* 2013;6(6):1142–1149.
48. Dunlop BW, Nemeroff CB. The role of dopamine in the pathophysiology of depression. *Arch Gen Psychiatry.* 2007;64(3):327–337.
49. Zhang Y, Wang Y, Wang L, Bai M, Zhang X, Zhu X. Dopamine receptor D2 and associated microRNAs are involved in stress susceptibility and resistance to escitalopram treatment. *Int J Neuropsychopharmacol.* 2015;18(8):pyv025.
50. Gray AL, Hyde TM, Deep-Soboslay A, Kleinman JE, Sodhi MS. Sex differences in glutamate receptor gene expression in major depression and suicide. *Mol Psychiatry.* 2015;20(9):1057–1068.
51. Cruceanu C, Lopez JP, Tsai WT, Turecki G. Dysregulation of the glutamatergic receptors after antidepressant treatment in human neural progenitor cells. *Mol Psychiatry.* 2017;22(9):1228–1229.
52. Pilc A, Chaki S, Nowak G, Witkin JM. Mood disorders: regulation by metabotropic glutamate receptors. *Biochem Pharmacol.* 2008;75(5):997–1006.
53. Davis MJ, Iancu OD, Acher FC, et al. Role of mGluR4 in acquisition of fear learning and memory. *Neuropharmacology.* 2013;66:365–372.
54. Dadkhah T, Rahimi-Aliabadi S, Jamshidi J, et al. A genetic variant in miRNA binding site of glutamate receptor 4, metabotropic (GRM4) is associated with increased risk of major depressive disorder. *J Affect Disord.* 2017;208:218–222.

55. Li J, Meng H, Cao W, Qiu T. MiR-335 is involved in major depression disorder and antidepressant treatment through targeting GRM4. *Neurosci Lett.* 2015;606:167–172.
56. Suderman M, Borghol N, Pappas JJ, et al. Childhood abuse is associated with methylation of multiple loci in adult DNA. *BMC Med Genomics.* 2014;7(1):13.
57. Labonté B, Suderman M, Maussion G, et al. Genome-wide epigenetic regulation by early-life trauma. *Arch Gen Psychiatry.* 2012;69(7):722–731.
58. Booij L, Szyf M, Carballedo A, et al. DNA methylation of the serotonin transporter gene in peripheral cells and stress-related changes in hippocampal volume: a study in depressed patients and healthy controls. *PLos One.* 2015;10(3):e0119061.
59. Domschke K, Tidow N, Schwarte K, et al. Serotonin transporter gene hypomethylation predicts impaired antidepressant treatment response. *Int J Neuropsychopharmacol.* 2014;17(8):1167–1176.
60. Tadic A, Muller-Engling L, Schlicht KF, et al. Methylation of the promoter of brain-derived neurotrophic factor exon IV and antidepressant response in major depression. *Mol Psychiatry.* 2014;19(3):281–283.
61. D'Addario C, Dell'Osso B, Galimberti D, et al. Epigenetic modulation of BDNF gene in patients with major depressive disorder. *Biol Psychiatry.* 2013;73(2):e6–e7.
62. Lopez JP, Mamdani F, Beaulieu MM, et al. Epigenetic regulation of BDNF expression according to antidepressant response. *Mol Psychiatry.* 2013;18(4):398–399.
63. Belzeaux R, Formisano-Treziny C, Loundou A, et al. Clinical variations modulate patterns of gene expression and define blood biomarkers in major depression. *J Psychiatr Res.* 2010;44(16):1205–1213.

CHAPTER TWELVE

Transgenerational Epigenetics of Traumatic Stress

Ali Jawaid[2], Martin Roszkowski[2], Isabelle M. Mansuy[1]
Laboratory of Neuroepigenetics, Medical Faculty of the University of Zurich and Department of Health Science and Technology of the Swiss Federal Institute of Technology, Neuroscience Center Zurich, Zurich, Switzerland
[1]Corresponding author: e-mail address: mansuy@hifo.uzh.ch

Contents

Abstract

Traumatic stress is a type of environmental experience that can modify behavior, cognition and physiological functions such as metabolism, in mammals. Many of the effects of traumatic stress can be transmitted to subsequent generations even when individuals from these generations are not exposed to any traumatic stressor. This book chapter discusses the concept of epigenetic/non-genomic inheritance of such traits involving the germline in mammals. It includes a comprehensive review of animal and human studies on inter- and transgenerational inheritance of the effects of traumatic stress, some

[2] Equal contribution.

Progress in Molecular Biology and Translational Science, Volume 158
ISSN 1877-1173
https://doi.org/10.1016/bs.pmbts.2018.03.003

of the epigenetic changes in the germline currently known to be associated with traumatic stress, and possible mechanisms for their induction and maintenance during development and adulthood. We also describe some experimental interventions that attempted to prevent the transmission of these effects, and consider the evolutionary importance of transgenerational inheritance and future outlook of the field.

1. INTRODUCTION

Life experiences can have a long-lasting impact on cognition and behavior of an individual. Cognitive and behavioral effects usually involve dynamic changes in gene expression and cellular signaling pathways in the brain. These molecular changes are complex and implicate epigenetic or non-genetic mechanisms such as DNA methylation, histone post-translational modifications (HPTMs) and non-coding RNAs (ncRNAs). Emerging evidence suggests that environment-induced epigenetic changes can persist for a long time and some of these changes can also be transmitted to subsequent generations in the absence of further exposure to the initial trigger (reviewed in Ref. 1).

Traumatic stress is a type of life experience that has transgenerational effects in mammals. Traumatic experiences, both in early and adult life, are major risk factors for behavioral dysfunctions and mental disorders. Importantly, there is increasing evidence for inter- and transgenerational cycles of behavioral adaptation in response to traumatic stress, which are largely mediated by epigenetic mechanisms. In most scenarios, stressful experiences negatively affect behaviors across generations and constitute heritable risk factors for neuropsychiatric disorders (reviewed in Ref. 2). For example, early life traumatic stress leads to a multitude of behavioral abnormalities including depressive-like behaviors across successive offspring in mice.[3] However, there is also evidence for transmission of adaptative behaviors expressed as behavioral flexibility and improved goal-directed responses, and for the involvement of some brain pathways in such adaptations in the offspring of traumatized mice.[4]

This chapter discusses the epigenetic mechanisms important for brain functions in the context of susceptibility to traumatic stress and its consequences. Current evidence for inter- and transgenerational inheritance of the effects of traumatic stress in animals and humans is reviewed and some of the potential underlying molecular mechanisms described. The evolutionary importance of transgenerational epigenetic inheritance, its potential clinical implications, and future outlook are also briefly discussed.

1.1 Traumatic Stress

In a medical or biological context, stress is a physical, mental or emotional factor that can be acute or chronic, and causes body and/or mental tension. Stress can be induced in response to external (environmental or psychological) or internal (physical illness or injury) stimuli. It typically leads to a "fight or flight" response, which involves broad activation of neurological and endocrinological processes. Acute stress is a strong psychological response to a novel and unpredictable threatening event. When exposure to such events is repetitive and strong enough to induce a persistent response, generally over a month in humans and more than a week in mice, it becomes chronic (reviewed in Refs. 5,6). Traumatic stress is a particular type of stress involving an acute or chronic reaction to shocking and emotionally overwhelming situations generally involving a threat to physical or personal integrity. Clinically, in humans, it exists in the form of acute stress disorder (ASD), which can be classified as post-traumatic stress disorder (PTSD) when persisting for more than a month. Both ASD and PTSD are characterized by intrusion symptoms involving recurrent memories, dreams, or flash backs of the traumatic event, dissociative symptoms including traumatic amnesia, avoidance symptoms, hyper-arousal and negative mood according to Diagnostic and Statistical Manual of Mental Disorders 5th edition (DSM V).[7] An important trigger of traumatic stress is early childhood trauma, which involves traumatic events in childhood, such as physical or sexual abuse, deprived parental care, natural disasters, or forced displacement. Similarly, experiencing or witnessing traumatic events during adulthood, such as military combat, physical or sexual abuse, disease, physical injury, etc. can lead to chronic traumatic stress in many individuals (reviewed in Ref. 8).

Long-term traumatic stress can lead to marked changes in brain functions, in particular mental health. Human studies suggest that PTSD is often comorbid with depression and anxiety[9] and is associated with a greater prevalence and incidence of dementia.[10] This could be partially attributed to a decrease in hippocampal volume observed in individuals with PTSD.[11] Beyond the brain, traumatic stress over prolonged periods also leads to harmful effects on other body functions. PTSD is a risk factor for cardiovascular and cerebrovascular diseases, gastrointestinal dysfunctions, rheumatoid arthritis, and cancer.[12]

At the molecular level, traumatic stress alters the integrity of the hypothalamic–pituitary–adrenal (HPA) axis. An important endocrinological circuit in HPA axis involves the release of corticotropin-releasing hormone (CRH) from the hypothalamus, which stimulates the release of adrenocorticotropin hormone (ACTH) from the pituitary. ACTH, in turn, induces the release of cortisol (CORT) from adrenal glands. Importantly, cortisol inhibits the release of ACTH as a physiological negative feedback, whereas ACTH has similar negative feedback on the release of CRH (reviewed in Ref. 8). PTSD patients have gross alterations in the HPA axis, including sustained release of CRH, blunted negative feedback of ACTH on CRH release, and hypercortisolism.[13] In contrast, other studies suggest a role for hypocortisolism in the development of PTSD and other stress-related disorders induced by traumatic stress.[14,15] This is combined with altered responses in hippocampus and amygdala, brain regions providing critical input to the HPA axis.[16]

Importantly, the effects of traumatic stress on the brain are pleiotropic and mediated by complex signaling pathways besides HPA axis. Notably, traumatic stress affects the expression and functioning of mineralocorticoids, which are implicated in the regulation of basal HPA axis activity, traumatic memories, and stress resilience (reviewed in Ref. 17). Traumatic stress can also dysregulate the balance between oxytocin and arginine vasopressin (avp), which have a critical role in the regulation of social and emotional responses, ranging from parenting to partner preference and stress-reactivity (reviewed in Ref. 18). Similarly, noradrenergic system and catecholamines are thought to underlie the development of PTSD after trauma exposure (reviewed in Ref. 19), and serotonergic signaling in the susceptibility to and long-term effects of traumatic stress.[20,21]

The effects of traumatic stress on the brain are more profound during certain periods of life. The brain is particularly sensitive to environmental stimuli during prenatal, early postnatal and pubertal periods due to higher plasticity of the brain, increased neurogenesis, relative immaturity of the stress-responsive circuitry and neuroendocrine dynamics.[22,23] Animal studies have reflected that there are sensitive periods for transgenerational transmission of environmental effects as well.[24] In the next section, we briefly review the epigenetic mechanisms that are relevant to the effects of traumatic stress on the brain and that are of potential importance in the transgenerational epigenetic inheritance of the effects of traumatic stress.

1.2 Epigenetic Mechanisms Relevant to the Effects of Traumatic Stress on the Brain

1.2.1 DNA Methylation

DNA methylation, which in mammals most commonly occurs at cytosine-guanine dinucleotides (CpG), has traditionally been viewed as a fairly stable epigenetic mark mostly responsible for gene silencing (reviewed in

Refs. 25,26). Silencing takes place by direct inhibition of the binding of transcription factors, or by recruitment of methyl-CpG binding proteins (MBPs) and associated repressive chromatin-remodeling components.[27] However, CpG methylation is also present at the promoter and coding region of actively transcribed genes, where it can associate with transcriptional activity.[28] Recent evidence has shown that in many cells including post-mitotic cells like adult neurons, DNA methylation is dynamically regulated and CpGs can be actively methylated and demethylated.[29,30] DNA demethylation involves a succession of biochemical steps implicating several enzymes. During this process, 5-methylcytosine (5-mC) is successively oxidized into 5-hydroxymethylcytosine (5-hmC), 5-formylcytosine (5-fC) and 5-carboxylcytosine (5-cAC) by ten-eleven translocation (TET) proteins. 5-fC and 5-cAC can ultimately be removed and repaired by terminal deoxytidyl transferases (tdTs) generating an unmodified cytosine.[31] In contrast to 5-fC and 5-cAC, 5-hmC is enriched in the adult brain, especially in the hippocampus, cortex and cerebellum.[32–34] The high abundance of 5-hmC in the brain and its stability[35] suggest that this modification is not just a transient by-product of 5-mC metabolism but also serves as an epigenetic mark thought to play an important role in brain processes.

Importantly, traumatic stress leads to long-term changes in global, as well as, locus-specific DNA methylation (reviewed in Ref. 36) making it a candidate epigenetic mechanism underlying long-term effects of traumatic stress. Mice subjected to an acute stress paradigm involving forced swim in cold water have altered methylation of several immediate early genes in the dentate gyrus, which modulate the subsequent behavioral response.[37] Further, DNA methylation is a critical regulator of fear memory, pointing to its importance in long-term consequences of traumatic stress in the brain.[38]

1.2.2 Histone Modifications

Covalent post-translational modifications of histone proteins are other important epigenetic marks that primarily involve acetylation, methylation, phosphorylation, ubiquitination, and SUMOylation (reviewed in Ref. 39). These modifications control the accessibility of the chromatin to the transcriptional machinery, in a sequence- and activity-dependent manner.[40] Generally, acetylation and phosphorylation, occurring on lysine (K) and serine (S), threonine (T) or tyrosine (Y) residues, respectively, are associated with transcriptional activation.[39] In contrast, histone methylation on lysine residues is associated with both actively transcribed and silenced genes.[40–42] Similarly, histone ubiquitination is associated with both transcriptional silencing and activation depending on whether it occurs on

H2A or H2B, respectively.[43] Finally, histone SUMOylation, which requires addition of small-ubiquitin like modifier (SUMO) to histones, negatively regulates gene transcription in the brain, often in association with other epigenetic regulators.[44,45]

Importantly, several HPTMs are affected by both acute and chronic stress.[46,47] Key stress-induced cascades, such as corticosterone and glutamate signaling may contribute to the regulation of HPTMs to modulate gene expression and behavior.[48] Pre-clinical studies suggesting the use of histone deacetylase inhibitors in trauma therapy and anxiety treatment provide further support for a role of histone regulation during and after traumatic stress (reviewed in Ref. 49).

1.2.3 Non-coding RNAs

Epigenetic regulation of gene expression can also be achieved by ncRNAs like microRNAs (miRNAs). MiRNAs are 20–22 nucleotides RNAs that act by degradation of mRNA targets or translational repression (reviewed in Ref. 50). MiRNA biogenesis involves a succession of cellular steps regulated by the RNAse Drosha and its partner DeGeorge syndrome critical region 8 in the nucleus, followed by further processing by another RNAse Dicer and its partner TAR RNA-binding protein in the cytoplasm. Once formed, miRNAs associate with Argonaute proteins in the cytoplasm to form the RNA-induced silencing complex (RISC). RNA-binding proteins can additionally modulate the efficiency of the microprocessor, Dicer or RISC (reviewed in Ref. 50). Many miRNAs and components of miRNA biogenesis machinery are highly expressed in the brain, mostly in a region-specific manner. Importantly, miRNAs have been implicated in the susceptibility of the brain to trauma and in the effects of trauma exposure.[51,52] A number of studies on PTSD patients highlight a potential involvement of miRNAs as mediators and/or moderators of the effects of traumatic stress (reviewed in Ref. 53).

While other ncRNAs such as lncRNA and piRNAs have a demonstrated role in various molecular pathways, their contribution to the effects of traumatic stress has been less well studied. Similar to miRNAs, lncRNAs are widely abundant in the brain.[54] and in some cases, they share the miRNA biogenesis machinery.[55] LncRNAs can interfere and compete with the functions of miRNAs. They can also influence mRNA stability and splicing, and transcriptional regulation (reviewed in Ref. 56). The lncRNA Gomafu was recently shown to be down regulated after fear conditioning, providing a lncRNA candidate possibly involved in traumatic stress.[57]

piRNAs are mainly known for their role in genomic defense against mobile elements in the germ line.[58] However, they may have a similar role in the CNS by restricting the activity of mobile elements in the developing and mature brain.[59] Novel insight in the role of piRNAs in the epigenetic regulation of memory also points to their potential functions in traumatic stress.[60]

1.2.4 RNA Modifications

Most RNAs carry a large variety of chemical modifications such as 1-methyladenosine (m^1A), 1-methylguanosine (m^1G), N^6-methyladenosine (m^6A), 5-methylcytidine (m^5C) or pseudouridine (Ψ) methylation.[61,62] Although the role of such modifications in RNA is less well understood than in DNA, their main effect is to alter the binding affinity for the complementary strand and modify the structure of the folded RNA.[63] This in turn can drastically alter the binding affinity of RNA to other RNAs, and to RNA-binding proteins, thus influencing RNA stability and sub-cellular localization.[64,65] Currently, little is known about the potential role of the epitranscriptome in traumatic stress. Recent findings showed that the RNA demethylase FTO is affected by contextual fear conditioning.[66] It is therefore likely that the epitranscriptome contributes to the epigenetic effects of traumatic stress.

Most epigenetic mechanisms have so far been studied individually, but a combination of different epigenetic mechanisms is likely responsible for the effects of traumatic stress at the molecular level. Thus, it is important to take into account the complex interplay of the combined cellular machinery that shapes and regulates the epigenome. Although there is currently no study providing such integrated view of the epigenome and epitranscriptome after traumatic stress, developmental studies have examined HPTMs in the context of DNA methylation and showed that they can guide de novo methylation.[67]

2. TRANSGENERATIONAL INHERITANCE OF TRAUMATIC STRESS

A growing body of evidence suggests that the effects of traumatic stress can be transmitted to subsequent generations, even when these generations are not exposed to the initial traumatic event. When these effects are transferred from parents to the direct offspring, it is termed intergenerational inheritance while transmission to at least two generations of offspring is transgenerational.[2] In this chapter, we will review evidence for inter- and transgenerational effects of traumatic stress, which are both important although transgenerational inheritance is more relevant from an evolutionary point of view.[68] The chapter will focus on the inheritance of the effects of postnatal trauma and their transmission to subsequent generations.[1] The effects of prenatal stress will be discussed only briefly as evidence is still sparse and the mediating mechanisms have been less well studied.[69,70]

Effects of postnatal traumatic stress can be transmitted to the offspring independently of the germline, and in this case, involve social transmission and acquisition of behavioral or physiological patterns from parent to offspring. True inheritance of the effects of traumatic stress involves transmission through germ cells and implicates epigenetic mechanisms.[1] In rodents, traumatic stress has been shown to affect behavior, cognitive functioning, and metabolism of subsequent generations through complex and not fully understood epigenetic mechanisms. Similar effects exist in humans, although mechanistic explanations are even more limited. This section summarizes pertinent evidence in support of inter- and transgenerational inheritance of traumatic stress in animals and humans, and results are summarized in Tables 1 and 2.

2.1 Animal Models

Recent animal studies have reported inter- and transgenerational effects of traumatic stress using rodents and other species. Observations in birds and reptiles are interesting to consider because these animals are precocial (relatively mature at birth), unlike rodents which are relatively immature.[71] Animal studies on transgenerational inheritance typically use the nomenclature adapted from studies of genetic inheritance in which the offspring produced from the first genetic cross is labeled as F1 (filial 1) and subsequent generations involving breeding of affected F1 to naïve breeders or affected F1 are F2, F3, etc. (Fig. 1).

2.1.1 Rodent Studies

Rodent studies on inter- and transgenerational effects of traumatic stress have provided important mechanistic insight into the way these effects can be transmitted across generations. These models have used different paradigms including interrupted or impaired maternal care, maternal separation or fear conditioning.[8] Babb et al. used chronic social stress (CSS) during lactation to impair maternal care in a cohort of mice (F0). The deprived male offspring (F1) was then mated with naïve females to produce an F2 generation, which had decreased social behaviors during puberty and adulthood.

Table 1 Studies on Inter- and Transgenerational Inheritance of the Effects of Traumatic Stress in Animals

Species	Acute or Chronic Traumatic Stress	Early Life or Adult Stress	Main Effect(s)	Mechanism of Inter-/Transgenerational Inheritance	Generations Affected	First Author, Year
Domesticated chickens (*Gallus gallus*)	Chronic social isolation	Early	Altered gene expression in the offspring brain	Not tested	Two	Goerlich, 2012[81]
	Chronic food frustration, physical restrain, social isolation	Early, puberty, adult	Increased CORT, higher body weight in the offspring	Not tested	Two	Erricson, 2016[82]
Mice (*Mus musculus*)	Chronic (impaired maternal care)	Early	Decreased social behaviors, and alterations in cortisol, and oxytocin in the offspring	Not tested	Two	Babb, 2014[72]
	Chronic (impaired maternal care)	Early	Pro-depressive behaviors, and altered mTOR-Akt signaling in the offspring	Not tested	Two	Wu, 2016[73]
	Chronic (impaired maternal care)	Early	Pro-depressive behaviors, decreased social anxiety, and decreased social recognition memory in the offspring	Not tested	Three	Franklin, 2010[3]
	Chronic (impaired maternal care)	Early	Pro-depressive behaviors, and altered glucose metabolism in the offspring	Sperm RNA as vector of transgenerational inheritance	Three	Gapp, 2014[74]

Continued

Table 1 Studies on Inter- and Transgenerational Inheritance of the Effects of Traumatic Stress in Animals—cont'd

Species	Acute or Chronic Traumatic Stress	Early Life or Adult Stress	Main Effect(s)	Mechanism of Inter-/Transgenerational Inheritance	Generations Affected	First Author, Year
	Chronic (impaired maternal care)	Early	Impaired spatial memory, and synaptic plasticity in the offspring	Not tested	Two	Bohacek, 2015[75]
	Chronic (impaired maternal care)	Early	Improved goal-directed behavior in the offspring	Changes in sperm MR	Two	Gapp, 2014[4]
	Chronic (impaired maternal care)	Early	Changes in serotonergic circuitry in the brain of the offspring	Not tested	Two	Razoux, 2017[76]
	Chronic (impaired maternal care)	Early	Brain metabolic alterations in the offspring	Not tested	Two	Gapp, 2017[125]
	Chronic social defeat	Adult	Pro-depressive behaviors in the offspring	Not tested (however germline transmission of effect was confirmed by in vitro fertilization)	Three	Saavedra-Rodriquez, 2013[77]
	Acute odor fear conditioning	Adult	Behavioral sensitivity to odor in the offspring	Olfactory pathways, germline transmission of effect was confirmed by in vitro fertilization	Three	Dias, 2013[79]

Table 2 Studies on Intergenerational Inheritance of the Effects of Traumatic Stress in Humans

Acute or Chronic Traumatic Stress	Early or Late Life Stress	Main Effect(s)	First Author, Year
Chronic (low maternal bonding)	Early	Increased risk of BPD in the offspring	Infurna, 2016[84]
Chronic (paternal death)	Early	Prematurity and low birth weight in the offspring	Vàgerö, 2017[85]
Chronic (holocaust survivors)		Increased psychopathology, increased susceptibility to PTSD, and low cortisol levels in the offspring	Yehuda, 1998[88] Yehuda, 2000[90] Yehuda, 2008[89]
Chronic (genocide exposure in Cambodia)		Increased anxiety in the offspring	Field, 2013[91]
Chronic (genocide exposure in Rwanda)		Increased anxiety and depression in the offspring	Rieder, 2013[92]
Chronic (combat exposure during Vietnam war)		Increased hostility in the offspring	Glenn, 2002[97]
Chronic (combat exposure during Bosnia–Serbia war)		Increased emotional and developmental problems in the offspring Increased social and behavioral problems in the offspring	Klaric, 2008[98] Ćorić, 2016[99]
Chronic (evacuation during the World War II in Finland)		Increase psychiatric hospitalizations in women whose mothers were evacuated	Santavirta, 2018[100]
Chronic (immigrants exposed to torture and forced displacement)		Increased depressive and somatization symptoms in the offspring	Daud, 2005[95]

The behavioral anomalies were accompanied by alterations in corticosterone and oxytocin in both F2 females and males and prolactin in F2 females only.[72] Similarly, a study on mice with impaired maternal care (F1) due to postpartum depression induced by chronic stress exposure (daily restraint and night time illumination) in mothers (F0) prior to gestation showed

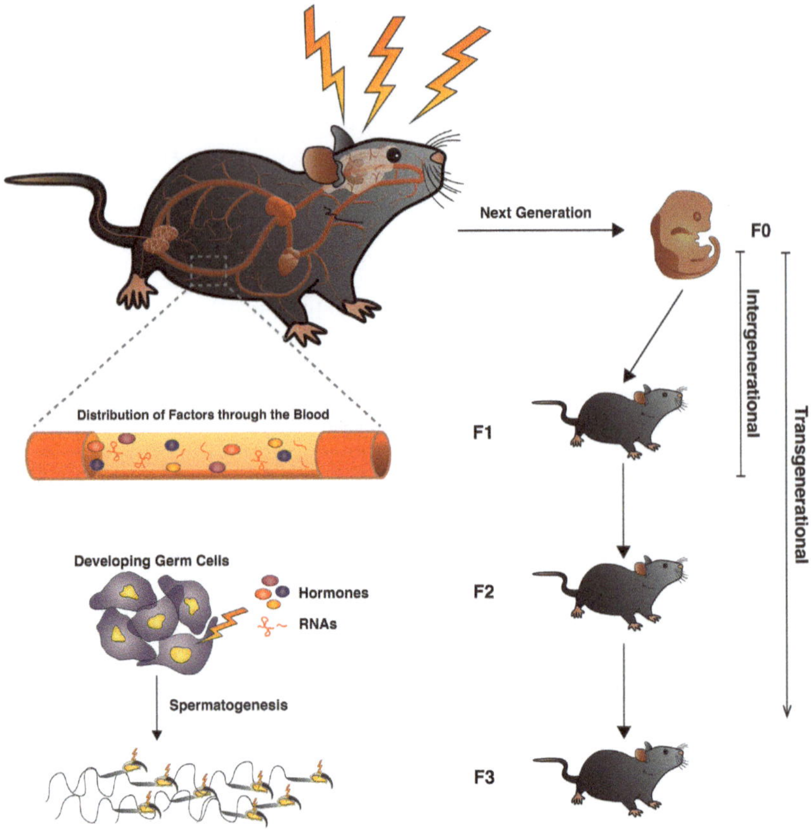

Fig. 1 Potential mechanisms of induction and persistence of transgenerational effects of traumatic stress. Traumatic stress may alter circulating factors (hormones, cytokines, and/or circulating ncRNAs) in trauma-exposed animals. These changes have the potential to affect the epigenome of the developing sperm. This could be either through direct uptake of circulating ncRNAs by developing gametes, or through the effect of circulating cytokines and hormones on enzymes regulating DNA methylation, HPTMs or ncRNAs. These changes may serve as vectors of transmission of the effects of traumatic stress to the next generation. Similar or related mechanisms may then be responsible for persistence of transgenerational phenotypes across generations.

transgenerational inheritance of depressive behaviors and altered mTOR-Akt signaling in the F2 offspring.[73]

Unpredictable maternal separation combined with unpredictable maternal stress (MSUS) is a solid model of transgenerational inheritance in mice, and one of the rare to affect several successive generations. This model uses a combination of daily unpredictable separation of newborn pups (F1) from their mother (F0), with unpredictable stress exposure of mothers

(F0) during separation.[3] These repetitive and unpredictable traumatic experiences during postnatal day 1–14 alter behavior of the offspring in up to three generations. F2 and F3 MSUS mice show depressive-like behaviors, which are not due to any alteration in maternal care as they persist after cross fostering.[3] Further, F2 and F3 mice have altered behavioral response to aversive environment, more risk-taking behaviors and impaired social recognition memory.[3,74] Hippocampus-dependent spatial memory and synaptic plasticity are also affected in F1 and F2 animals and molecular pathways related to learning and memory are altered.[75] Further, both F1 and F2 MSUS mice have disrupted functional connectivity of brain serotonergic pathways, including limbic circuits implicated in mood disorders.[76] Interestingly, MSUS also confers some adaptive advantages to the offspring, especially increased behavioral flexibility and improved goal-directed behaviors.[4] The transgenerational effects of MSUS are not only limited to the brain, but also affect glucose and insulin metabolism. MSUS F2 generation have a faster glucose clearance on the glucose tolerance test despite lower level of fasting insulin. They also weigh less than controls despite a higher dietary intake.[74]

Besides early life adversity, traumatic stressors during postnatal or adult life also have transmissible behavioral effects. Chronic stress involving social instability and disrupted social hierarchy from postnatal day 27 to 76 enhances anxiety and induces social deficits in mice in up to three generations.[77] Chronic social defeat can also induce depressive behaviors in female and male offspring (F1) of stressed male mice (F0).[78] Some of these effects could be transmitted by in vitro fertilization (IVF), suggesting their dependence on germ cells. Odor fear conditioning in mice (F0) was reported to induce behavioral sensitivity to an odor in F2 and F3 offspring, and be accompanied by changes in the olfactory system in the brain of F2 and F3 that could be recapitulated by IVF.[79] These results overall suggest that the effects of traumatic stress can be passed to subsequent generations in rodents.

2.1.2 Non-rodent Studies

In domesticated chickens (*Gallus gallus*), gene expression and feeding behavior are altered in the offspring (F1) of animals exposed to unpredictable food access.[80] While the study did not focus on behavioral or molecular aspects of traumatic stress, it was the first to suggest intergenerational inheritance in response to stress in birds. Another study using chronic stress by intermittent social isolation for the first 3 months of life showed that male offspring (F1) of stressed chickens (F0) have altered expression of over

40 genes in the brain. However, there was no difference in fear-related behavior in the F1 offspring of stressed versus unstressed F0.[81] A more recent study from the same group exposed chickens (F0) to a stress paradigm involving food frustration, physical restraint and social isolation on alternating days for a total of 6 days during early life, early puberty or late puberty. Behavioral and hormonal assessment of the progeny (F1) showed a heightened corticosterone response to acute restraint stress in F1 offspring of chickens exposed to stress (F0) during puberty. Similarly, F1 offspring of F0 exposed to stress during late puberty and early life had higher body weight at hatching and at 1 month of age, respectively.[82] Finally, a recent study on eastern fence lizards (*Sceloporus undulatus*) showed that ancestral exposure to stress (in the form of predator invasion for many generations) showed a more robust corticosterone response to restraint and ACTH injection. These results indicate that ancestral stress exposure can reprogram the offspring HPA axis.[83]

2.2 Human Studies

Literature about the inheritance of the effects of traumatic stress in humans has slowly accumulated in the past decade (Table 2). However, it remains thin and studies in humans also generally lack clear "cause and effect" association, mechanistic explanations or germline assessment, which are admittedly less tractable in humans than in animals. Like in animals, parental care seems to be a strong determinant of transgenerational inheritance of the effects of traumatic stress in humans. A recent study in Germany showed that low maternal bonding in mothers and high level of paternal stress are associated with a clinical diagnosis of borderline personality disorder (BPD) in the offspring.[84] Similarly, based on a large Swedish population-based cohort, a study showed that parental death during "slow growth period" (time before the start of puberty, generally 9–12 years of age for boys) in boys leads to prematurity and low birth weight in their offspring.[85] Evidence from small-scale studies from the 1960s and 1970s, indicates signs of psychopathology in children of holocaust survivors.[86] More recently, trauma exposure in holocaust survivors was shown to induce neuropsychiatric changes in their children and be associated with alterations in some epigenetic marks, in particular DNA methylation in FK506 binding protein 5 gene.[87] The children of holocaust survivors had increased susceptibility to PTSD, primarily women survivors,[88,89] which could be related to low cortisol level.[90] Increased anxiety symptoms have also been reported in children of individuals who developed hyperarousal related to genocide exposure in Cambodia[91] and increased depression and anxiety in children of genocide exposed people in Rwanda.[92]

Intergenerational inheritance of behavioral dysfunctions has also been reported in families subjected to forced displacement or immigration. A study on adult descendants of individuals exposed to forced displacement with or without PTSD showed implicit avoidance of displacement-related stimuli in relation to PTSD.[93] The effects of traumatic stress have also been suggested to be transmitted to the offspring in Latin American immigrants[94] and increased depressive and somatization symptoms in the offspring of immigrants exposed to torture and forced displacement was reported in a Swedish study.[95] Emerging evidence further shows intergenerational inheritance in relation to combat trauma. The descendants of World War II or Vietnam war veterans exhibit various psychological impairments,[96] violence and hostility behaviors, which positively correlated with the intensity and duration of exposure to combat trauma in their parents.[97] Further, veterans of the Serbian–Bosnian conflict with PTSD have children with increased developmental, behavioral and emotional problems.[98,99] Finally, a recent study on offspring of Finnish evacuees during World War II showed a two times increased risk of psychiatric hospitalization in women whose mothers were evacuated compared to those whose mothers were not evacuated.[100]

In conclusion, although less studied from a mechanistic point of view, inter- and possibly transgenerational inheritance of the effects of traumatic stress is supported by empirical evidence in humans.

3. MECHANISMS OF INTER- AND TRANSGENERATIONAL INHERITANCE

Inheritance across generations can occur through both matrilineal and patrilineal lineages.[1] Mechanistically, evidence for transmission through males is characteristically dominant, and will therefore be described here. Four major questions are addressed (1) Which epigenetic mechanisms serve as vectors of inheritance in the context of traumatic stress, (2) How does traumatic stress induce these epigenetic changes in the germline, (3) How epigenetic changes induced in the germline are transmitted to the offspring, and (4) How do these effects persist across generations.

3.1 Epigenetic Mechanisms in the Germline

Like somatic cells, germ cells also have DNA methylation, HPTMs and ncRNAs. In sperm, DNA can be differentially methylated by environmental factors at different loci. Traumatic stress induced by MSUS in mice induces hypomethylation at promoter CpGs of corticotropin-releasing factor 2 and protein kinase C gamma, and hypermethylation at the promoter of methyl CpG-binding protein 2 (*Mecp2*) and *Nr3c2* (the gene encoding for mineralocorticoid receptor (MR)). Similar alterations in DNA methylation are also observed in the brain of the offspring.[3,75] Environmental stressors can also modulate HPTMs and related proteins such as Polycomb repressive complex 2 for instance in *D. melanogaster* and *C. elegans*.[101,102] In rodents, the contribution of histones to transgenerational inheritance is more limited since most histones (95%–99%) are replaced by protamines in sperm.[1] ncRNAs have also been implicated in the effects of traumatic stress across generations, in this case with some causal evidence. MSUS was shown to alter the ncRNA content in sperm, in particular miRNAs and piRNAs.[74,103,104] Injection of sperm RNAs from MSUS-exposed males into wild-type fertilized oocytes could recapitulate behavioral and metabolic impairments in the resulting mice and their progeny, providing strong evidence for a direct link between sperm RNA and transmission of symptoms.[74] It was suggested that nine miRNAs, shown to be upregulated in sperm by adult paternal stress may recapitulate corticosterone level alteration, but further effects have not been addressed, limiting the conclusions of this study.[103]

3.2 Induction of Epigenetic Mechanisms in the Germline by Traumatic Stress

While changes in the germline epigenome have been proposed as vectors of transgenerational inheritance of the effects of traumatic stress, how traumatic stress affects the germline epigenome remains unknown. Circulating factors may be involved, as they can access germ cells and other tissues like the brain and possibly mediate the effects of traumatic stress. Interestingly, just like the brain, germ cells are in a immunologically privileged tissue protected by the blood–testis barrier. To access germ cells in adult gonads, circulating factors, therefore, need to be lipid soluble and easily transportable such as hormones, cytokines or circulating RNAs (Fig. 1). These factors may be released by stress pathways and act via receptors like adrenergic receptors,[105] MR,[106] gonadotropin-releasing hormone receptor[107] or arginine vasopressin receptors,[108] which are present on sperm cells. Hormonal treatment has indeed been shown to change sperm ncRNA content.[109] Cytokines make other important circulating factors to consider as they can cross the blood–testis and blood–brain barrier, and are even implicated in the regulation of the barrier.[110] Cytokine receptors are also present on sperm.[107]

Circulating RNAs could also be carriers of the impact of traumatic stress to germ cells. In mammals, circulating RNAs are mainly transported in extracellular vesicles called exosomes. Exosomes enriched in ncRNAs are generated from parent cells, and can cross blood–testes or blood–brain barrier to affect cells behind the barrier. The developing sperm can acquire ncRNAs from prostasomes and epidydimosomes, two specialized types of exosomes, during maturation.[111] It is possible that the ncRNA content of such exosomes is altered, e.g., from distal somatic cells, and then alter the ncRNA content of sperm. Evidence from *C. elegans* shows that extracellular dsRNA can reach the developing embryo through vesicles, and cause gene silencing in the progeny. These results suggest a possible mechanism for the transmission of experience-dependent effects between generations.[112]

3.3 How Are These Effects Passed to the Offspring?

Another important question to address is how the effects of traumatic stress are passed from the germline to the brain of the offspring. One possibility is that changes induced by traumatic stress in germ cells are maintained during development and escape reprogramming to be passed to all three embryonic layers. Evidence in the MSUS model shows that similar epigenetic alterations affect the brain and sperm of trauma-exposed mice (F1) and the brain and sperm of their offspring (F2).[74,75] Alternately, epigenetic modifications in germ cells may in turn regulate the mechanisms controlling brain development. For instance, beta catenin, which regulates neuronal differentiation in embryonic stem cells,[113] is regulated by miR-375, a microRNA altered in the sperm of trauma-exposed mice.[74] Parallel investigations on germ cells of trauma-exposed animals, as well as, the developing embryo, postnatal offspring brain and adult offspring brain, are likely to provide valuable insight about the mechanisms underlying the persistence of the transgenerational effects of traumatic stress.

3.4 Persistence of Epigenetic Marks Related to Trauma Across Generations

It is currently not known how persistent epigenetic modifications induced by trauma are maintained in brain and sperm. However, it has been observed that while the offspring of traumatized individuals exhibit similar behavioral symptoms, some epigenetics changes are not present in this offspring, e.g., miRNAs in sperm. It is possible that initial changes directly induced by trauma are transferred to other epigenetic marks to be perpetuated, consistent with evidence of strong cross-talks between epigenetic mechanisms, e.g., miRNAs regulating DNA methyl transferases (DNMTs),[114] or MeCP2 regulating miRNAs biogenesis.[115] It could also be that epigenetic changes induced by traumatic stress in germ cells are somehow transformed into genomic changes in subsequent generations due to decreased genomic stability, and lead to copy number variations (CNVs). Exposure of the endocrine disruptor vinclozolin was shown to increase CNVs in the F3 generation,[116] suggesting that it could also be the case after traumatic stress. Future studies on transgenerational inheritance of traumatic stress may benefit from whole germline genome, epigenome, as well as, RNA sequencing to ascertain how epigenetic marks of traumatic stress persist across generations.

3.5 Potential of Anti-stress Therapies to Prevent Transmission of the Effects of Traumatic Stress

While traumatic stress may lead to inherited effects in the offspring, transmission may be prevented by behavioral paradigms or pharmacological manipulations of trauma-exposed animals. Exposure of fathers to environmental enrichment (EE) for instance, was shown to prevent the transmission of the effects of traumatic stress and to reverse epigenetic alterations, in particular DNA methylation alterations in the hippocampus, in the offspring.[117] In Ras-GRF knock out mice, 2 weeks of EE at postnatal 15 day EE could correct long-term potentiation (LTP) deficits and fear memory impairment, and even enhanced LTP in the offspring of control mice.[118] EE can also reverse increased anxiety-like behaviors and prevent against seizures in a genetic absence epilepsy rat from Strasbourg (GAERS) model of absence seizures, an effect transmitted across generation.[119] It is however important to note that while EE has shown potential to prevent the transmission of effects of traumatic stress or other behaviors, it has shown only limited effects on anxiety- and depression-related behaviors when used alone. A recent study involving parental EE failed to improve anxiety or depression-related behaviors and did not normalize serum corticosterone levels post-stress in the offspring.[104]

Not surprisingly, oxytocin (OXT) and vasopressin (AVP) have shown a potential to reverse the effects of CSS in mice. F2 female offspring of dams exposed to CSS have increased repetitive/perseverative and anxiety behaviors and deficits in social behavior. Treatment of F1 dams with OXT following CSS prevented transmission of repetitive and anxiety behaviors in F2 offspring, whereas intranasal AVP treatment in F1 dams showed positive effects on behavior deficits and anxiety. Interferon gamma (IFN-gamma) was proposed as a mediator of these effects.[120]

4. TRANSGENERATIONAL INHERITANCE OF TRAUMATIC STRESS FROM AN EVOLUTIONARY POINT OF VIEW

The evolutionary significance of transgenerational epigenetic inheritance is an extremely interesting aspect that challenges the Darwinian view of evolution.[121] With respect to the effects of traumatic stress, one can question whether their transmission has any evolutionary benefit, since many of these effects are negative. However, these effects may help the offspring better cope in certain conditions, as shown for instance in MSUS mice which have better goal-directed behaviors and behavioral flexibility in challenging situations.[3] However; most of the effects are maladaptive in normal and insult-free conditions suggesting that this could provide a way of "epigenetic natural selection" in which selection is driven by an organism's ability to adapt to a changing environment rather than survive the environmental adversity based on innate abilities. Epigenetic inheritance may also contribute to genome evolution since DNA methylation can lead to the formation of epialleles then to genomic changes.[121] For example, increased CNVs in response to changes in methylation could be a means through which epigenetic inheritance could contribute to evolution.[116] The emerging role of RNAs in transgenerational epigenetic inheritance could be a supplement to the RNA theory of evolution (reviewed in Ref. 122). Therefore, cumulating evidence and theoretical hypotheses do suggest that transgenerational epigenetic inheritance may play a role in evolution. Expanding transgenerational studies to more than two to three generations, delineating transgenerational studies based on how well the animals cope with trauma and exposing the offspring to both adverse and non-adverse conditions are important considerations to assess the evolutionary importance of transgenerational inheritance of traumatic stress.

5. CONCLUSIONS AND OUTLOOK

Evidence that the effects of traumatic stress can be passed to subsequent generations has accumulated in the past years. The epigenetic factors acting as potential vectors of such transmission and the molecular components responsible for their induction in germ cells are currently being studied. However, a number of questions and challenges remain in this field. First, it is important to have more comprehensive analyses of transmitted phenotypes. Considering that chronic stress can have wide physiological consequences, it is important to determine if the cardiovascular, immunological, gastrointestinal system, and endocrine systems are also affected in models of trauma. It is also important to delineate the transgenerational effects in individuals in addition to large cohorts. Increasing evidence from human studies suggest that only a fraction of individuals develop PTSD after exposure to trauma while a few individuals become more resilient; a phenomenon described as post-traumatic growth.[123] It is possible that a transgenerational phenotype is different in trauma-susceptible versus resilient animals/individuals. Third, it is also important to identify the factors responsible for the induction of epigenetic modifications in the germline and the signaling pathways sensitive to such factors. More direct evidence for the relevance of epigenetic modifications, is also needed, using for instance gene-editing tools such as CRISPR-dCas9 (reviewed in Ref. 124). The epigenetic machinery in germ and/or embryonic cells may be edited to determine the role of some of its components in transmission. Finally, it is important that accessible biological fluids such as saliva, blood, and semen be analyzed in traumatized human populations and ideally also their offspring to assess if mechanisms of transmission identified in animal models are relevant for humans. This step is also likely to provide valuable information to screen populations with parental trauma for phenotypic atypicalities. In the future, one could envision translation of research on transgenerational epigenetic inheritance to the bed-side. A clinical world offering "epigenetic counseling," "epigenetic prophylaxis" or "epigenetic therapy" in populations exposed to traumatic stress and/or their offspring might eventually become reality.

REFERENCES

1. Bohacek J, Mansuy IM. Molecular insights into transgenerational non-genetic inheritance of acquired behaviours. *Nat Rev Genet*. 2015;16(11):641–652.
2. Klengel T, Dias BG, Ressler KJ. Models of intergenerational and transgenerational transmission of risk for psychopathology in mice. *Neuropsychopharmacology*. 2016;41(1):219–231.
3. Franklin TB, Russig H, Weiss IC, et al. Epigenetic transmission of the impact of early stress across generations. *Biol Psychiatry*. 2010;68(5):408–415.
4. Gapp K, Soldado-Magraner S, Alvarez-Sánchez M, et al. Early life stress in fathers improves behavioural flexibility in their offspring. *Nat Commun*. 2014;5:5466.
5. Schneiderman N, Ironson G, Siegel SD. Stress and health: psychological, behavioral, and biological determinants. *Annu Rev Clin Psychol*. 2005;1(1):607–628.

6. Beery AK, Kaufer D. Stress, social behavior, and resilience: insights from rodents. *Neurobiol Stress*. 2015;1(1):116–127.
7. American Psychiatric Association. *Diagnostic and Statistical Manual of Mental Disorders*. 5th ed. Washington, DC: American Psychiatric Association; 2013.
8. Lupien SJ, McEwen BS, Gunnar MR, Heim C. Effects of stress throughout the lifespan on the brain, behaviour and cognition. *Nat Rev Neurosci*. 2009;10(6):434–445.
9. Ginzburg K, Ein-Dor T, Solomon Z. Comorbidity of posttraumatic stress disorder, anxiety and depression: a 20-year longitudinal study of war veterans. *J Affect Disord*. 2010;123(1–3):249–257.
10. Qureshi SU, Kimbrell T, Pyne JM, et al. Greater prevalence and incidence of dementia in older veterans with posttraumatic stress disorder. *J Am Geriatr Soc*. 2010;58(9):1627–1633.
11. Childress JE, McDowell EJ, Dalai VVK, et al. Hippocampal volumes in patients with chronic combat-related posttraumatic stress disorder: a systematic review. *J Neuropsychiatry Clin Neurosci*. 2013;25(1):170.
12. Qureshi SU, Pyne JM, Magruder KM, Schulz PE, Kunik ME. The link between post-traumatic stress disorder and physical comorbidities: a systematic review. *Psychiatry Q*. 2009;80(2):87–97.
13. de Kloet CS, Vermetten E, Geuze E, Kavelaars A, Heijnen CJ, Westenberg HGM. Assessment of HPA-axis function in posttraumatic stress disorder: pharmacological and non-pharmacological challenge tests, a review. *J Psychiatr Res*. 2006;40(6):550–567.
14. Heim C, Ehlert U, Hellhammer DH. The potential role of hypocortisolism in the pathophysiology of stress-related bodily disorders. *Psychoneuroendocrinology*. 2000;25(1):1–35.
15. Rohleder N, Joksimovic L, Wolf JM, Kirschbaum C. Hypocortisolism and increased glucocorticoid sensitivity of pro-inflammatory cytokine production in Bosnian war refugees with posttraumatic stress disorder. *Biol Psychiatry*. 2004;55(7):745–751.
16. Sherin JE, Nemeroff CB. Post-traumatic stress disorder: the neurobiological impact of psychological trauma. *Dialogues Clin Neurosci*. 2011;13(3):263–278.
17. ter Heegde F, De Rijk RH, Vinkers CH. The brain mineralocorticoid receptor and stress resilience. *Psychoneuroendocrinology*. 2015;52(1):92–110.
18. Neumann ID, Landgraf R. Balance of brain oxytocin and vasopressin: implications for anxiety, depression, and social behaviors. *Trends Neurosci*. 2012;35(11):649–659.
19. Hendrickson RC, Raskind MA. Noradrenergic dysregulation in the pathophysiology of PTSD. *Exp Neurol*. 2016;284:181–195.
20. Lee HJ, Lee MS, Kang RH, et al. Influence of the serotonin transporter promoter gene polymorphism on susceptibility to posttraumatic stress disorder. *Depress Anxiety* 2005;21(3):135–139.
21. Bryant RA, Felmingham KL, Falconer EM, et al. Preliminary evidence of the short allele of the serotonin transporter gene predicting poor response to cognitive behavior therapy in posttraumatic stress disorder. *Biol Psychiatry*. 2010;67(12):1217–1219.
22. Fuhrmann D, Knoll LJ, Blakemore SJ. Adolescence as a sensitive period of brain development. *Trends Cogn Sci*. 2015;19(10):558–566.
23. Hensch TK. Critical period plasticity in local cortical circuits. *Nat Rev Neurosci*. 2005;6(11):877–888.
24. Gapp K, von Ziegler L, Tweedie-Cullen RY, Mansuy IM. Early life epigenetic programming and transmission of stress-induced traits in mammals: how and when can environmental factors influence traits and their transgenerational inheritance? *Bioessays*. 2014;36(5):491–502.
25. Bird A. DNA methylation patterns and epigenetic memory. *Genes Dev*. 2002;16(1):6–21.
26. Klose RJ, Bird AP. Genomic DNA methylation: the mark and its mediators. *Trends Biochem Sci*. 2006;31(2):89–97.

27. Jones PL, Veenstra GJC, Wade PA, et al. Methylated DNA and MeCP2 recruit histone deacetylase to repress transcription. *Nat Genet.* 1998;19:187–191.
28. Bahar Halpern K, Vana T, Walker MD. Paradoxical role of DNA methylation in activation of FoxA2 gene expression during endoderm development. *J Biol Chem.* 2014;289(34):23882–23892.
29. Tognini P, Napoli D, Tola J, et al. Experience-dependent DNA methylation regulates plasticity in the developing visual cortex. *Nat Neurosci.* 2015;18(7):956–958.
30. Rudenko A, Dawlaty M, Seo J, et al. Tet1 is critical for neuronal activity-regulated gene expression and memory extinction. *Neuron.* 2013;79(6):1109–1122.
31. Kohli RM, Zhang Y. TET enzymes, TDG and the dynamics of DNA demethylation. *Nature.* 2013;502(7472):472–479.
32. Globisch D, Münzel M, Müller M, et al. Tissue distribution of 5-hydroxymethylcytosine and search for active demethylation intermediates. Croft AK, ed. *PLoS One.* 2010;5(12): e15367.
33. Münzel M, Globisch D, Brückl T, et al. Quantification of the sixth DNA base hydroxymethylcytosine in the brain. *Angew Chem Int Ed Engl.* 2010;49(31):5375–5377.
34. Song C-X, Szulwach KE, Fu Y, et al. Selective chemical labeling reveals the genome-wide distribution of 5-hydroxymethylcytosine. *Nat Biotechnol.* 2011;29(1):68–72.
35. Bachman M, Uribe-Lewis S, Yang X, Williams M, Murrell A, Balasubramanian S. 5-Hydroxymethylcytosine is a predominantly stable DNA modification. *Nat Chem.* 2014;6(12):1049–1055.
36. Jawahar MC, Murgatroyd C, Harrison EL, Baune BT. Epigenetic alterations following early postnatal stress: a review on novel aetiological mechanisms of common psychiatric disorders. *Clin Epigenetics.* 2015;7(1):122.
37. Saunderson EA, Spiers H, Mifsud KR, et al. Stress-induced gene expression and behavior are controlled by DNA methylation and methyl donor availability in the dentate gyrus. *Proc Natl Acad Sci USA.* 2016;113(17):4830–4835.
38. Zovkic IB, Sweatt JD. Epigenetic mechanisms in learned fear: implications for PTSD. *Neuropsychopharmacology.* 2013;38(1):77–93.
39. Bannister AJ, Kouzarides T. Regulation of chromatin by histone modifications. *Cell Res.* 2011;21(3):381–395.
40. Li B, Carey M, Workman JL. The role of chromatin during transcription. *Cell.* 2007;128(4):707–719.
41. Peters AH, Schübeler D. Methylation of histones: playing memory with DNA. *Curr Opin Cell Biol.* 2005;17(2):230–238.
42. Klose RJ, Zhang Y. Regulation of histone methylation by demethylimination and demethylation. *Nat Rev Mol Cell Biol.* 2007;8(4):307–318.
43. Cao J, Yan Q. Histone ubiquitination and deubiquitination in transcription, DNA damage response, and cancer. *Front Oncol.* 2012;2:26.
44. Cheng J, Huang M, Zhu Y, et al. SUMOylation of MeCP2 is essential for transcriptional repression and hippocampal synapse development. *J Neurochem.* 2014;128(6):798–806.
45. Stielow C, Stielow B, Finkernagel F, Scharfe M, Jarek M, Suske G. SUMOylation of the polycomb group protein L3MBTL2 facilitates repression of its target genes. *Nucleic Acids Res.* 2014;42(5):3044–3058.
46. Reed B, Fang N, Mayer-Blackwell B, et al. Chromatin alterations in response to forced swimming underlie increased prodynorphin transcription. *Neuroscience.* 2012;220:109–118.
47. Hunter RG, McCarthy KJ, Milne TA, Pfaff DW, McEwen BS. Regulation of hippocampal H3 histone methylation by acute and chronic stress. *Proc Natl Acad Sci USA.* 2009;106(49):20912–20917.
48. Reul JMHM. Making memories of stressful events: a journey along epigenetic, gene transcription, and signaling pathways. *Front Psychiatry.* 2014;5(Jan):1–11.

49. Whittle N, Singewald N. HDAC inhibitors as cognitive enhancers in fear, anxiety and trauma therapy: where do we stand? *Biochem Soc Trans*. 2014;42(2):569–581.
50. Ha M, Kim VN. Regulation of microRNA biogenesis. *Nat Rev Mol Cell Biol*. 2014;15(8):509–524.
51. Mannironi C, Camon J, De Vito F, et al. Acute stress alters amygdala microRNA miR-135a and miR-124 expression: inferences for corticosteroid dependent stress response. *PLoS One*. 2013;8(9):1–11.
52. Pan-Vazquez A, Rye N, Ameri M, et al. Impact of voluntary exercise and housing conditions on hippocampal glucocorticoid receptor, miR-124 and anxiety. *Mol Brain*. 2015;8(1):40.
53. Snijders C, de Nijs L, Baker DG, et al. MicroRNAs in post-traumatic stress disorder. In: Geyer MA, Ellenbroek BA, Marsden CA, Barnes TRE, Andersen SL, eds. *Current Topics in Behavioral Neurosciences*. Berlin, Heidelberg: Springer; 2017:1–24.
54. Mercer TR, Dinger ME, Sunkin SM, Mehler MF, Mattick JS. Specific expression of long noncoding RNAs in the mouse brain. *Proc Natl Acad Sci USA*. 2007;105(2):716–721.
55. Quinn JJ, Chang HY. Unique features of long non-coding RNA biogenesis and function. *Nat Rev Genet*. 2016;17(1):47–62.
56. Kung JTY, Colognori D, Lee JT. Long noncoding RNAs: past, present, and future. *Genetics*. 2013;193(3):651–669.
57. Spadaro PA, Flavell CR, Widagdo J, et al. Long noncoding RNA-directed epigenetic regulation of gene expression is associated with anxiety-like behavior in mice. *Biol Psychiatry*. 2015;78(12):848–859.
58. Aravin AA, Sachidanandam R, Bourc'his D, et al. A piRNA pathway primed by individual transposons is linked to de novo DNA methylation in mice. *Mol Cell*. 2008;31(6):785–799.
59. Rajan KS, Ramasamy S. Retrotransposons and piRNA: the missing link in central nervous system. *Neurochem Int*. 2014;77:94–102.
60. Landry CD, Kandel ER, Rajasethupathy P. New mechanisms in memory storage: piRNAs and epigenetics. *Trends Neurosci*. 2013;36(9):535–542.
61. Harcourt EM, Kietrys AM, Kool ET. Chemical and structural effects of base modifications in messenger RNA. *Nature*. 2017;541(7637):339–346.
62. Basanta-Sanchez M, Temple S, Ansari SA, D'Amico A, Agris PF. Attomole quantification and global profile of RNA modifications: epitranscriptome of human neural stem cells. *Nucleic Acids Res*. 2015;44(3):1–10.
63. Lewis CJT, Pan T, Kalsotra A. RNA modifications and structures cooperate to guide RNA–protein interactions. *Nat Rev Mol Cell Biol*. 2017;18(3):202–210.
64. Zhao BS, Roundtree IA, He C. Post-transcriptional gene regulation by mRNA modifications. *Nat Rev Mol Cell Biol*. 2016;18(1):31–42.
65. Liu N, Pan T. N6-methyladenosine–encoded epitranscriptomics. *Nat Struct Mol Biol*. 2016;23(2):98–102.
66. Walters BJ, Mercaldo V, Gillon CJ, et al. The role of the RNA demethylase FTO (fat mass and obesity-associated) and mRNA methylation in hippocampal memory formation. *Neuropsychopharmacology*. 2017;42:1502–1510.
67. Ambrosi C, Manzo M, Baubec T. Dynamics and context-dependent roles of DNA methylation. *J Mol Biol*. 2016;429(10):1459–1475.
68. Burggren W. Epigenetic inheritance and its role in evolutionary biology: re-evaluation and new perspectives. *Biology (Basel)*. 2016;5(2):24.
69. Ward ID, Zucchi FCR, Robbins JC, et al. Transgenerational programming of maternal behaviour by prenatal stress. *BMC Pregnancy Childbirth*. 2013;13(suppl 1):S9.
70. Weber-Stadlbauer U, Richetto J, Labouesse MA, Bohacek J, Mansuy IM, Meyer U. Transgenerational transmission and modification of pathological traits induced by prenatal immune activation. *Mol Psychiatry*. 2017;22(1):102–112.

71. Scheiber IBR, Weiß BM, Kingma SA, Komdeur J. The importance of the altricial–precocial spectrum for social complexity in mammals and birds—a review. *Front Zool* 2017;14(1):3.
72. Babb JA, Carini LM, Spears SL, Nephew BC. Transgenerational effects of social stress on social behavior, corticosterone, oxytocin, and prolactin in rats. *Horm Behav*. 2014;65(4):386–393.
73. Wu R, Zhang H, Xue W, et al. Transgenerational impairment of hippocampal Akt-mTOR signaling and behavioral deficits in the offspring of mice that experience postpartum depression-like illness. *Prog Neuropsychopharmacol Biol Psychiatry*. 2016;73: 11–18 [1878–4216 (Electronic)].
74. Gapp K, Jawaid A, Sarkies P, et al. Implication of sperm RNAs in transgenerational inheritance of the effects of early trauma in mice. *Nat Neurosci*. 2014;17(5):667–669.
75. Bohacek J, Farinelli M, Mirante O, et al. Pathological brain plasticity and cognition in the offspring of males subjected to postnatal traumatic stress. *Mol Psychiatry*. 2015;20(5):621–631.
76. Razoux F, Russig H, Mueggler T, et al. Transgenerational disruption of functional 5-HT1AR-induced connectivity in the adult mouse brain by traumatic stress in early life. *Mol Psychiatry*. 2017;22(4):519–526.
77. Saavedra-Rodríguez L, Feig LA. Chronic social instability induces anxiety and defective social interactions across generations. *Biol Psychiatry*. 2013;73(1):44–53.
78. Dietz DM, Laplant Q, Watts EL, et al. Paternal transmission of stress-induced pathologies. *Biol Psychiatry*. 2011;70(5):408–414.
79. Dias BG, Ressler KJ. Parental olfactory experience influences behavior and neural structure in subsequent generations. *Nat Neurosci*. 2013;17(1):89–96.
80. Nätt D. Stress and the offspring: adaptive transgenerational effects of unpredictability on behaviour and gene expression in chickens (Gallus gallus). *Linköping University Electronic Press 1385*. 2008;1–22.
81. Goerlich VC, Nätt D, Elfwing M, Macdonald B, Jensen P. Transgenerational effects of early experience on behavioral, hormonal and gene expression responses to acute stress in the precocial chicken. *Horm Behav*. 2012;61(5):711–718.
82. Ericsson M, Henriksen R, Bélteky J, Sundman A-S, Shionoya K, Jensen P. Long-term and transgenerational effects of stress experienced during different life phases in chickens (*Gallus gallus*). van den Bos R, ed. *PLoS One*. 2016;11(4):e0153879.
83. McCormick GL, Robbins TR, Cavigelli SA, Langkilde T. Ancestry trumps experience: transgenerational but not early life stress affects the adult physiological stress response. *Horm Behav*. 2017;87:115–121.
84. Infurna MR, Fuchs A, Fischer-Waldschmidt G, et al. Parents' childhood experiences of bonding and parental psychopathology predict borderline personality disorder during adolescence in offspring. *Psychiatry Res*. 2016;246:373–378.
85. Vågerö D, Rajaleid K. Does childhood trauma influence offspring's birth characteristics? *Int J Epidemiol*. 2017;46(1):219–229.
86. Barocas HA, Barocas CB. Wounds of the fathers: the next generation of holocaust victims. *Int Rev Psychoanal*. 1979;6:331–340.
87. Yehuda R, Daskalakis NP, Bierer LM, et al. Holocaust exposure induced intergenerational effects on FKBP5 methylation. *Biol Psychiatry*. 2016;80(5):372–380.
88. Yehuda R, Schmeidler J, Giller EL, Siever LJ, Binder-Brynes K. Relationship between posttraumatic stress disorder characteristics of holocaust survivors and their adult offspring. *Am J Psychiatry*. 1998;155(6):841–843.
89. Yehuda R, Bell A, Bierer LM, Schmeidler J. Maternal, not paternal, PTSD is related to increased risk for PTSD in offspring of Holocaust survivors. *J Psychiatr Res*. 2008;42(13):1104–1111.

90. Yehuda R, Bierer LM, Schmeidler J, Aferiat DH, Breslau I, Dolan S. Low cortisol and risk for PTSD in adult offspring of Holocaust survivors. *Am J Psychiatry*. 2000;157(8):1252–1259.
91. Field NP, Muong S, Sochanvimean V. Parental styles in the intergenerational transmission of trauma stemming from the Khmer Rouge regime in Cambodia. *Am J Orthopsychiatry*. 2013;83(4):483–494.
92. Rieder H, Elbert T. The relationship between organized violence, family violence and mental health: findings from a community-based survey in Muhanga, Southern Rwanda. *Eur J Psychotraumatol*. 2013;4(1). 21329.
93. Wittekind CE, Jelinek L, Kellner M, Moritz S, Muhtz C. Intergenerational transmission of biased information processing in posttraumatic stress disorder (PTSD) following displacement after World War II. *J Anxiety Disord*. 2010;24(8):953–957.
94. Phipps RM, Degges-White S. A new look at transgenerational trauma transmission: second-generation Latino immigrant youth. *J Multicult Couns Devel*. 2014;42(3): 174–187.
95. Daud A, Skoglund E, Rydelius P-A. Children in families of torture victims: transgenerational transmission of parents' traumatic experiences to their children. *Int J Soc Welf*. 2005;14(1):23–32.
96. Rosenheck R. Impact of posttraumatic stress disorder of World War II on the next generation. *J Nerv Ment Dis*. 1986;174(6):319–327.
97. Glenn DM, Beckham JC, Feldman ME, Kirby AC, Hertzberg MA, Moore SD. Violence and hostility among families of Vietnam veterans with combat-related posttraumatic stress disorder. *Violence Vict*. 2002;17(4):473–489.
98. Klaric M, Franciskovic T, Klaric B, et al. Psychological problems in children of war veterans with posttraumatic stress disorder in Bosnia and Herzegovina: cross-sectional study. *Croat Med J*. 2008;49(4):491–498.
99. Ćorić MK, Klarić M, Petrov B, Mihić N. Psychological and behavioral problems in children of war veterans with Post Traumatic Stress Disorder. *Eur J Psychiatry*. 2016;30(3):219–230.
100. Santavirta T, Santavirta N, Gilman SE. Association of the World War II Finnish evacuation of children with psychiatric hospitalization in the next generation. *JAMA Psychiat*. 2018;75(1):21.
101. Ciabrelli F, Comoglio F, Fellous S, et al. Stable Polycomb-dependent transgenerational inheritance of chromatin states in Drosophila. *Nat Genet*. 2017;49(6):876–886.
102. Gaydos LJ, Wang W, Strome S. H3K27me and PRC2 transmit a memory of repression across generations and during development. *Science*. 2014;345(6203): 1515–1518.
103. Rodgers AB, Morgan CP, Leu NA, Bale TL. Transgenerational epigenetic programming via sperm microRNA recapitulates effects of paternal stress. *Proc Natl Acad Sci USA*. 2015;112(44):13699–13704.
104. Short AK, Yeshurun S, Powell R, et al. Exercise alters mouse sperm small noncoding RNAs and induces a transgenerational modification of male offspring conditioned fear and anxiety. *Transl Psychiatry*. 2017;7(5):e1114.
105. Adeoya-Osiguwa SA, Gibbons R, Fraser LR. Identification of functional alpha2- and beta-adrenergic receptors in mammalian spermatozoa. *Hum Reprod*. 2006;21(6): 1555–1563.
106. Fiore C, Sticchi D, Pellati D, et al. Identification of the mineralocorticoid receptor in human spermatozoa. *Int J Mol Med*. 2006;18(4):649–652.
107. Naz RK. Receptors in spermatozoa: are they real? *J Androl*. 2006;27(5):627–636.
108. Kwon W-S, Park Y-J, Kim Y-H, You Y-A, Kim IC, Pang M-G. Vasopressin effectively suppresses male fertility. *PLoS One*. 2013;8(1):e54192.

109. Short AK, Fennell KA, Perreau VM, et al. Elevated paternal glucocorticoid exposure alters the small noncoding RNA profile in sperm and modifies anxiety and depressive phenotypes in the offspring. *Transl Psychiatry*. 2016;6(6):e837.
110. Pan W, Stone KP, Hsuchou H, Manda VK, Zhang Y, Kastin AJ. Cytokine signaling modulates blood-brain barrier function. *Curr Pharm Des*. 2011;17(33):3729–3740.
111. Sullivan R, Saez F, Girouard J, Frenette G. Role of exosomes in sperm maturation during the transit along the male reproductive tract. *Blood Cells Mol Dis*. 2005;35(1):1–10.
112. Marré J, Traver EC, Jose AM. Extracellular RNA is transported from one generation to the next in Caenorhabditis elegans. *Proc Natl Acad Sci USA*. 2016;113(44): 12496–12501.
113. Otero JJ, Fu W, Kan L, Cuadra AE, Kessler JA. Beta-catenin signaling is required for neural differentiation of embryonic stem cells. *Development*. 2004;131(15):3545–3557.
114. Denis H, Ndlovu MN, Fuks F. Regulation of mammalian DNA methyltransferases: a route to new mechanisms. *EMBO Rep*. 2011;12(7):647–656.
115. Cheng TL, Wang Z, Liao Q, et al. MeCP2 suppresses nuclear microRNA processing and dendritic growth by regulating the DGCR8/Drosha complex. *Dev Cell*. 2014;28(5):547–560.
116. Skinner MK, Guerrero-Bosagna C, Haque MM. Environmentally induced epigenetic transgenerational inheritance of sperm epimutations promote genetic mutations. *Epigenetics*. 2015;10(8):762–771.
117. Gapp K, Bohacek J, Grossmann J, et al. Potential of environmental enrichment to prevent transgenerational effects of paternal trauma. *Neuropsychopharmacology*. 2016;41(11):2749–2758.
118. Arai JA, Li S, Hartley DM, Feig LA. Transgenerational rescue of a genetic defect in long-term potentiation and memory formation by juvenile enrichment. *J Neurosci*. 2009;29(5):1496–1502.
119. Dezsi G, Ozturk E, Salzberg MR, Morris M, O'Brien TJ, Jones NC. Environmental enrichment imparts disease-modifying and transgenerational effects on genetically-determined epilepsy and anxiety. *Neurobiol Dis*. 2016;93:129–136.
120. Murgatroyd CA, Hicks-Nelson A, Fink A, et al. Effects of chronic social stress and maternal intranasal oxytocin and vasopressin on offspring interferon-γ and behavior. *Front Endocrinol (Lausanne)*. 2016;7:155.
121. Sharma A. Transgenerational epigenetic inheritance: resolving uncertainty and evolving biology. *Biomol Concepts*. 2015;6(2):87–103.
122. Kun Á, Szilágyi A, Könnyu B, Boza G, Zachar I, Szathmáry E. The dynamics of the RNA world: insights and challenges. *Ann NY Acad Sci*. 2015;1341(1):75–95.
123. Horn SR, Charney DS, Feder A. Understanding resilience: new approaches for preventing and treating PTSD. *Exp Neurol*. 2016;284:119–132.
124. Sander JD, Joung JK. CRISPR-Cas systems for editing, regulating and targeting genomes. *Nat Biotechnol*. 2014;32(4):347–355.
125. Gapp K, Corcoba A, van Steenwyk G, Mansuy IM, Duarte JM. Brain metabolic alterations in mice subjected to postnatal traumatic stress and in their offspring. *J Cereb Blood Flow Metab*. 2017;37(7):2423–2432.

CHAPTER THIRTEEN

From Epigenetic Associations to Biological and Psychosocial Explanations in Mental Health

Chiara Renzi*, Nadine Provencal†,‡, Katherine C. Bassil§, Kathinka Evers¶, Ulrik Kihlbom¶, Elizabeth J. Radford||, Ilona Koupil#,, Bertram Mueller-Myhsok††,‡‡,§§, Mats G. Hansson¶, Bart P.F. Rutten¶¶,1**

*Applied Research Division for Cognitive and Psychological Science, European Institute of Oncology, Milan, Italy
†Faculty of Health Sciences, Simon Fraser University, Burnaby, BC, Canada
‡Healthy Starts Theme, BC Children's Hospital Research Institute, Vancouver, BC, Canada
§School for Mental Health and Neuroscience (MHeNS), Department of Psychiatry and Neuropsychology, Faculty of Health, Medicine and Life Sciences, Maastricht University, Maastricht, The Netherlands
¶Centre for Research Ethics & Bioethics, Department of Public Health and Caring Sciences, Uppsala University, Uppsala, Sweden
||Department of Pediatrics, University of Cambridge, Cambridge, United Kingdom
#Department of Public Health Sciences, Stockholm University, Stockholm, Sweden
**Department of Public Health Sciences, Karolinska Institutet, Stockholm, Sweden
††Department of Translational Research in Psychiatry, Max Planck Institute of Psychiatry, Munich, Germany
‡‡Munich Cluster for Systems Neurology (SyNergy), Munich, Germany
§§University of Liverpool, Liverpool, United Kingdom
¶¶Department of Psychiatry and Neuropsychology, School for Mental Health and Neuroscience (MHeNs), European Graduate School of Neuroscience (EURON), Faculty of Health, Medicine and Life Sciences, Maastricht University Medical Centre +, Maastricht, The Netherlands
[1]Corresponding author: e-mail address: b.rutten@maastrichtuniversity.nl

Contents

Progress in Molecular Biology and Translational Science, Volume 158
ISSN 1877-1173
https://doi.org/10.1016/bs.pmbts.2018.04.011

Abstract

The development of mental disorders constitutes a complex phenomenon driven by unique social, psychological and biological factors such as genetics and epigenetics, throughout an individual's life course. Both environmental and genetic factors have an impact on mental health phenotypes and act simultaneously to induce changes in brain and behavior. Here, we describe and critically evaluate the current literature on gene–environment interactions and epigenetics on mental health by highlighting recent human and animal studies. We furthermore review some of the main ethical and social implications concerning gene–environment interactions and epigenetics and provide explanations and suggestions on how to move from statistical and epigenetic associations to biological and psychological explanations within a multi-disciplinary and integrative approach of understanding mental health.

1. INTRODUCTION

A major challenge of current research in the field of mental health is to understand how different factors related to the social environment, psychological characteristics, genetic make-up, transcription of genes (epigenetics) and cerebral architecture are involved in well-being and in the development of mental disorders in an individual. In order to be more effective, treatment and prevention need to be adjusted to individual differences related to all these factors.

There is abundant evidence in social epidemiological research regarding social determinants of mental illness during one individual's life course, e.g., unemployment, financial strain, living on social welfare, work environment with lack of decision latitude, job strain and bullying.[1–3] Social and psychological programs are available to help population groups and affected individuals cope with stressful conditions, e.g., community-based interventions for parents (parenting support). However, the effect of these interventions and programs may also depend on biological parameters that are unique for each individual. Some of these interventions and programs may at least work better for some individuals than others, possibly due to differences in their biological make-up.

Inter-disciplinary studies of the nature of these relationships and their effects on mental health conducted from genetic, neuroscientific and social perspectives are key to deepening our understanding of mental disorders.

There is growing evidence of associations between non-genetic variables and gene expression in regard to nutrition, maternal care/behavior, psychosocial stress, adversity and neglect in early life indicating that the environment plays an important role in the evolving neuronal networks of the brain and in shaping the proteome and the metabolome.[4,5] Molecular epigenetic mechanisms in the cell may explain how the exposure to environmental factors influences the phenotypic outcome and variability both between individuals and within an individual at different points in time. As an example, the organizational complexity of the brain is not only a result of gene expression patterns but is also affected by the strength, selection and stabilization of synapses in a critical reciprocal interaction between the brain and its physical, psychosocial and cultural environment. Similarly, the homeostasis and the complex interplays of proteins' expression and metabolic pathways can be modified by these factors. Notably, the plasticity of gene expression, brain maturation, and metabolic signatures in response to postnatal and adult social life events, both with increasing evidence of trans-generational effects, suggest the evolutionary significance of these mechanisms in enabling organisms to adapt to changing environmental conditions. In this sense, the subjective experience, or even the experiences of ancestors, may model the equilibria of human systems.

An interesting feature of epigenetics as well as of neuronal epigenesis is the putative reversibility of epigenetic changes, e.g., due to demethylation or synaptic plasticity, opening new avenues for prevention. The reversible nature of epigenetic pathways illustrates the potential of developing and bringing epigenetic drugs into therapeutics,[6,7] accompanied by the possibility of developing approaches focusing on the understanding and improvement of psychosocial, cultural and political environments for the purpose of reversing unfavorable processes toward being epigenetically proactive.[8,9]

2. MENTAL HEALTH: STILL AWAITING A MULTI-DISCIPLINARY APPROACH

Mental health involves emotional, psychological and social well-being. It affects directly how individuals manage challenges in everyday life. The World Health Organization (WHO) reports from systematic reviews in some European countries indicate that as much as 27% of the adult population (aged 18–65) has experienced at least one of a series of mental disorders in the past year, including problems arising from substance use, psychoses, depression, anxiety, and eating disorders. Figures like these represent an enormous human toll of ill health. Yet, as WHO acknowledges, even these figures are likely to underestimate the scale of the problem, as only a limited number of disorders were included and it did not collect data on those aged over 65, a group that is at particular risk.[10] In addition, there is also data suggesting increasing prevalence of anxiety disorders in young children, with some even given medication in pre-school.[11,12] Mental disorders are by far the largest contributor to chronic conditions afflicting the population of Europe. They rank as the first cause of years lived with disability in Europe, accounting for 36.1% of those attributable to all causes.[13] WHO estimates that three out of four patients with major depression in Europe, do not receive adequate treatment.[14] The treatment gap extends to other disorders like schizophrenia (32.2%), bipolar disorder (50.2%) and panic disorder (55.9%), where individuals remain untreated despite the availability of effective treatments.[15]

Therefore, it is no surprise that since the last century, research has devoted particular attention in establishing the determinants of mental health and mental disorders. However, deterministic models ranging from theories of biochemical imbalance to genetic approaches are not sufficient in explaining the phenotypic variability in the mental illness severity and in the response of individuals to treatment.[16]

Some studies indicate that the simple use of medication is not sufficient to obtain a stable remission for disorders like psychosis, and that minimal or time limited use of antipsychotic medication seems to be associated with better long-term outcomes.[17–19] Similarly, although genetic loci have been identified which confer risk to mental disorders, the absolute risk conferred by individual loci is small, similar to other complex traits. Genetic association studies have demonstrated the importance of large cohorts, rigorous statistical analyses and independent replication in order to avoid confounding.[20]

Diagnosis based solely on signs and symptoms like the diagnostic and statistical manual do not reflect the nature of mental phenotypes which cut across the traditional diagnostic boundaries. This may lead to a number of fallacies including misdiagnosis but also categorization of conditions and as such overlooking valuable information. After all, depression, bipolar disorder, and posttraumatic stress disorder (PTSD) may share more common underlying mechanisms than what was originally expected. It has also been observed that the possibility to adapt to a changing environment is linked to mental health (see, e.g., Ref. 21). Evolutionary approaches have suggested that some behavioral syndromes or mental conditions which are classically considered mental disorders actually represent normal evolutionary attempts to overcome challenging situations and increase fitness to the environment (e.g., Refs. 22–25).

In an attempt to provide a broader approach to health, the biopsychosocial model[26,27] had the merit to re-propose a holistic view of the individual as part of a complex network of factors where circularity rather than causality prevails.[28] Yet, it has been argued that this model has some limitations as research framework to investigate the etiology, diagnosis and treatment of mental disorders such as depression[29,30] and in implementing effective preventative policies.[31] It has been suggested that the biopsychosocial model needs to be revisited in light of the systems theory,[30] in order to bridge better and to integrate biological, psychological, and social components of mental disorders.

For example, an individual that has experienced a highly stressful situation may go through a phase of depression characterized by the predominance of (a) unpleasant emotions such as sadness, despair, anxiety; (b) thoughts related to sense of uselessness, hopelessness, self-pity, self-blaming, unworthiness; (c) lack of sleep, chronic fatigue, difficulties focusing, and reduction of pleasant activities. In some cases, the person may enter into a circuit where these emotions, thoughts, and behaviors constitute the premises to confirm the negative image that the individual has of himself, associated with deeper negative emotions and with the adoption of behaviors that tend self-perpetuate and constantly reinforce the feeling of being depressed.[32,33]

Garcia-Toro and Aguirre[30] identified and selected 10 relevant empirical findings about depression from different research domains: (1) genetic predisposition; (2) predisposition due to early cerebral damage; (3) precipitating biological factors (such as somatic illnesses or drugs); (4) structural and functional neuroimaging changes; (5) effective biological treatment (such as the use of certain pharmaceutical drugs or of brain stimulation techniques); (6) predisposing early psychological stress; (7) predisposing personality traits (such as neuroticism); (8) predisposition due to social maladjustment (such as low socio-economic level and lack of social support); (9) precipitating psychological stress; and (10) effective psychosocial treatments (such as interpersonal therapy and cognitive-behavioral therapy). Further, they propose possible interactions between these evidences in the attempt to provide a more integrated non-causal view of the possible generative mechanisms related to depression. For instance, a parallel could be draw between reinforce of hyperactive neurons and atrophy of hypoactive neurons at the biological level and reinforce of the most used (depressive) cognitive-emotional schemas and segregation of the others (pleasure). This, in turn, calls for the implementation of early interventions, which could intervene before these schemas become rigid.

Network medicine, a newly emerging field, is based on the use of large data from different contributing aspects to a better understanding of disease development.[34–37] The integrative network approach bridges findings from the clinic (signs and symptoms), biological findings in the laboratory (metabolomics and genomics), increasing the understanding of mental disorders, even in the context of comorbidity between disorders of the brain like depression and medical conditions like inflammation. It has been argued that network-based models have been more successful in the diagnosis of non-mental diseases characterized by less "complexity" and more straightforward distinctive symptomatic relations. Nevertheless, it is important to note that this observation could be attributed to nosological complexities or the lack of information concerning the fundamental etiology of mental disorders such as depression or schizophrenia compared to diseases such as lung cancer, and not due to "fuzzy symptom networks" in mental disorders. In principle, mental disorders like depression constitute a complex dynamic network of symptoms constantly creating a reinforcing loop of the disorder itself.

While the network approach allows the clinicians a closer approach to personalized medicine, several fundamental concepts of psychiatry are being overlooked and hence improvements in diagnosis and treatment using this approach are being stalled. For instance, network methods have reinstalled division between the two co-existing school of thoughts—biological and psychological—instead of complementing one another to improve diagnosis and treatment. Therefore, how can we rely on studies based on the network approach using tools unqualified to make accurate conclusions? To truly benefit from the network theory, replication studies for transparency and reproducibility, and longitudinal data for statements on temporal and causal relationships are needed. This comes hand-in-hand with increasing population size and heterogeneity, and relying on a combination of criteria and fundamental functional domains, like the Research Domain Criteria (RDoC).[38]

3. ENVIRONMENTAL INFLUENCES ON MENTAL HEALTH PHENOTYPES

Associations between a range of environmental exposures and mental health phenotypes have been well-established.[39–44] One should keep in mind that the nature of these associations may act through a variety of pathways, which may contribute differently depending on the period of life, the tissues analyzed or on the type of phenotype considered. The classical perspective is that genetic variants may modulate how an individual respond to environmental stressors, and so represent intrinsic susceptibilities to mental health or disease. An alternative perspective is that genetic variants impact on behavioral patterns that shape or select the surrounding environment and thereby modulate the likelihood of being exposed to adversity, stress or protective environments.[45] Another framework is represented by structural effects such that population groups with certain genetic profiles are exposed to, e.g., poorer living conditions or welfare policies as well as to racial discrimination or refugee status. In addition, of course, personal or familiar experiences may trigger chains of events that represent risk or protective factors.[43,46] In the first phases of development, the family represents the main environment to which a child is exposed. Since it represents one of the pivotal vectors through which the influence of the surrounding environment is exerted, it is important to take this perspective into account as it may give important insights. An important challenge is to understand how environmental stressors or life events may have pervasive effects such that they alter mental health in the long term. Environmental factors have been shown to affect brain development,[47–49] neuroendocrine functioning,[50] the immune system,[51] and possibly disease progression.[52] In turn, patterns of cognitive and affective processing which result from the complex interplay between different body systems may affect the selection of environments by the individual as well as how the information from the environment or from interactions with other individuals is conceptualized and mentally represented.

In this vein, a series of studies by the group of Helen Neville reported that in pre-school children low socio-economic status was associated to a delay in selective attentional patterns.[53–55] Furthermore, the genotype of the serotonin transporter linked polymorphic region (5-HTTLPR) was also linked to a difference in the effects of selective attention on neural processing in this population of children. Interestingly, a family based training involving pre-school children and their parents was shown to reverse the effects of the socio-economic status and of the genotype on children' selective attention.[56]

Importantly, whatever the perspective assumed, it is crucial to consider that these associations may not represent causal mechanisms. Particular caution should be taken in distinguishing between risk indicators (e.g., parental divorce) and risk mediators (e.g., family discord and conflict which often precede or accompany parental divorce), and between distal risks (e.g., poverty or parental loss) and proximal risks (impaired parenting often experienced in poor economic conditions or in the absence of one parent).[46] Erroneous assumptions on causality or risk type may partially arise from the use of single frameworks or level of analysis that may be overcome by adopting a multi-level, complex approach. Complex approaches increase the likelihood of including simultaneously risk indicators and risk mediators as well as proximal and distal risks.

An important factor to be considered is that studies often assumed that environmental influences would be active in extreme conditions (e.g., parental loss) while they may well operate with a gradient of effect across the distribution (e.g., parenting difficulties). Another assumption concerns the period of plasticity for such influences. It is often supposed that the prenatal and early postnatal period is a unique window of opportunity. However, one should keep in mind that different environmental factors might have a variable degree of influence depending on the developmental stage, and that the period of plasticity for the influence of some environmental factors may be wider than expected ranging from prenatally to adulthood. Furthermore, individual characteristics or prior experiences may concur in determining greater vulnerability or protection from environmental risks.

4. GENETIC INFLUENCES ON MENTAL HEALTH

The relative importance of genetic and environmental influences on mental health has for long been studied in twin cohorts.[57] Comparisons of similarity between genetically identical (monozygotic, MZ) pairs and fraternal (dizygotic, DZ) pairs have been made for many decades in order to establish genetic contributions to normal human variation and the risk of clinically significant outcomes. The average heritability across different phenotypes is estimated to 49%.[58] Twin study designs have also been able to demonstrate causal interrelationships between brain structure and neuropsychological performance.[59,60] Comparing MZ with DZ pairs with shared environmental exposures will give information on the relative contribution of genetic and shared environmental effects. Of particular interest is to identify gene–environment interactions, such as the findings of Hicks et al.[61] that the genetic contribution to mental disorders expressing externalizing behaviors (e.g., attention-deficit/hyperactivity disorder, alcohol- and substance-related disorders) was especially pertinent in those individuals that had experienced high environmental adversity, and as described for psychotic disorders.[62,63]

There is abundant evidence in terms of association between genotypes and mental disorders and health, e.g., genetic overlaps with schizophrenia and gender differences,[64] genomic predictors of combat stress vulnerability and resilience[65] and posttraumatic stress.[66] In order to determine the causal influence of environmental factors such as early stressful life events and trauma, family relationships, socio-economic factors or school factors on neuropsychological development and mental health, one needs to assess genetic influence and gene–environment interactions as well as features that are correlated both with the environment, the brain and gene expression.[67]

5. EFFECTS OF ENVIRONMENTAL EXPOSURES ON BEHAVIOR AND MENTAL HEALTH MEDIATED VIA EPIGENETIC MECHANISMS

There is growing evidence of the impact of non-genetic variables on gene expression, e.g., nutrition, maternal care/behavior, psychosocial stress, adversity and neglect in early life, hormones and drugs.[68–70] For instance, the Developmental Origins of Health and Disease (DOHaD[71,72]) paradigm states that environmental exposures in critical developmental periods may contribute to long-term modifications, programming our functioning at the biological level. For instance, deprived environments early in life were associated with changes in the stress response system and with changes in brain architecture.[73,74]

Molecular epigenetic mechanisms (DNA methylation and hydroxymethylation, histone modifications, non-coding RNAs and the three-dimensional organization of chromatin within the nucleus[75]) may explain how the exposure of environmental factors influence the phenotypic outcome and variability both between individuals and within an individual at different times.[63,68,76–78] The possibility of maternal behavior to influence epigenetic programming has been demonstrated in rat models. For example, increased pup licking and grooming by rat mothers altered the offspring epigenome, and was associated with differences in stress reactivity across generations.[79] These effects were reversed through cross fostering, indicating reversibility in programming through changes in behavior. Moreover, Murgatroyd and Spengler recently demonstrated that adversity in early life might shape the experience-dependent maturation of stress-regulating pathways underlying emotional functions.[80] Similar social experience-dependent memory has been suggested to have an epigenetic basis, induced by the social and/or physical environment, via intracellular pathways.[81] It has therefore been hypothesized that exposures to adversities during early life may substantially affect stress sensitivity and immunity trajectories later in life by modifying DNA methylation during critical periods earlier in life.

Since the mid-20th century, several studies have also analyzed the regulatory stress response in both children and adults as one of the most important mediating mechanisms for the influence of poverty on cognitive, emotional, and social functioning.[82] For example, threats, negative events, exposure to environmental hazards, family and community violence, changes in the dynamics of family life, job loss, instability and economic deprivation—which are most likely to occur under poverty conditions,[83] are all phenomena that activate differently the systems of stress regulation.[84,85] The physiological responses to stress can also be manifested in different ways: vagal tone, allostatic load and neuroendocrine activity. In rodent models, stress has been observed to result in intergenerational transmission of phenotypes. For instance, repeated unpredictable separations of pups from the dam results in depressive-like behaviors and altered methylation profiles up to the third generation with a complex gender-dependent profile.[86] Family history of stress is associated also with altered miRNA expression and behavioral and metabolic signatures in the progeny, such as altered sensory-motor development, with some evidence for the strongest impact on the third generation.[86,87] A growing body of literature indicates that traumatic prenatal experiences in the mother, such as wars, invasions or bereavement may be linked to a higher likelihood of experiencing mental ill health in the offspring.[88,89] At a less extreme level, maternal stress or depression during pregnancy (a condition that is more likely to co-exist with low socio-economic status) was found to have an effect on diurnal cortisol pattern and to be transmitted to the offspring through biological mechanisms such as "diurnal cortisol coupling."[90,91]

Importantly, the psychosocial environment does not only exert negative influences on development. Positive early experiences and enriched environments can have a protective or mitigating effect on stress pathways[52,92,93] via epigenetic mechanisms. For instance, the results of the follow up of the Bucharest Early Intervention Project indicate that positive caregiving can reduce the effects of early life institutionalization on the stress pathway.[93] Furthermore, even in later phases of life, individuals can use (brain) plasticity to mitigate the negative effects of traumatic experiences.[94]

Interestingly, Van den Bergh[39] proposed to extend the DOHaD into the "Developmental Origins of Behavior, Health, and Disease" (DOBHaD) hypothesis, in order to integrate also aspects of the behavioral development and to adopt a preventive approach to diseases and disorders. Starting from the evidence that many disorders have specific architectural or electric signatures before the individual shows clinical symptoms,[53,95] he argues that more research is needed in order to increase our comprehension of the complex interplay between genes, gene expression, brain and behavior and the biological mechanisms involved.

6. EFFECTS OF ENVIRONMENTAL EXPOSURES ON BRAIN NETWORKS: THE NEURONAL EPIGENESIS

The environment plays an important role also in the evolving neuronal network of the brain. The organizational complexity of the brain is not only a result of gene expression but affected by the selection and stabilization of synapses in a critical reciprocal interaction between the brain and its physical, social and cultural environment. As previously described,[9] the word "epigenesis" can be traced back to William Harvey (1651), who stated in contrast to contemporary preformationist views that the embryo arises by "the addition of parts budding out from one another."[96] It was subsequently used by Conrad Waddington (1942) to specify the relationship between the genes and their environment to produce a phenotype.[96] This is also the meaning adopted in the theory of the epigenesis of neuronal networks by selective stabilization of synapses, according to which the environment affects the organization of connections in an evolving neuronal network through the stabilization or elimination (pruning) of labile synapses, under the control of the state of activity of the network.[97] This meaning contrast with the more recent and biochemically distinct meaning of the word *epigenetic*, as described above, which refers to the molecular mechanisms through which altered genomic activity states are achieved. The modulatory role of chromatin modifications in long-term memory has already been described (see, e.g., Ref. 98), but the informational content involved—which relies upon cell bodies—is expected to be in orders of magnitude smaller that of synaptic epigenesis, based upon the combinatorial power of individual synapses.

During embryonic and postnatal development, the million billion (1015) synapses that form the human brain network do not assemble like the parts of a computer, that is, according to a plan that precisely defines the disposition of all the individual components. If this were the case, the slightest error in the instructions for carrying out this program could have catastrophic consequences. On the contrary, the mechanism appears to rely on the progressive setting of robust interneuronal connections through trial-and-error mechanisms that formally resemble an evolutionary process by variation selection.[97,99,100] At sensitive periods of brain development, the phenotypic variability of nerve cell distribution and position, as well as the exuberant spreading and the multiple figures of transiently-formed connections originating from the erratic wandering of growth cone behavior, introduce a maximal diversity of synaptic connections. This variability is then reduced by the selective stabilization of some of the labile contacts and the elimination (or retraction) of others. The crucial hypothesis of the model is that the evolution of the connective state of each synaptic contact is governed globally, and within a given time window, by the overall "message" of signals experienced by the cell on which it terminates.[97] One consequence of this is that particular electrical and chemical spatiotemporal patterns of activity in developing neuronal networks are liable to be inscribed under the form of defined and stable topologies of connections within the frame of the genetic envelope. In humans, about half of all adult connections are formed after birth at a very fast rate.[101] The nesting of these multiple traces directly contributes to forming and shaping the micro- and macroscopic architecture of the wiring network of the adult human brain.

Both epigenetic regulation and neuronal epigenesis point to the importance of multi-disciplinary approaches in order to understand the effects of epigenetic mechanisms and the role of psychological, social and cultural factors in shaping behavior, susceptibility to disease and effects of treatment interventions. Furthermore, a significant feature of epigenetic mechanisms is their reversibility. Unlike genetic mutations, if a phenotype is caused by epigenetic phenomena, such as DNA methylation and histone acetylation, they can be chemically reversed. The reversible nature of epigenetic pathways indicates a potential of developing therapeutics and social interventions, which modulate epigenetics state. However, in order to be safely applied, therapeutic drugs will need to be extremely targeted, which remains a major challenge in the field. Reversibility is also a feature related to synaptic plasticity. Both epigenetic and neuronal epigenesis mechanisms indicate the importance of the environment and the need to understand and improve psychosocial and cultural environments in order to reverse an unfavorable process.

The neuronal organization of the adult brain develops over a 25 year period following birth, during which it is subject to cultural influence, both on the individual level and at the social group level.[102] Synaptic epigenesis theories of cultural and social imprinting on our brain architecture (which differ from less discriminative epigenetic modifications of nuclear chromatin) suggest that there is an interesting possibility, namely, that one could potentially be *epigenetically proactive*[8,9,103,104] and adapt social structures, in both the short and the long term, to benefit, influence, and constructively interact with the ever-developing neuronal architecture of the brains. For example, with the aim of preventing the development of mental disorders.

7. ETHICAL AND SOCIAL IMPLICATIONS REGARDING GENE–BRAIN–ENVIRONMENT INTERACTIONS AND PROACTIVITY

Social practices, local cultures and family patterns of lifestyle clearly also play a role in creating a milieu for the developing conceptus or infant. That in turn can induce parental effects that may or may not involve epigenetic processes but nevertheless have clear trans-generational effects. That being said, when non-contagious diseases become "contagious" to future generations, the distinction between communicable and non-communicable diseases in the area of public health may become less relevant from an ethical point of view. Links between individual liberties and health conditions for large population groups may change over time, and may lead to arguments for strengthening the social context, especially when the health effects of strengthening the social context would be positive for both current and future generations. However, social prevention programs need to be carefully considered from an ethical point of view. Mental illness and disorders have been connected with social stigma, and explanatory hypotheses that connects biological and psychosocial data with mental disorders and illness will most likely give rise to a number of ethical issues.

As mentioned above, the use of large data that underlies network medicine give rise to issues around privacy and integrity. Comprehensive sets of health care information can be expected to be easier to collect, handle and transfer, even though the question if epigenetic information will as sensitive as genetic information remains open. Since environmental factors may be perceived to contribute more to the phenotype than genetic factors,[105] also the issue around "the right not to know" may be less controversial in epigenetics than in relation to genetic information, see, e.g., UNESCO's Universal Declaration on the Human Genome and Human Rights, article 5c.

One issue relates to distribution of responsibilities and justice. There is an interrelationship between individual and social responsibility where the balancing point may shift due to developments in genetics and epigenetics.

Arribas-Ayllon, Featherstone and Atkinson[106] placed the balance close to individuals and their families when suggesting that there is a genetic responsibility whereby "individuals and families have a right, a duty, even a compulsion, to choose in relation to managing the risks of themselves and others." Relating to the insights of how important parental behavior may be to the health of future generations, one might argue that this is a question of intergenerational equity.[107]

That individuals should be held responsible for their health as well as for the health of their offspring, is controversial since genetic risks are involuntary and may rather be seen as part of an individual's identity. However, when knowledge on how the environment, e.g., traumatic events in childhood, psychological stress in a family, poor living conditions, may affect the reading of a genetic make-up, both the individual and his/her family will have at least a theoretical possibility to affect the phenotype, and *eo ipso* a greater responsibility, at least to try to manage the environmental factors. However, from a public health and social justice perspective the extent to which individuals are free to choose their lives is to a significant extent depending on how opportunities are distributed and made available to them.

Public health research makes clear that justice in relation to health does not only involve the distribution of health care. It reaches into the fabric of society as a whole. A large number of studies show that social factors affect health, such as education, socio-economic status, relative wealth/poverty, gender, ethnicity, and individual and group behavior. For instance, it was shown in the so-called Black Report that life expectancy varied strictly with social class position.[108–111] Other studies have shown that the prevalence of early deaths due to smoking and unhealthy food consumption is greater among the least well off.[109,112–115] Furthermore, the importance of early life socio-economic environment for healthy development and life long health has been well documented.[116] Inequalities in health across the life course have been studied from various aspects: socio-economic, socio-demographic, gender, immigration status and ethnicity and these factors should be understood as intersecting categories.[109,117,118]

Arguably individuals and their families may have responsibilities for that which is in their capacity to change. However, from a social justice perspective the responsibilities fall heavily also on society in order to provide equal access and conditions and on this basis give extra support to those who are "worst off," due either to genetic make-up or to environmental factors affecting the transcription. Justice was explicitly and extensively addressed by John Rawls,[119] who also presented the idea of giving priority to the worst off. According to Rawls' theory, inequalities can be justified, but only to the extent that they exist in the alternative that leaves the worst off as well off as possible. Special concern for the worst off is now a standard feature of most theories of justice. Norman Daniels adjusted Rawls' theory to accommodate health care.[120]

Amartya Sen raised the question "Equality of what?," a central theme in discussions on justice ever since. Equality in the distribution of functions, opportunities, capabilities, and primary goods are some suggestions found in research, all relevant for sorting out the ethical quandaries related to accommodating epigenetics of mental disorders in a just society.[121–125]

Some of the theoretical assumptions and suggested social implications in this field may also be ethically controversial. Mental conditions may be problematic for the individual per se—intrinsic problems—while others only arise in a social context, e.g., when the individual encounters others: relational problems. Theories about causes of mental disorders include references to diverse factors: genetic, neurobiological, social and cultural. In addition to the scientific challenge of assessing suggested possible causal factors of mental disorders there is a challenge to avoid hype, over-interpretations, unjustified generalizations and ideological bias particularly for the field of neuroepigenetics and mental health.

As stated above, epigenetic variability may also be related to variation in results from empirical analyses and in the understanding of central concepts used to define phenotypic expressions, such as "stress," "traumatic stress," and "resilience." Indeed, different conceptualizations may also rely on different theoretical underpinnings, in particular psychological or social theory. One needs also understand the effects of environmental cues related to risk, resilience at both the molecular and the synaptic levels. Reversibility is a potential feature of epigenetic and synaptic mechanisms, thus making some of these amenable to both pharmaceutical and nutritional interventions. But even so, to what extent may social changes have the same effect, such as reversing or compensating for the effects of childhood deprivation for children less than 3 years of age? Linked to this are pertinent ethical issues related to the policy implications and the need to understand how social perceptions and norms of stress, resilience and aggression affect the social context of individuals and groups. Most importantly, evidence on gene–environment interactions across the life course of individuals and families calls for stronger and better coordinated efforts in improving the environments and living conditions during sensitive periods of development accompanied by interventions that will attempt to reverse and compensate for earlier disadvantage in the current aging populations.

8. GOING FROM ASSOCIATIONS TO EXPLANATIONS

As described by Frances Champagne and in the beginning of this chapter, decades of longitudinal, preclinical and clinical based studies have highlighted the relationship between the social environment and behavioral/health outcomes.[126] There are several epidemiological registry-based studies also pointing in the same direction.[127–131] Continuous similar studies are coming from the Stockholm Public Health Cohorts,[132] as well as other study cohorts such as a prospective military cohorts.[133] The potential role of epigenetic mechanisms in these processes has so far been based primarily on animal models. Public health, population registries and biobanks are vital in order to apply what has been learnt from animal models to a human context. There are several examples of environmental factors, both during prenatal and childhood/adolescence developmental stages, that have been proposed to interact with genetic factors in major psychotic disorders.[77] However, as noted by Rutten and Mill,[77] these factors often represent only statistical interactions, and despite the existence of several epigenetic mediators identified in animal models, there remains a lack of knowledge about the underlying etiological mechanisms in human populations. Psychotic syndromes may be understood as disorders of adaption to a social context, implying that one needs to understand both the social context in addition to the genetic, molecular and cellular factors. A lesson from epigenetics is that heritability estimates from classical twin studies reflect both the genetic influence and the underlying gene–environment interactions.[63] As concluded by Champagne, "epigenetic mechanisms play a critical role in development and may serve both to shape development in response to social experiences and to induce variation in social behaviour."[126]

As noted in a recent review by Vinkers et al.[134] the seminal observations by Weaver et al.[79] about the influence of the environment on the significant methylation effects related to stress has resulted in fairly consistent replication. It has also triggered numerous observational studies suggesting associations between DNA methylation and environmental impact. However, as Vinkers et al.[134] conclude there is a need to move beyond simply collecting large volumes of observational data of associations and design research projects that may provide plausible explanations of how epigenetic variability is related to meaningful biologic and psychosocial differences. A recent review by Provençal and Binder[68] concludes that exposure to stress in early life may prime the stress related system toward future responses to environmental challenges, being silent until the appropriate challenge occurs. As such, the consequences may be beneficial or detrimental depending on the challenge. However, as Provençal and Binder conclude, a full understanding of these effects will require longitudinal studies starting before conception with repeated sampling of different tissues, outcomes and environmental exposure.[68]

In order to move beyond mere associations to plausible explanations of the relation between environmental impact and gene transcription or synaptic plasticity a number of questions must be addressed. For instance, a methodological shortcoming is that most studies in the field are cross-sectional, preventing conclusions on causal relationships, when in fact longitudinal and intergenerational studies are needed. One recent example of the first upcoming wave of longitudinal studies is a genome-wide, blood-based DNA methylation profiles in relation to the development of PTSD symptoms in two prospective military cohorts. This study comprised discovery analyses in combat-trauma exposed Dutch military soldiers and replication analyses in U.S. marines, and indicated that the emergence of PTSD symptoms over a deployment period to Afghanistan was significantly associated with alterations in DNA methylation levels at several genomic positions regions.[36]

Besides, DNA methylation may also have a protective function for the organism rather than being a detrimental causal factor. Furthermore, epigenetic marks can be the consequence of phenotypic development, instead of causal,[135] and therefore longitudinal study designs and techniques aiming at understanding causality such as Mendelian randomization[136] or additive noise models[137] are required. There is therefore a great need to specify and understand confounding factors, including biological, psychosocial and cultural aspects. One needs also to tackle the fact that despite similar environmental exposures some individuals adapt and survive with a strong sense of identity while others suffer from poor health or low self-esteem. Advanced statistical methods are needed in order to understand the relative importance of different factors. Although researchers are now able to collect a wealth of clinical and molecular information in a large number of individuals, most of the discoveries to date in mental health, originate mainly from isolated analysis using only few selected factors. This results in loss of valuable information on how disease associated factors act in synergy to cause complex phenotypes. The lack of data integration using all available information is mainly due to the high complexity of bioinformatics and statistical analysis required.

Advances in bioinformatics are beginning to provide such tools using a network perspective to integrate multi-factor analysis in the study of complex phenotypes. In network models, correlations between variables, e.g., symptoms, are no longer explained by the common latent factor, the mental disorders, but rather mental disorders are conceptualized as complex systems where psychological, biological and environmental factors have autonomous causal power to influence each other.[138] There are a range of network analysis methods available with varying complexity from simple correlation networks to undirected graphical networks, such as Gaussian graphical networks, which are created on partial correlations, to Bayesian networks, which represent directed a-cyclical graphs and thus estimates of causal relations in a probabilistic way, again exploiting conditional dependencies between the observed variables. For an early example of a Bayesian network analysis see Curiac et al.[139] Further research into causality methods in networks within this very general framework has come from additive noise models, where the structure of the noise is used to infer causality also in high dimensional data. Some of these methods have been shown to be quite robust to violations of the original underlying assumptions, which make them practical in applications. The use of longitudinal registries on environmental exposure combined with individual studies in genetics, brain development and epigenetics may provide the multi-dimensional data of sufficiently large sample size required for these novel integrative analysis techniques. While recent advances have suggested that mechanistic reasoning paradigms may be applied using formalized computable knowledge, much research still being warranted in this domain.[140,141]

Thus, taken together, it is evident that there is a current, great need, promise and challenge in embracing multi-disciplinary research approaches in moving from epigenetic associations to biological and psychosocial explanations in mental health and illness.

ACKNOWLEDGMENT

Part of this research is supported by funding from the European Union's Horizon 2020 research and innovation program under grant agreement 720270 (HBP SGA1-2).

REFERENCES

1. World Health Organization. *Social Determinants of Mental Health*. World Health Organization; 2014.
2. Stansfeld S, Candy B. Psychosocial work environment and mental health—a meta-analytic review. *Scand J Work Environ Health*. 2006;32(6):443–462.

3. LaMontagne AD, Keegel T, Louie AM, Ostry A. Job stress as a preventable upstream determinant of common mental disorders: a review for practitioners and policy-makers. *Adv Ment Health*. 2010;9(1):17–35.
4. Meaney MJ. Maternal care, gene expression, and the transmission of individual differences in stress reactivity across generations. *Annu Rev Neurosci*. 2001;24(1):1161–1192.
5. Schanberg SM, Evoniuk G, Kuhn CM. Tactile and nutritional aspects of maternal care: specific regulators of neuroendocrine function and cellular development. *Proc Soc Exp Biol Med*. 1984;175(2):135–146.
6. Boks MP, de Jong NM, Kas MJ, et al. Current status and future prospects for epigenetic psychopharmacology. *Epigenetics*. 2012;7(1):20–28.
7. Seo S, Grzenda A, Lomberk G, Ou X-M, Cruciani RA, Urrutia R. Epigenetics: a promising paradigm for better understanding and managing pain. *J Pain*. 2013;14(6):549–557.
8. Evers K. Can we be epigenetically proactive? In: *Open Mind*. Frankfurt am Main: MIND Group; 2014.
9. Evers K, Changeux JP. Proactive epigenesis and ethical innovation: a neuronal hypothesis for the genesis of ethical rules. *EMBO Rep*. 2016;17(10):1361–1364.
10. World Health Organization. Data and Statistics. Mental Health. http://www.euro.who.int/en/health-topics/noncommunicable-diseases/mental-health/data-and-statistics; March 2018.
11. Zito JM, Safer DJ, Gardner JF, Boles M, Lynch F. Trends in the prescribing of psychotropic medications to preschoolers. *JAMA*. 2000;283(8):1025–1030.
12. Costello EJ, Mustillo S, Erkanli A, Keeler G, Angold A. Prevalence and development of psychiatric disorders in childhood and adolescence. *Arch Gen Psychiatry*. 2003;60(8):837–844.
13. World Health Organization. Global Health Estimates 2000-2015. Disease Burden. http://www.who.int/healthinfo/global_burden_disease/estimates/en/index2.html.
14. World Health Organization. 3 Out of 4 People Suffering From Major Depression Do Not Receive Adequate Treatment. Media Centre. Press release. Available at: http://www.euro.who.int/en/media-centre/sections/press-releases/2017/3-out-of-4-people-suffering-from-major-depression-do-not-receive-adequate-treatment; 2017.
15. Kohn R, Saxena S, Levav I, Saraceno B. The treatment gap in mental health care. *Bull World Health Organ*. 2004;82(11):858–866.
16. Moffitt TE, Caspi A, Rutter M. Measured gene-environment interactions in psychopathology: concepts, research strategies, and implications for research, intervention, and public understanding of genetics. *Perspect Psychol Sci*. 2006;1(1):5–27.
17. Carpenter WT, McGlashan TH, Strauss JS. The treatment of acute schizophrenia without drugs: an investigation of some current assumptions. *Am J Psychiatry*. 1977;134(1):14–20.
18. Rappaport M, Hopkins HK, Hall K, Belleza T, Silverman J. Are there schizophrenics for whom drugs may be unnecessary or contraindicated? *Int Pharmacopsychiatry*. 1978;13:100–111.
19. Bola JR, Mosher LR. Treatment of acute psychosis without neuroleptics: two-year outcomes from the Soteria project. *J Nerv Ment Dis*. 2003;191(4):219–229.
20. Balderston NL, Mathur A, Adu-Brimpong J, Hale EA, Ernst M, Grillon C. Effect of anxiety on behavioural pattern separation in humans. *Cognit Emot*. 2017;31(2):238–248.
21. Durisko Z, Mulsant BH, McKenzie K, Andrews PW. Using evolutionary theory to guide mental health research. *Can J Psychiatry*. 2016;61(3):159–165.
22. Durisko Z, Mulsant BH, Andrews PW. An adaptationist perspective on the etiology of depression. *J Affect Disord*. 2015;172:315–323.
23. Joness I, Blackshaw JK. An evolutionary approach to psychiatry. *Aust N Z J Psychiatry*. 2000;34(1):8–13.
24. Nesse RM. Is depression an adaptation? *Arch Gen Psychiatry*. 2000;57(1):14–20.

25. Baptista T, Rangel N, Fernández V, et al. Metformin as an adjunctive treatment to control body weight and metabolic dysfunction during olanzapine administration: a multicentric, double-blind, placebo-controlled trial. *Schizophr Res.* 2007;93(1): 99–108.
26. Engel GL. The need for a new medical model: a challenge for biomedicine. *Science.* 1977;196(4286):129–136.
27. Engel GL. The clinical application of the biopsychosocial model. *Am J Psychiatry.* 1980;137(5):535–544.
28. Bateson G. *Steps to an Ecology of Mind: Collected Essays in Anthropology, Psychiatry, Evolution, and Epistemology.* University of Chicago Press; 1972.
29. Reiser MF. Implications of a biopsychosocial model for research in psychiatry. *Psychosom Med.* 1980;42(1):141–151.
30. Garcia-Toro M, Aguirre I. Biopsychosocial model in depression revisited. *Med Hypotheses.* 2007;68(3):683–691.
31. Ghaemi SN. *The Rise and Fall of the Biopsychosocial Model.* RCP; 2009.
32. Leader JB, Klein DN. Social adjustment in dysthymia, double depression and episodic major depression. *J Affect Disord.* 1996;37(2):91–101.
33. Angst J, Kupfer D, Rosenbaum J. Recovery from depression: risk or reality? *Acta Psychiatr Scand.* 1996;93(6):413–419.
34. Barabási A-L, Gulbahce N, Loscalzo J. Network medicine: a network-based approach to human disease. *Nat Rev Genet.* 2011;12(1):56.
35. Barabási A-L. *Network Medicine—From Obesity to the "Diseasome".* Massachusetts Medical Society; 2007.
36. Pawson T, Linding R. Network medicine. *FEBS Lett.* 2008;582(8):1266–1270.
37. Zanzoni A, Soler-López M, Aloy P. A network medicine approach to human disease. *FEBS Lett.* 2009;583(11):1759–1765.
38. Guloksuz S, Pries L, van Os J. Application of network methods for understanding mental disorders: pitfalls and promise. *Psychol Med.* 2017;47(16):1–10.
39. Van den Bergh BR. Developmental programming of early brain and behaviour development and mental health: a conceptual framework. *Dev Med Child Neurol.* 2011;53(s4):19–23.
40. Heim C, Binder EB. Current research trends in early life stress and depression: review of human studies on sensitive periods, gene–environment interactions, and epigenetics. *Exp Neurol.* 2012;233(1):102–111.
41. McGowan PO, Szyf M. The epigenetics of social adversity in early life: implications for mental health outcomes. *Neurobiol Dis.* 2010;39(1):66–72.
42. Rasic D, Hajek T, Alda M, Uher R. Risk of mental illness in offspring of parents with schizophrenia, bipolar disorder, and major depressive disorder: a meta-analysis of family high-risk studies. *Schizophr Bull.* 2013;40(1):28–38.
43. Rutter M. Environmentally mediated risks for psychopathology: research strategies and findings. *J Am Acad Child Adolesc Psychiatry.* 2005;44(1):3–18.
44. Schmitt A, Malchow B, Hasan A, Falkai P. The impact of environmental factors in severe psychiatric disorders. *Front Neurosci.* 2014;8:19.
45. Krapohl E, Patel H, Newhouse S, et al. Multi-polygenic score approach to trait prediction. *Mol Psychiatry.* 2017;23(5):1368–1374.
46. Rutter M. *How the Environment Affects Mental Health.* RCP; 2005.
47. Schore AN. Effects of a secure attachment relationship on right brain development, affect regulation, and infant mental health. *Infant Ment Health J.* 2001;22(1–2):7–66.
48. Schore AN. The effects of early relational trauma on right brain development, affect regulation, and infant mental health. *Infant Ment Health J.* 2001;22(1–2):201–269.
49. Schore AN. Attachment, affect regulation, and the developing right brain: linking developmental neuroscience to pediatrics. *Pediatr Rev.* 2005;26(6):204–217.

50. Pierrehumbert B, Torrisi R, Ansermet F, Borghini A, Halfon O. Adult attachment representations predict cortisol and oxytocin responses to stress. *Attach Hum Dev.* 2012;14(5):453–476.
51. Lubach GR, Coe CL, Ershler WB. Effects of early rearing environment on immune-responses of infant rhesus monkeys. *Brain Behav Immun.* 1995;9(1):31–46.
52. Renzi C, Vadilonga V, Gandini S, et al. Stress exposure in significant relationships is associated with lymph node status in breast cancer. *PLoS One.* 2016;11(2):e0149443.
53. Wray AH, Stevens C, Pakulak E, Isbell E, Bell T, Neville H. Development of selective attention in preschool-age children from lower socioeconomic status backgrounds. *Dev Cogn Neurosci.* 2017;26:101–111.
54. Stevens C, Paulsen D, Yasen A, Neville H. Atypical auditory refractory periods in children from lower socio-economic status backgrounds: ERP evidence for a role of selective attention. *Int J Psychophysiol.* 2015;95(2):156–166.
55. Stevens C, Lauinger B, Neville H. Differences in the neural mechanisms of selective attention in children from different socioeconomic backgrounds: an event-related brain potential study. *Dev Sci.* 2009;12(4):634–646.
56. Isbell E, Stevens C, Pakulak E, Wray AH, Bell TA, Neville HJ. Neuroplasticity of selective attention: research foundations and preliminary evidence for a gene by intervention interaction. *Proc Natl Acad Sci USA.* 2017;114(35):9247–9254.
57. Stringer S, Minică C, Verweij KJ, et al. Genome-wide association study of lifetime cannabis use based on a large meta-analytic sample of 32 330 subjects from the International Cannabis Consortium. *Transl Psychiatry.* 2017;6(3):e769.
58. Polderman TJ, Benyamin B, De Leeuw CA, et al. Meta-analysis of the heritability of human traits based on fifty years of twin studies. *Nat Genet.* 2015;47(7):702–709.
59. Hopfer CJ, Crowley TJ, Hewitt JK. Review of twin and adoption studies of adolescent substance use. *J Am Acad Child Adolesc Psychiatry.* 2003;42(6):710–719.
60. Lynskey MT, Agrawal A, Heath AC. Genetically informative research on adolescent substance use: methods, findings, and challenges. *J Am Acad Child Adolesc Psychiatry.* 2010;49(12):1202–1214.
61. Hicks BM, DiRago AC, Iacono WG, McGue M. Gene–environment interplay in internalizing disorders: consistent findings across six environmental risk factors. *J Child Psychol Psychiatry.* 2009;50(10):1309–1317.
62. Van Os J, Rutten BP. *Gene-Environment-Wide Interaction Studies in Psychiatry.* American Psychiatric Association; 2009.
63. van Os J, Kenis G, Rutten BP. The environment and schizophrenia. *Nature.* 2010;468(7321):203–212.
64. Duncan L, Ratanatharathorn A, Aiello A, et al. Largest GWAS of PTSD (N=20 070) yields genetic overlap with schizophrenia and sex differences in heritability. *Mol Psychiatry.* 2018;23(3):666–673.
65. Nievergelt CM, Maihofer AX, Mustapic M, et al. Genomic predictors of combat stress vulnerability and resilience in US Marines: a genome-wide association study across multiple ancestries implicates PRTFDC1 as a potential PTSD gene. *Psychoneuroendocrinology.* 2015;51:459–471.
66. Stein MB, Chen C-Y, Ursano RJ, et al. Genome-wide association studies of posttraumatic stress disorder in 2 cohorts of US Army soldiers. *JAMA Psychiatry.* 2016;73(7):695–704.
67. Hanson JL, Chandra A, Wolfe BL, Pollak SD. Association between income and the hippocampus. *PLoS One.* 2011;6(5):e18712.
68. Provençal N, Binder EB. The effects of early life stress on the epigenome: from the womb to adulthood and even before. *Exp Neurol.* 2015;268:10–20.
69. Andlauer TF, Buck D, Antony G, et al. Novel multiple sclerosis susceptibility loci implicated in epigenetic regulation. *Sci Adv.* 2016;2(6):e1501678.

70. Stuffrein-Roberts S, Joyce PR, Kennedy MA. Role of epigenetics in mental disorders. *Aust N Z J Psychiatr.* 2008;42(2):97–107.
71. Barker D. *Mothers, Babies and Health in Later Life.* Edinburgh: Churchill Livingston; 1998.
72. Barker DJ. The fetal and infant origins of adult disease. *BMJ.* 1990;301(6761):1111.
73. McLaughlin KA, Sheridan MA, Tibu F, Fox NA, Zeanah CH, Nelson CA. Causal effects of the early caregiving environment on development of stress response systems in children. *Proc Natl Acad Sci USA.* 2015;112(18):5637–5642.
74. Keding TJ, Herringa RJ. Abnormal structure of fear circuitry in pediatric post-traumatic stress disorder. *Neuropsychopharmacology.* 2015;40(3):537.
75. Bird A. Perceptions of epigenetics. *Nature.* 2007;447(7143):396–398.
76. Pishva E, Kenis G, van den Hove D, et al. The epigenome and postnatal environmental influences in psychotic disorders. *Soc Psychiatry Psychiatr Epidemiol.* 2014;49(3): 337–348.
77. Rutten BP, Mill J. Epigenetic mediation of environmental influences in major psychotic disorders. *Schizophr Bull.* 2009;35(6):1045–1056.
78. Rutten BP, Hammels C, Geschwind N, et al. Resilience in mental health: linking psychological and neurobiological perspectives. *Acta Psychiatr Scand.* 2013;128(1):3–20.
79. Weaver IC, Cervoni N, Champagne FA, et al. Epigenetic programming by maternal behavior. *Nat Neurosci.* 2004;7(8):847–854.
80. Murgatroyd C, Spengler D. Epigenetics of early child development. *Front Psychiatry.* 2011;2:16.
81. Hoffmann A, Spengler D. DNA memories of early social life. *Neuroscience.* 2014;264:64–75.
82. Doom JR, Gunnar MR. Stress physiology and developmental psychopathology: past, present, and future. *Dev Psychopathol.* 2013;25(4 pt 2):1359–1373.
83. Evans GW. The environment of childhood poverty. *Am Psychol.* 2004;59(2):77.
84. Bradley RH, Corwyn RF. Socioeconomic status and child development. *Annu Rev Psychol.* 2002;53(1):371–399.
85. Maholmes V, King RB. *The Oxford Handbook of Poverty and Child Development.* USA: OUP; 2012.
86. Franklin TB, Russig H, Weiss IC, et al. Epigenetic transmission of the impact of early stress across generations. *Biol Psychiatry.* 2010;68(5):408–415.
87. Gapp K, Jawaid A, Sarkies P, et al. Implication of sperm RNAs in transgenerational inheritance of the effects of early trauma in mice. *Nat Neurosci.* 2014;17(5):667.
88. Santavirta T, Santavirta N, Gilman SE. Association of the World War II Finnish evacuation of children with psychiatric hospitalization in the next generation. *JAMA Psychiatry.* 2018;75(1):21–27.
89. Babenko O, Kovalchuk I, Metz GA. Stress-induced perinatal and transgenerational epigenetic programming of brain development and mental health. *Neurosci Biobehav Rev.* 2015;48:70–91.
90. LeMoult J, Chen MC, Foland-Ross LC, Burley HW, Gotlib IH. Concordance of mother–daughter diurnal cortisol production: understanding the intergenerational transmission of risk for depression. *Biol Psychol.* 2015;108:98–104.
91. Sandman CA, Davis EP, Buss C, Glynn LM. Exposure to prenatal psychobiological stress exerts programming influences on the mother and her fetus. *Neuroendocrinology.* 2012;95(1):8–21.
92. Laviola G, Hannan AJ, Macrì S, Solinas M, Jaber M. Effects of enriched environment on animal models of neurodegenerative diseases and psychiatric disorders. *Neurobiol Dis.* 2008;31(2):159–168.
93. Nelson CA, Zeanah CH, Fox NA, Marshall PJ, Smyke AT, Guthrie D. Cognitive recovery in socially deprived young children: the Bucharest Early Intervention Project. *Science.* 2007;318(5858):1937–1940.

94. Karatsoreos IN, McEwen BS. Annual research review: the neurobiology and physiology of resilience and adaptation across the life course. *J Child Psychol Psychiatry.* 2013;54(4):337–347.
95. Ben-Ari Y. Neuro-archaeology: pre-symptomatic architecture and signature of neurological disorders. *Trends Neurosci.* 2008;31(12):626–636.
96. Changeux J-P. Synaptic epigenesis and the evolution of higher brain functions. In: *Epigenetics, Brain and Behavior.* Springer; 2012:11–22.
97. Changeux J-P, Courrége P, Danchin A. A theory of the epigenesis of neuronal networks by selective stabilization of synapses. *Proc Natl Acad Sci USA.* 1973;70(10): 2974–2978.
98. Levenson JM, Sweatt JD. Epigenetic mechanisms in memory formation. *Nat Rev Neurosci.* 2005;6(2):108–118.
99. Changeux J-P, Danchin A. Selective stabilisation of developing synapses as a mechanism for the specification of neuronal networks. *Nature.* 1976;264(5588):705–712.
100. Edelman GM. *Neural Darwinism: The Theory of Neuronal Group Selection.* Basic Books; 1987.
101. Bruer JT. Neural connections: some you use, some you lose. *Phi Delta Kappan.* 1999;81(4):264–277.
102. Collin G, van den Heuvel MP. The ontogeny of the human connectome: development and dynamic changes of brain connectivity across the life span. *Neuroscientist.* 2013;19(6):616–628.
103. Evers K, Changeux JP. Response by the authors. *EMBO Rep.* 2017;e201744696, Epub ahead of print.
104. Evers K. Personalized medicine in psychiatry: ethical challenges and opportunities. *Dialogues Clin Neurosci.* 2009;11(4):427.
105. Rothstein MA, Cai Y, Marchant GE. The ghost in our genes: legal and ethical implications of epigenetics. *Health Matrix Clevel.* 2009;19(1):1.
106. Arribas-Ayllon M, Featherstone K, Atkinson P. The practical ethics of genetic responsibility: non-disclosure and the autonomy of affect. *Soc Theory Health.* 2011;9(1):3–23.
107. Rothstein MA, Cai Y, Marchant GE. Ethical implications of epigenetics research. *Nat Rev Genet.* 2009;10(4):224.
108. Hofrichter R. *Health and Social Justice: Politics, Ideology, and Inequality in the Distribution of Disease.* San Francisco: Jossey-Bass A Wiley Imprint; 2003.
109. Bartley M. *Health Inequality: An Introduction to Theories, Concepts and Methods.* Malden, USA: Polity Press Google Scholar; 2004.
110. Smith JP. Healthy bodies and thick wallets: the dual relation between health and economic status. *J Econ Perspect.* 1999;13(2):145–166.
111. Wamala SP, Lynch JP. *Gender and Social Inequities in Health: A Public Health Issue.* Studentlitteratur; 2002.
112. Marmot MG, Allen J, Goldblatt P, et al. *Fair Society, Healthy Lives: Strategic Review of Health Inequalities in England Post-2010.* Department for International Development, Gov. UK; 2010.
113. Marmot M, Goldblatt P, Allen J, et al. *Fair Society Healthy Lives (The Marmot Review).* Department for International Development, Gov. UK; 2010.
114. World Health Organization. *Closing the Gap in a Generation: Health Equity Through Action on the Social Determinants of Health;* 2009.
115. Marmot SM. Closing the health gap in a generation: the work of the Commission on Social Determinants of Health and its recommendations. *Glob Health Promot.* 2009; 16(suppl 1):23–27.
116. Maggi S, Irwin LJ, Siddiqi A, Hertzman C. The social determinants of early child development: an overview. *J Paediatr Child Health.* 2010;46(11):627–635.
117. Östlin P, George A, Sen G. Gender, health, and equity: the intersections. In: *Challenging Inequities in Health: From Ethics to Action.* Oxford University Press; 2001:174–189.

118. Whitehead M, Evans T, Diderichsen F, Bhuiya A. *Challenging Inequities in Health: From Ethics to Action.* Oxford University Press; 2001.
119. Rawls J. *A Theory of Justice.* Cambridge, MA: Oxford University Press; 1972.
120. Daniels N. *Just Health Care.* Cambridge University Press; 1985.
121. Callahan D. *Setting Limits: Medical Goals in an Aging Society With "A Response to My Critics".* Georgetown University Press; 1995.
122. Callahan D. *False Hopes: Overcoming the Obstacles to a Sustainable, Affordable Medicine.* Rutgers University Press; 1999.
123. Shotton L. *Health Care Law and Ethics.* Social Science Press; 1997.
124. Gruskin S. Hofrichter R, ed. *Health and Social Justice: Politics, Ideology and Inequity in the Distribution of Disease.* San Francisco, CA: Jossey-Bass, John Wiley and Sons; 2003. ISBN: 0787967335:688. £ 34.50 (PB). *Int J Epidemiol.* 2004;33(5):1159–1160.
125. Sen A. *Inequality Reexamined.* Clarendon Press; 1992.
126. Champagne FA. Interplay between social experiences and the genome: epigenetic consequences for behavior. *Adv Genet.* 2012;77:33.
127. Modin B, Vågerö D, Hallqvist J, Koupil I. The contribution of parental and grandparental childhood social disadvantage to circulatory disease diagnosis in young Swedish men. *Soc Sci Med.* 2008;66(4):822–834.
128. Faris REL, Dunham HW. *Mental Disorders in Urban Areas: An Ecological Study of Schizophrenia and Other Psychoses.* Chicago: University of Chicago Press; 1939. https://searchworks.stanford.edu/view/10842446.
129. Jackson JS, Brown TN, Williams DR, Torres M, Sellers SL, Brown K. Racism and the physical and mental health status of African Americans: a thirteen year national panel study. *Ethn Dis.* 1996;6(1–2):132–147.
130. Williams DR, Yu Y, Jackson JS, Anderson NB. Racial differences in physical and mental health: socio-economic status, stress and discrimination. *J Health Psychol.* 1997;2(3):335–351.
131. Yen IH, Syme SL. The social environment and health: a discussion of the epidemiologic literature. *Annu Rev Public Health.* 1999;20(1):287–308.
132. Svensson AC, Fredlund P, Laflamme L, et al. Cohort profile: the Stockholm public health cohort. *Int J Epidemiol.* 2012;42(5):1263–1272.
133. Rutten BP, Vermetten E, Vinkers CH, et al. Longitudinal analyses of the DNA methylome in deployed military servicemen identify susceptibility loci for post-traumatic stress disorder. *Mol Psychiatry.* 2017;23(5):1145–1156.
134. Vinkers CH, Kalafateli AL, Rutten BP, et al. Traumatic stress and human DNA methylation: a critical review. *Epigenomics.* 2015;7(4):593–608.
135. Wahl S, Drong A, Lehne B, et al. Epigenome-wide association study of body mass index, and the adverse outcomes of adiposity. *Nature.* 2017;541(7635):81–86.
136. Caramaschi D, Sharp GC, Nohr EA, et al. Exploring a causal role of DNA methylation in the relationship between maternal vitamin B12 during pregnancy and child's IQ at age 8, cognitive performance and educational attainment: a two-step Mendelian randomization study. *Hum Mol Genet.* 2017;26(15):3001–3013.
137. Peters J, Mooij JM, Janzing D, Schölkopf B. Causal discovery with continuous additive noise models. *J Mach Learn Res.* 2014;15(1):2009–2053.
138. Silbersweig D. Integrating models of neurologic and psychiatric disease. *JAMA Neurol.* 2017;74(7):759–760.
139. D.-I. Curiac, G. Vasile, O. Banias, C. Volosencu and A. Albu, In: *Bayesian network model for diagnosis of psychiatric diseases*, Paper presented at: Information Technology Interfaces, 2009. ITI'09. Proceedings of the ITI 2009 31st International Conference on Information Technology Interfaces.

140. Catlett NL, Bargnesi AJ, Ungerer S, et al. Reverse causal reasoning: applying qualitative causal knowledge to the interpretation of high-throughput data. *BMC Bioinformatics*. 2013;14(1):340.
141. Hofmann-Apitius M, Ball G, Gebel S, et al. Bioinformatics mining and modeling methods for the identification of disease mechanisms in neurodegenerative disorders. *Int J Mol Sci*. 2015;16(12):29179–29206.

INDEX

Note: Page numbers followed by "*f*" indicate figures, "*t*" indicate tables, and "*ge*" indicate glossary terms.

W

Z

FSC
www.fsc.org
MIX
Papier | Fördert gute Waldnutzung
FSC® C083411

Kolibri 360 GmbH
Ferdinand-Jühlke-Straße 7
99095 Erfurt, Deutschland
produktsicherheit@kolibri360.de